ADVANCES IN
Chromatography

VOLUME 38

ADVANCES IN
Chromatography

VOLUME 38

A Tribute to J. Calvin Giddings

Edited by

Phyllis R. Brown
UNIVERSITY OF RHODE ISLAND
KINGSTON, RHODE ISLAND

Eli Grushka
THE HEBREW OF JERUSALEM
JERUSALEM, ISRAEL

MARCEL DEKKER, INC.　　NEW YORK · BASEL · HONG KONG

Library of Congress Cataloging-in-Publication Data
Main entry under title:

Advances in Chromatography. v. 1-
1965-
New York, M. Dekker
  v. illus. 24 cm.
  Editors: v.1-J.C. Giddings and R.A. Keller
  1. Chromatographic analysis-Addresses, essays, lectures.
I. Giddings, John Calvin, [date] ed.   II. Keller, Roy A., [date] ed.
QD271.A23 544.92 65-27435
ISBN: 0-8247-9999-2

The publisher offers discounts on this book when ordered in bulk quantities. For more information, write to Special Sales/Professional Marketing at the address below.

This book is printed on acid-free paper.

Copyright © 1998 by MARCEL DEKKER, INC. All Rights Reserved.

Neither this book nor any part may be reproduced or transmitted in any form or by any means, electronic or mechanical, including photocopying, microfilming, and recording, or by any information storage and retrieval system, without permission in writing from the publisher.

MARCEL DEKKER, INC.
270 Madison Avenue, New York, New York 10016
http://www.dekker.com

Current printing (last digit):
10 9 8 7 6 5 4 3 2 1

PRINTED IN THE UNITED STATES OF AMERICA

A Tribute to J. Calvin Giddings

On October 24, 1996, at the age of 66, my father, J. Calvin Giddings, lost a prolonged and courageous battle with cancer. He leaves behind a broad legacy in science, in exploration, in environmental preservation, and most of all in the many lives on which he has had a positive influence.

Throughout his life Dad maintained a great passion for and curiosity about the natural world. In chemistry he contributed a great deal to the field of separation science, including the invention of field-flow fractionation and other related techniques. His remarkable publication record included authorship on more than 400 publications, editing of 32 books, and authorship of two others. The quality and depth of his research were recognized by numerous awards, and its utility led to the founding of a company, FFFractionation, to make his techniques widely available. There is much more that could be listed here, but what was, perhaps, most important to him was not the numerous hon-

---

Reprinted with permission from *Journal of Microcolumn Separations.*

ors his work received but rather just the joy of seeing his separation methods applied to the many problems for which they could be useful, in a wide range of areas extending beyond industry to medicine and environmental studies.

His passion and curiosity extended far beyond science, and combined with a zest for adventure and a love of the outdoors led him to a long career of outdoor exploration. His outdoor enthusiasm was kindled through trips with his own father to the mountains near their American Fork home. Later in life, he began to further explore these and other mountains on his own. His climbing career included numerous first and early ascents, including the first ascent of the west face of Lone Peak, prominently visible from the Salt Lake Valley floor. Other notable ascents include Zion Canyon's Great White Throne, Devil's Tower in Wyoming, and climbs in the Tetons. He also was one of the pioneers in discovering numerous back country ski-touring routes in Utah's unique Wasatch Mountains.

Still later, he turned to a long career of river exploration. Beginning with homemade kayaks, the first in Utah, and techniques learned from a book, his enthusiasm led him to explore many western rivers. The list of his pioneering descents is too long to give here, but river runners will recognize among them a number of respectable runs such as Cross Mountain Canyon on the Yampas, the Black Boxes of the San Rafael, Idaho's Big Creek, and the South Fork of the Salmon. Even today these runs are considered significantly challenging.

Earlier generations of explorers such as John Wesley Powell had already navigated the West's major rivers such as the Colorado, so in looking for even greater challenges, Dad had to look elsewhere. His focus shifted to the longest and largest river in the world: the Amazon. The longest tributary (hence the recognized source) of the Amazon is the Apurimac River in Peru. In 1974 he and a partner mounted an expedition to kayak some of the upper canyons of this river, but were soon turned back by the recognition that their resources did not allow them to continue with a sufficient safety margin. He returned with a larger expedition in 1975, and despite enormous hazard and toil managed to descend a large fraction of the canyons of the Apurimac. This adventure is chronicled in his book, *Demon River Apurimac*, which he was fortunate enough to see published just before his death.

Beyond these remarkable achievements, Dad's love of exploration and adventure, and joy in sharing it with friends and family, pervaded his life. In later years his focus shifted to mountain biking, which he began at age 56 with a ride on Moab's challenging Slickrock trail. He found that mountain biking opened up a whole new way of seeing new sights, discovering new surprises, and simply having fun. Even on hikes near his canyon home he found happiness exploring, always looking for a new path to tread—and frequently leading family and friends through thick brush and enduring the consequent joke that he loved bushwacking.

Out of my father's love for the natural world came a strong vision that it needed protection. He early realized the fragility of the environment, and the many disasters both real and potential that unbridled development and population growth was visiting on it. He became one of the pioneers of the Utah environmental movement. Soon after the passage of the Wilderness Act, he proposed Utah's first wilderness, the Lone Peak Wilderness Area, which won protection several years later. He was a leader in several local environmental organizations, and also participated in the founding of the American Rivers Conservation Council (now American Rivers). Realizing the crucial importance of arresting unlimited population growth, he also served as a member of the board of directors of Negative Population Growth.

He also recognized the importance of education and of bringing a scientific perspective to the environmental debate, and authored the text *Chemistry, Man, and Environmental Change*, another of his pioneering works which has been widely used as a textbook to educate students about the relation between chemistry and the health of the planet. This book received an award for Outstanding Environmental Achievement in Education from the Rocky Mountain Council on the Environment in 1973.

But beyond this list of accomplishments, and many more not listed, there is another equally important legacy left in the influence that he had on others. He had a remarkable ability to inspire others to achieve beyond themselves. In all that I observed, this was almost universally done through gentle encouragement rather than through intimidation or other means. He was truly a leader, with a talent for not making you feel like you were being led. He was able to inspire colleagues, students, friends, and family solely through his understated enthusiasm.

As a father, he has been an inspiration both in outdoor adventure and in science. He would almost always include the family in outdoor trips. He introduced my brother Mike and me to hiking and skiing in the Wasatch at an early age, and slightly later to kayaking on Utah and Idaho rivers. He took great pride and delight in seeing our development, and was always there to rescue us when we took the inevitable fall while skiing or swim through the rapids while kayaking. We were able to share in some exploratory descents of rivers, although others were beyond us. Later, when I took up climbing myself, Dad gave guidance, sometimes too firm for the rebelliousness of a teenager but possibly keeping me alive nonetheless. Despite the fact that he had given up climbing himself, he joined me on several occasions so that he could share in his son's growth in the mountains. He planned family trips to kayak other exotic rivers such as the Nahanni in the Northwest Territories, the Tatshenshini in Alaska, and the Colorado in the Grand Canyon. He always watched over us, yet let us develop independence.

Beginning at an early age he taught me to always apply logic and reason to the mysteries of nature. I learned that most things could be explained by science, and that those that could not were just interesting puzzles waiting to be solved. He recognized the great importance of mathematical learning, and introduced me to calculus long before it would ordinarily be taught. This mathematical background served as the foundation for my career as a theoretical physicist. My inspiration in physics came later, when he explained to me some of the truly weird properties of quantum mechanics and I decided that if the world of physics was that different from ordinary experience I had to learn about it. He helped arrange for me to enter college early, and constantly gave me encouragement in my studies. In particular, he urged me to take as much math as possible—an extremely valuable suggestion. As my knowledge in my area of physics gradually surpassed his, he always asked me to explain the latest ideas, who the players were, and what I was working on in my research.

Although he allowed himself to take pride in his sons, he was nearly the definition of humility, generosity, and tolerance. I don't recall him ever speaking in self-aggrandizing terms. In fact, even as his son it was difficult to learn about his many accomplishments—only rarely and reluctantly did he speak of them. He was generous almost beyond reason, and always wanted to lend a helping hand to those who needed it. He was tolerant of other's views, even if they were widely divergent from his own, and he respected people for who they were and accepted them despite their flaws. This did not mean, however, that he accepted untruths or flawed reasoning: he was always patiently insistent on ferreting out the truth, guided by logic and reason.

Gone in body, he leaves behind his wife, partner, and best friend Leslie, his two sons, Steven and Michael, two brothers, and two sisters. However, Dad will long live on in the imprint that he has made on so many lives. His impact extends over a wide range of human endeavor. His scientific work continues to grow. His vision has helped preserve a small part of the natural world, and continues to influence others to fight to save us from the destruction of our environment. His adventures and explorations will be widely remembered both by those who were lucky enough to share them and by many others inspired by them. Finally, his ability to lead family, friends, and colleagues to achieve beyond their own expectations will live on, although perhaps his leadership will be even better than it was in life.

Steven B. Giddings
University of California
Santa Barbara, California

Preface

Volumes 38 and 39 of *Advances in Chromatography* are dedicated to the memory of Dr. J. Calvin Giddings, the founder and executive editor of the series for 28 years. As executive editor, he set the policy and the high standards that successfully guided the series, bringing the latest advances in chromatographic methods to both the theoreticians and the users. Cal Giddings had the foresight to see the need for a yearly volume of "critical, stimulating, and relevant reviews," which supplied scientists from all disciplines with cutting-edge information in an easily readable form. That the series has continued since 1965 is a tribute to Cal's understanding that separations are essential not only in analytical chemistry but also in almost every scientific discipline. In fact, the development of good separation techniques made possible the advances in the biological sciences that led to biotechnology.

J. Calvin Giddings died on October 24, 1996, at the age of 66. He received his B.Sc. degree at Brigham Young University and his Ph.D. at the University of Utah in 1954 under Henry Eyring. He did postdoctoral work at the University of Wisconsin with Joseph Herschfelder before returning to the University of Utah as a faculty member. Although he was trained as a physical chemist, Cal was the consummate scientist and became a giant in the field of separation science. He made truly outstanding contributions to the development of chromatographic theory. He was a brilliant pioneer who blazed the trail with his precise mathematical treatment of mechanisms involved in chemical separations. Cal was able to picture in his mind a complex process, describe it mathematically, and then paint a word picture clearly and concisely. He was mainly interested in understanding chromatography and describing it with physical and mathematical models. In 1958, together with Dr. Eyring, he developed the random walk theory of chromatography. Later on, he developed the quasi-equilibrium model that elegantly simplified the mathematical description of the complex processes occurring in the chromatographic column during the separation. He used the models he developed to optimize separations. He proposed that the optimal velocity of a separation was based on a compromise between diffusion and rate processes. He also realized that liquid chromatography was not a different system from gas chromatography, and that both techniques are described using the same theoretical models.

Theory was important to Cal because it provided "the power of prediction, control, correlation, and calculation." With theory, he found that mechanisms of physical and chemical phenomena could be elucidated and simplified. He strongly believed that understanding the theoretical basis of a process was fundamental and he showed how theory could guide the development and optimization of a separation system. To him, theoretical work was "like laying the bricks of a satisfying edifice, tying diverse chromatographic phenomena to dynamical roots which lead to predictions of efficiency." His famous book *The Dynamics of Chromatography, Volume I* was published by Marcel Dekker, Inc. in 1965 and is a classic in the field. It is as relevant today as it was on the day of its publication.

In the past three decades, Cal concentrated his efforts on a new series of techniques that he termed field flow fractionation (FFF), in which retention is controlled by a force field perpendicular to the flow of the sample. These techniques extended the range of separations to macromolecules and particles of every size and shape. He was a tireless innovator and made unique contributions to the development and understanding of these techniques.

He had published over 400 articles, mainly on separations but he also wrote scientific articles on diverse subjects such as flame and nuclear kinetics, quantum mechanics, snow and avalanche physics, and the probability factors in nuclear holocausts. His two books in the separation areas are classics in the field. *The Dynamics of Chromatography* defined the theoretical basis of chromatography, and his more recent book (published in 1991) *Unified Separation Science* is an essential text to all scientists who deal, directly or indirectly, with separations. In his clear and concise manner, he tied together many separation techniques, showing their similarities and differences.

Cal was a man of great vision, a scientist who not only developed theoretical concepts, but also applied them to the real world. He warned us about the environmental problems we see today. Back in 1972, in his textbook for undergraduates *Man, Chemistry and Environmental Change*, he wrote "Spaceship Earth is the only home we will ever have, a home worth being concerned and educated about. The issues involved here transcend political differences, social barriers and cultural gaps, for without such a home and its life support-system, all mankind will perish."

Cal received many honors for his pioneering work. Among his awards were the ACS Awards in Chromatography, in Separation Science, and in Analytical Chemistry; the Stephan dal Nogare Award in Chromatography; the Tswett Chromatography Medal; Fulbright Fellowship; and the Romcoe Award for Outstanding Environmental Achievement in Education.

Although Cal was an outstanding scientist, his interests were not narrowly focused on science. He was a "Renaissance man," a man of great depth, breadth, and intellectual curiosity. He loved nature, and was a renowned explorer, dedicated environmentalist, and avid sportsman. Most importantly, he was a wonderful human being who cared deeply about his students, his colleagues, his friends, and his family.

Both of the present editors of the *Advances in Chromatography* series had the honor of working with Cal on this series for about 20 years and have learned a great deal from his leadership, his insight, and his knowledge. He was a superb role model and had a tremendous influence on our professional lives. He was a gentleman in every respect of the word, and his willingness to help us and other colleagues with professional advice was invaluable. Eli was fortunate to have been a postdoctoral fellow in Cal's laboratory about 30 years ago. The results of that experience are evident to this day. In addition, he was our mentor, our colleague, and our friend. Cal was a very special person who will be missed greatly by us and everyone who knew him.

His son, Steven, a physicist at the University of California at Santa

Barbara, wrote a beautiful memorial to his father for the *Journal of Microcolumn Separations,* which recently published two issues in memory of Cal. Steven and the *Journal* very kindly gave us the permission to reprint Steven's tribute, which lovingly depicts his father and shows us the side of Cal that many scientists did not know: that of devoted father, family man, explorer, sportsman, and avid environmentalist.

Phyllis R. Brown
Eli Grushka

Contributors

Alberto Cavazzini, Ph.D.   University of Ferrara, Ferrara, Italy

Catherine O. Dasenbrock, Ph.D.   Department of Chemistry and Biochemistry, University of Maryland Baltimore County, Baltimore, Maryland

J. W. Dolan, Ph.D.   LC Resources Inc., Walnut Creek, California

Franceso Dondi, Ph.D.   University of Ferrara, Ferrara, Italy

András Guttman, Ph.D.   Genetic BioSystems, Inc., San Diego, California

John H. Knox, D.Sc.   Department of Chemistry, University of Edinburgh, Edinburgh, United Kingdom

William R. LaCourse, Ph.D.   Department of Chemistry and Biochemistry, University of Maryland Baltimore County, Baltimore, Maryland

Milton L. Lee, Ph.D. Department of Chemistry and Biochemistry, Brigham Young University, Provo, Utah

H. Poppe, Ph.D. Laboratory for Analytical Chemistry, Amsterdam Institute for Molecular Studies (AIMS), University of Amsterdam, Amsterdam, The Netherlands

Maurizio Remelli, Ph.D. University of Ferrarra, Ferrarra, Italy

Yufeng Shen Department of Chemistry and Biochemistry, Brigham Young University, Provo, Utah

L. R. Snyder, Ph.D. LC Resources Inc., Walnut Creek, California

Kathi J. Ulfelder, Ph.D.* Beckman Instruments, Inc., Fullerton, California

---

**Current affiliation*: Caliper Technologies Corp., Palo Alto, California

Contents of Volume 38

A Tribute to J. Calvin Giddings   *Steven B. Giddings*   iii
Preface   ix
Contributors to Volume 38   xiii
Contents of Other Volumes   xix

1. **Band Spreading in Chromatography: A Personal View**   1

   John H. Knox

   I. Column Structure
   II. Early Theories
   III. First Meeting with Calvin Giddings—Reduced Parameters
   IV. Obstructed Diffusion and the *B*-Term
   V. Eddy Diffusion, Coupling, Radial Dispersion, and the Infinite Diameter Effect
   VI. Nonequilibrium Between the Stationary and Moving Zones or Phases
   VII. Separation of the *A*- and *C*-Terms
   VIII. Exclusion Chromatography
   IX. Slow Chemical Equilibration and Wide Peaks
   X. Band Spreading in Electroseparation Systems
   XI. Concluding Remarks
   References

2. **The Stochastic Theory of Chromatography**   **51**

   Francesco Dondi, Alberto Cavazzini, and Maurizio Remelli

   I. Introduction
   II. Theoretical Approaches in Chromatography
   III. Level One Theory: Structural Stochastic Concepts
   IV. Level Two Theory: Fundamentals and Basic Achievements of the Stochastic Theory of Chromatography
   V. Level Three Theory: Some Applications of the Stochastic Theory to Experimental Peak Shape Analysis
   VI. The Characteristic Function Method in the Stochastic Theory of Chromatography
   References

3. **Solvating Gas Chromatography Using Packed Capillary Columns**   **75**

   Yufeng Shen and Milton L. Lee

   I. Introduction
   II. Packed-Column HPGC, SGC, SFC, and LC
   III. Flow Transport in SGC
   IV. Efficiency in SGC
   V. Separation Speed in SGC
   VI. Polarity and Solvating Power in SGC
   References

4. **The Linear-Solvent-Strength Model of Gradient Elution**   **115**

   L. R. Snyder and J. W. Dolan

   I. Introduction
   II. Liquid Chromatography Basics
   III. The Linear-Solvent-Strength (LSS) Model

IV. Optimizing Gradient Separation
V. Nonideal Effects in Gradient Elution
VI. Application of the LSS Model to Gas Chromatography
VII. Conclusion
References
Appendix: Isocratic Versus Gradient Resolution for Case of Unequal S-Values

5. **High-Performance Liquid Chromatography-Pulsed Electrochemical Detection for the Analysis of Antibiotics**   **189**

   William R. LaCourse and Catherine O. Dasenbrock

   I. Introduction
   II. Aminoglycosides
   III. Penicillins and Cephalosporins
   IV. Future Research
   V. Conclusions
   References

6. **Theory of Capillary Zone Electrophoresis**   **233**

   H. Poppe

   I. In Memory of J. Calvin Giddings
   II. Introduction
   III. Mobilities
   IV. Ions in Equilibrium
   V. Electrophoresis in Capillaries
   VI. Electroosmotic Flow
   VII. Dispersion
   VIII. Interaction of Ions
   IX. Effect of Electrolysis on Buffer Composition
   X. Peak Integrals
   References

7. **Separation of DNA by Capillary Electrophoresis**   **301**

András Guttman and Kathi J. Ulfelder

    I. Introduction
   II. Theory
  III. Methods
  IV. Selected Applications
   V. Future Directions
      References

Index   341

Contents of Other Volumes

Volumes 1-10 out of print

Volume 11

Quantitative Analysis by Gas Chromatography *Josef Novák*
Polyamide Layer Chromatography *Kung-Tsung Wang, Yau-Tang Lin, and Iris S. Y. Wang*
Specifically Adsorbing Silica Gels *H. Bartels and P. Prijs*
Nondestructive Detection Methods in Paper and Thin-Layer Chromatography *G. C. Barrett*

Volume 12

The Use of High-Pressure Liquid Chromatography in Pharmacology and Toxicology *Phyllis R. Brown*
Chromatographic Separation and Molecular-Weight Distributions of Cellulose and Its Derivatives *Leon Segal*
Practical Methods of High-Speed Liquid Chromatography *Gary J. Fallick*
Measurement of Diffusion Coefficients by Gas-Chromatography Broadening Techniques: A Review *Virgil R. Maynard and Eli Grushka*
Gas-Chromatography Analysis of Polychlorinated Diphenyls and Other Nonpesticide Organic Pollutants *Joseph Sherma*
High-Performance Electrometer Systems for Gas Chromatography *Douglas H. Smith*
Steam Carrier Gas-Solid Chromatography *Akira Nonaka*

Volume 13 Out of Print

Volume 14

Nutrition: An Inviting Field to High-Pressure Liquid Chromatography
    Andrew J. Clifford
Polyelectrolyte Effects in Gel Chromatography *Bengt Stenlund*
Chemically Bonded Phases in Chromatography *Imrich Sebestian and István Halász*
Physicochemical Measurements Using Chromatography *David C. Locke*
Gas-Liquid Chromatography in Drug Analysis *W. J. A. VandenHeuvel and
    A. G. Zacchei*
The Investigation of Complex Association by Gas Chromatography and Related
    Chromatographic and Electrophoretic Methods *C. L. de Ligny*
Gas-Liquid-Solid Chromatography *Antonio De Corcia and Arnaldo Liberti*
Retention Indices in Gas Chromatography *J. K. Haken*

Volume 15

Detection of Bacterial Metabolites in Spent Culture Media and Body Fluids by
    Electron Capture Gas-Liquid Chromatography *John B. Brooks*
Signal and Resolution Enhancement Techniques in Chromatography
    Raymond Annino
The Analysis of Organic Water Pollutants by Gas Chromatography and Gas Chroma-
    tography-Mass Spectrometry *Ronald A. Hites*
Hydrodynamic Chromatography and Flow-Induced Separations *Hamish Small*
The Determination of Anticonvulsants in Biological Samples by Use of High-
    Pressure Liquid Chromatography *Reginald F. Adams*
The Use of Microparticulate Reversed-Phase Packing in High-Pressure Liquid
    Chromatography of Compounds of Biological Interest *John A. Montgomery,
    Thomas P. Johnston, H. Jeanette Thomas, James R. Piper, and
    Carroll Temple Jr.*
Gas-Chromatographic Analysis of the Soil Atmosphere *K. A. Smith*
Kinematics of Gel Permeation Chromatography *A. C. Ouano*
Some Clinical and Pharmacological Applications of High-Speed Liquid Chroma-
    tography *J. Arly Nelson*

Volume 16 Out of Print

Volume 17

Progress in Photometric Methods of Quantitative Evaluation in TLO *V. Pollak*
Ion-Exchange Packings for HPLC Separations: Care and Use *Fredric M. Rabel*

Micropacked Columns in Gas Chromatography: An Evaluation *C. A. Cramers and J. A. Rijks*
Reversed-Phase Gas Chromatography and Emulsifier Characterization *J. K. Haken*
Template Chromatography *Herbert Schott and Ernst Bayer*
Recent Usage of Liquid Crystal Stationary Phases in Gas Chromatography *George M. Janini*
Current State of the Art in the Analysis of Catecholamines *Anté M. Krstulovic*

Volume 18

The Characterization of Long-Chain Fatty Acids and Their Derivatives by Chromatography *Marcel S. F. Lie Ken Jie*
Ion-Pair Chromatography on Normal- and Reversed-Phase Systems *Milton T. W. Hearn*
Current State of the Art in HPLC Analyses of Free Nucleotides, Nucleosides, and Bases in Biological Fluids *Phyllis R. Brown, Anté M. Krstulovic, and Richard A. Hartwick*
Resolution of Racemates by Ligand-Exchange Chromatography *Vadim A. Danankov*
The Analysis of Marijuana Cannabinoids and Their Metabolites in Biological Media by GC and/or GC-MS Techniques *Benjamin J. Gudzinowicz, Michael J. Gudzinowicz, Joanne Hologgitas, and James L. Driscoll*

Volume 19

Roles of High-Performance Liquid Chromatography in Nuclear Medicine *Steven How-Yan Wong*
Calibration of Separation Systems in Gel Permeation Chromatography for Polymer Characterization *Josef Janča*
Isomer-Specific Assay of 2,4-D Herbicide Products by HPLC: Regulaboratory Methodology *Timothy S. Stevens*
Hydrophobic Interaction Chromatography *Stellan Hjertén*
Liquid Chromatography with Programmed Composition of the Mobile Phase *Pavel Jandera and Jaroslav Churaček*
Chromatographic Separation of Aldosterone and Its Metabolites *David J. Morris and Ritsuko Tsai*

Volume 20

High-Performance Liquid Chromatography and Its Application to Protein Chemistry *Milton T. W. Hearn*
Chromatography of Vitamin D_3 and Metabolites *K. Thomas Koshy*

High-Performance Liquid Chromatography: Applications in a Children's Hospital *Steven J. Soldin*

The Silica Gel Surface and Its Interactions with Solvent and Solute in Liquid Chromatography *R. P. W. Scott*

New Developments in Capillary Columns for Gas Chromatography *Walter Jennings*

Analysis of Fundamental Obstacles to the Size Exclusion Chromatography of Polymers of Ultrahigh Molecular Weight *J. Calvin Giddings*

Volume 21

High-Performance Liquid Chromatography/Mass Spectrometry (HPLC/MS) *David E. Grimes*

High-Performance Liquid Affinity Chromatography *Per-Olof Larsson, Magnus Glad, Lennart Hansson, Mats-Olle Månsson, Sten Ohlson, and Klaus Mosbach*

Dynamic Anion-Exchange Chromatography *Roger H. A. Sorel and Abram Hulshoff*

Capillary Columns in Liquid Chromatography *Daido Ishii and Toyohide Takeuchi*

Droplet Counter-Current Chromatography *Kurt Hostettmann*

Chromatographic Determination of Copolymer Composition *Sadao Mori*

High-Performance Liquid Chromatography of K Vitamins and Their Antagonists *Martin J. Shearer*

Problems of Quantitation in Trace Analysis by Gas Chromatography *Josef Novák*

Volume 22

High-Performance Liquid Chromatography and Mass Spectrometry of Neuropeptides in Biologic Tissue *Dominic M. Desiderio*

High-Performance Liquid Chromatography of Amino Acids: Ion-Exchange and Reversed-Phase Strategies *Robert F. Pfeifer and Dennis W. Hill*

Resolution of Racemates by High-Performance Liquid Chromatography *Vadium A. Davankov, Alexander A. Kurganov, and Alexander S. Bochkov*

High-Performance Liquid Chromatography of Metal Complexes *Hans Veening and Bennett R. Willeford*

Chromatography of Carotenoids and Retinoids *Richard F. Taylor*

High Performance Liquid Chromatography *Zbyslaw J. Petryka*

Small-Bore Columns in Liquid Chromatography *Raymond P. W. Scott*

Volume 23

Laser Spectroscopic Methods for Detection in Liquid Chromatography *Edward S. Yeung*

Low-Temperature High-Performance Liquid Chromatography for Separation of
    Thermally Labile Species   *David E. Henderson and Daniel J. O'Connor*
Kinetic Analysis of Enzymatic Reactions Using High-Performance Liquid
    Chromatography   *Donald L. Sloan*
Heparin-Sepharose Affinity Chromatography   *Akhlaq A. Farooqui and
    Lloyd A. Horrocks*
New Developments in Capillary Columns for Gas Chromatography
    Walter Jennings

Volume 24

Some Basic Statistical Methods for Chromatographic Data   *Karen Kafadar and
    Keith R. Eberhardt*
Multifactor Optimization of HPLC Conditions   *Stanley N. Deming, Julie G. Bower,
    and Keith D. Bower*
Statistical and Graphical Methods of Isocratic Solvent Selection for Optimal
    Separation in Liquid Chromatography   *Haleem J. Issaq*
Electrochemical Detectors for Liquid Chromatography   *Ante M. Krstulović
    Henri Colin, and Georges A. Guiochon*
Reversed-Flow Gas Chromatography Applied to Physicochemical Measurements
    Nicholas A. Katsanos and George Karaiskakis
Development of High-Speed Countercurrent Chromatography   *Yoichiro Ito*
Determination of the Solubility of Gases in Liquids by Gas-Liquid Chromatography
    Jon F. Parcher, Monica L. Bell, and Ping J. Lin
Multiple Detection in Gas Chromatography   *Ira S. Krull, Michael E. Swartz, and
    John N. Driscoll*

Volume 25

Estimation of Physicochemical Properties of Organic Solutes Using HPLC Retention
    Parameters   *Theo L. Hafkenscheid and Eric Tomlinson*
Mobile Phase Optimization in RPLC by an Iterative Regression Design
    Leo de Galan and Hugo A. H. Billiet
Solvent Elimination Techniques for HPLC/FT-IR   *Peter R. Griffiths and
    Christine M. Conroy*
Investigations of Selectivity in RPLC of Polycyclic Aromatic Hydrocarbons
    Lane C. Sander and Stephen A. Wise
Liquid Chromatographic Analysis of the Oxo Acids of Phosphorus
    Roswitha S. Ramsey
HPLC Analysis of Oxypurines and Related Compounds   *Katsuyuki Nakano*
HPLC of Glycosphingolipids and Phospholipids   *Robert H. McCluer,
    M. David Ullman, and Firoze B. Jungalwala*

Volume 26

RPLC Retention of Sulfur and Compounds Containing Divalent Sulfur
    Hermann J. Möckel
The Application of Fleuric Devices to Gas Chromatographic Instrumentation
    Raymond Annino
High Performance Hydrophobic Interaction Chromatography *Yoshio Kato*
HPLC for Therapeutic Drug Monitoring and Determination of Toxicity
    Ian D. Watson
Element Selective Plasma Emission Detectors for Gas Chromatography
    A. H. Mohamad and J. A. Caruso
The Use of Retention Data from Capillary GC for Qualitative Analysis: Current Aspects *Lars G. Blomberg*
Retention Indices in Reversed-Phase HPLC *Roger M. Smith*
HPLC of Neurotransmitters and Their Metabolites *Emilio Gelpi*

Volume 27

Physicochemical and Analytical Aspects of the Adsorption Phenomena Involved in GLC *Victor G. Berezkin*
HPLC in Endocrinology *Richard L. Patience and Elizabeth S. Penny*
Chiral Stationary Phases for the Direct LC Separation of Enantiomers *William H. Pirkle and Thomas C. Pochapsky*
The Use of Modified Silica Gels in TLC and HPTLC *Willi Jost and Heinz E. Hauck*
Micellar Liquid Chromatography *John G. Dorsey*
Derivatization in Liquid Chromatography *Kazuhiro Imai*
Analytical High-Performance Affinity Chromatography *Georgio Fassina and Irwin M. Chaiken*
Characterization of Unsaturated Aliphatic Compounds by GC/Mass Spectrometry
    Lawrence R. Hogge and Jocelyn G. Millar

Volume 28

Theoretical Aspects of Quantitative Affinity Chromatography: An Overview
    Alain Jaulmes and Claire Vidal-Madjar
Column Switching in Gas Chromatography *Donald E. Willis*
The Use and Properties of Mixed Stationary Phases in Gas Chromatography
    Gareth J. Price
On-Line Small-Bore Chromatography for Neurochemical Analysis in the Brain
    William H. Church and Joseph B. Justice, Jr.

The Use of Dynamically Modified Silica in HPLC as an Alternative to Chemically Bonded Materials *Per Helboe, Steen Honoré Hansen, and Mogens Thomsen*
Gas Chromatographic Analysis of Plasma Lipids *Arnis Kuksis and John J. Myher*
HPLC of Penicillin Antibiotics *Michel Margosis*

Volume 29

Capillary Electrophoresis *Ross A. Wallingford and Andrew G. Ewing*
Multidimensional Chromatography in Biotechnology *Daniel F. Samain*
High-Performance Immunoaffinity Chromatography *Terry M. Phillips*
Protein Purification by Multidimensional Chromatography *Stephen A. Berkowitz*
Fluorescence Derivitization in High-Performance Liquid Chromatography *Yosuke Ohkura and Hitoshi Nohta*

Volume 30

Mobile and Stationary Phases for Supercritical Fluid Chromatography *Peter J. Schoenmakers and Louis G. M. Uunk*
Polymer-Based Packing Materials for Reversed-Phase Liquid Chromatography *Nobuo Tanaka and Mikio Araki*
Retention Behavior of Large Polycyclic Aromatic Hydrocarbons in Reversed-Phase Liquid Chromatography *Kiyokatsu Jinno*
Miniaturization in High-Performance Liquid Chromatography *Masashi Goto, Toyohide Takeuchi, and Daido Ishii*
Sources of Errors in the Densitometric Evaluation of Thin-Layer Separations with Special Regard to Nonlinear Problems *Viktor A. Pollak*
Electronic Scanning for the Densitometric Analysis of Flat-Bed Separations *Viktor A. Pollak*

Volume 31

Fundamentals of Nonlinear Chromatography: Prediction of Experimental Profiles and Band Separation *Anita M. Katti and Georges A. Guiochon*
Problems in Aqueous Size Exclusion Chromatography *Paul L. Dubin*
Chromatography on Thin Layers Impregnated with Organic Stationary Phases *Jiri Gasparic*
Countercurrent Chromatography for the Purification of Peptides *Martha Knight*
Boronate Affinity Chromatography *Ram P. Singhal and S. Shyamali M. DeSilva*
Chromatographic Methods for Determining Carcinogenic Benz(c)-acridine *Noboru Motohashi, Kunihiro Kamata, and Roger Meyer*

Volume 32

Porous Graphitic Carbon in Biomedical Applications *Chang-Kee Lim*
Tryptic Mapping by Reversed Phase Liquid Chromatography *Michael W. Dong*
Determination of Dissolved Gases in Water by Gas Chromatography
    Kevin Robards, Vincent R. Kelly, and Emilios Patsalides
Separation of Polar Lipid Classes into Their Molecular Species Components by
    Planar and Column Liquid Chromatography *V. P. Pchelkin and
    A. G. Vereshchagin*
The Use of Chromatography in Forensic Science *Jack Hubball*
HPLC of Explosives Materials *John B. F. Lloyd*

Volume 33

Planar Chips Technology of Separation Systems: A Developing Perspective in
    Chemical Monitoring *Andreas Manz, D. Jed Harrison, Elizabeth Verpoorte,
    and H. Michael Widmer*
Molecular Biochromatography: An Approach to the Liquid Chromatographic
    Determination of Ligand–Biopolymer Interactions *Irving W. Wainer and
    Terence A. G. Noctor*
Expert Systems in Chromatography *Thierry Hamoir and D. Luc Massart*
Information Potential of Chromatographic Data for Pharmacological Classification
    and Drug Design *Roman Kaliszan*
Fusion Reaction Chromatography: A Powerful Analytical Technique for Condensation Polymers *John K. Haken*
The Role of Enantioselective Liquid Chromatographic Separations Using Chiral
    Stationary Phases in Pharmaceutical Analysis *Shulamit Levin and Saleh
    Abu-Lafi*

Volume 34

High-Performance Capillary Electrophoresis of Human Serum and Plasma Proteins
    Oscar W. Reif, Ralf Lausch, and Ruth Freitag
Analysis of Natural Products by Gas Chromatography/Matrix Isolation/Infrared
    Spectrometry *W. M. Coleman III and Bert M. Gordon*
Statistical Theories of Peak Overlap in Chromatography *Joe M. Davis*
Capillary Electrophoresis of Carbohydrates *Ziad El Rassi*
Environmental Applications of Supercritical Fluid Chromatography
    Leah J. Mulcahey, Christine L. Rankin, and Mary Ellen P. McNally
HPLC of Homologous Series of Simple Organic Anions and Cations
    Norman E. Hoffman
Uncertainty Structure, Information Theory, and Optimization of Quantitative Analysis in Separation Science *Yuzuru Hayashi and Rieko Matsuda*

Volume 35

Optical Detectors for Capillary Electrophoresis   *Edward S. Yeung*
Capillary Electrophoresis Coupled with Mass Spectrometry   *Kenneth B. Tomer, Leesa J. Deterding, and Carol E. Parker*
Approaches for the Optimization of Experimental Parameters in Capillary Zone Electrophoresis   *Haleem J. Issaq, George M. Janini, King C. Chan, and Ziad El Rassi*
Crawling Out of the Chiral Pool: The Evolution of Pirkle-Type Chiral Stationary Phases   *Christopher J. Welch*
Pharmaceutical Analysis by Capillary Electrophoresis   *Sam F. Y. Li, Choon Lan Ng, and Chye Peng Ong*
Chromatographic Characterization of Gasolines   *Richard E. Pauls*
Reversed-Phase Ion-Pair and Ion-Interaction Chromatography   *M. C. Gennaro*
Error Sources in the Determination of Chromatographic Peak Size Ratios   *Veronika R. Meyer*

Volume 36

Use of Multivariate Mathematical Methods for the Evaluation of Retention Data Matrices   *Tibor Cserháti and Esther Forgács*
Separation of Fullerenes by Liquid Chromatography: Molecular Recognition Mechanism in Liquid Chromatographic Separation   *Kiyokatsu Jinno and Yoshihiro Saito*
Emerging Technologies for Sequencing Antisense Oligonucleotides: Capillary Electrophoresis and Mass Spectrometry   *Aharon S. Cohen, David L. Smisek, and Bing H. Wang*
Capillary Electrophoretic Analysis of Glycoproteins and Glycoprotein-Derived Oligosaccharides   *Robert P. Oda, Benjamin J. Madden, and James P. Landers*
Analysis of Drugs of Abuse in Biological Fluids by Liquid Chromatography   *Steven R. Binder*
Electrochemical Detection of Biomolecules in Liquid Chromatography and Capillary Electrophoresis   *Jian-Ge Chen, Steven J. Woltman, and Steven G. Weber*
The Development and Application of Coupled HPLC-NMR Spectroscopy   *John C. Lindon, Jeremy K. Nicholson, and Ian D. Wilson*
Microdialysis Sampling for Pharmacological Studies: HPLC and CE Analysis   *Susan M. Lunte and Craig E. Lunte*

Volume 37

Assessment of Chromatographic Peak Purity   *Muhammad A. Sharaf*
Fluorescence Detectors in HPLC   *Maria Brak Smalley and Linda B. McGown*
Carbon-Based Packing Materials for Liquid Chromatography: Structure, Performance, and Retention Mechanisms   *John H. Knox and Paul Ross*

Carbon-Based Packing Materials for Liquid Chromatography: Applications
  Paul Ross and John H. Knox
Directly Coupled (On-Line) SFE-GC: Instrumentation and Applications *Mark D. Burford, Steven B. Hawthorne, and Keith D. Bartle*
Sample Preparation for Gas Chromatography with Solid-Phase Extraction and Solid-Phase Microextraction *Zelda E. Penton*
Capillary Electrophoresis of Proteins *Tim Wehr, Robert Rodriguez-Diaz, and Cheng-Ming Liu*
Chiral Micelle Polymers for Chiral Separations in Capillary Electrophoresis
  Crystal C. Williams, Shahab A. Shamsi, and Isiah M. Warner
Analysis of Derivatized Peptides Using High-Performance Liquid Chromatography and Capillary Electrophoresis *Kathryn M. De Antonis and Phyllis R. Brown*

Volume 39

Theory of Field Flow Fractionation *Michel Martin*
Particle Simulation Methods in Separation Science *Mark R. Schure*
Mathematical Analysis of Multicomponent Chromatograms *Attila Felinger*
Determination of Association Constants by Chromatography and Electrophoresis
  Daniel W. Armstrong
Method Development and Selectivity Optimization in High Performance Liquid Chromatography *H.A.H. Billiet and G. Rippel*
Chemical Equilibria in Ion Chromatography: Theory and Applications
  Péter Hajós, Ottó Horváth, and Gabriella Révész
Fundamentals of Simulated Moving Bed Chromatography Under Linear Conditions
  Guoming Zhong and Georges Guiochon

ADVANCES IN Chromatography

VOLUME 38

1
Band Spreading in Chromatography: A Personal View

John H. Knox *University of Edinburgh, Edinburgh, United Kingdom*

I.	COLUMN STRUCTURE	2
II.	EARLY THEORIES	5
III.	FIRST MEETING WITH CALVIN GIDDINGS—REDUCED PARAMETERS	11
IV.	OBSTRUCTED DIFFUSION AND THE B-TERM	12
V.	EDDY DIFFUSION, COUPLING, RADIAL DISPERSION, AND THE INFINITE DIAMETER EFFECT	15
VI.	NONEQUILIBRIUM BETWEEN THE STATIONARY AND MOVING ZONES OR PHASES	21
VII.	SEPARATION OF THE A- AND C-TERMS	31
VIII.	EXCLUSION CHROMATOGRAPHY	36
IX.	SLOW CHEMICAL EQUILIBRATION AND WIDE PEAKS	39
X.	BAND SPREADING IN ELECTROSEPARATION SYSTEMS	41
XI.	CONCLUDING REMARKS	45
	REFERENCES	47

Calvin Giddings made some of the most important contributions to the field of band spreading in chromatography. His simple random walk theory set out the basis of band spreading in terms which everyone could understand, while his nonequilibrium theory elucidated the underlying phenomena in clear mathematical terms and provided rigorous equations which could be tested experimentally. This major contribution to chromatography is set out in his book *Dynamics of Chromatography—Part 1* [1].

There is no question that Cal Giddings has been the major influence on my thinking about chromatography, although it was with Howard Purnell that I first became involved in gas chromatography in 1953. Howard and I were research students together in Cambridge, and in 1953 we constructed a rudimentary gas chromatograph with which we were able to surprise our fellow research students by correctly determining the quantities of butane and isobutane in mixtures that they had made up for us. One of our colleagues was so astonished that he suggested that we must have seen has lab notebook, and we had to do the experiment again to convince him. We were greatly excited by the potential of gas chromatography and almost immediately applied it with effect in our own researches into gas-phase hydrocarbon pyrolysis (Howard) and hydrocarbon combusiton and chlorination (myself). This was some years before Cal made his first theoretical contributions around 1957 [2]. I would like to pay tribute to his influence in this review, which is not intended to be comprehensive but will, I hope, give a useful, if ideosyncratic, overview of the field of band dispersion in chromatography, particularly liquid chromatography.

I. COLUMN STRUCTURE

Before proceeding with the main part of this review it is useful to clarify some basic ideas regarding the structure of the chromatography column. Gas chromatography (GC) was the first chromatographic method to which the term "high performance" could be applied, in that the operating conditions were carefully controlled (temperature, flow rate, pressure drop, particle size, etc.). In GC it is a simple matter to distinguish between the mobile and stationary phases. Retention is governed by the relative volumes of the phases, and band spreading, as shown by van Deemter, Zuiderweg, and Klinkenberg [3], comes from three main sources: (1) eddy diffusion, (2) axial diffusion, and (3) slow mass transfer within the stationary phase. These ideas were taken over by high-performance liquid chromatography (HPLC) when it was introduced in the late 1960s [4,5].

The situation in LC is more complex than in GC, largely because both the mobile and stationary phases are liquid or quasi-liquid, and therefore less clearly distinguished in their physical properties than a gas and a liquid. In particular, the diffusion rates in both phases are rather similar. It now becomes necessary to distinguish between the true thermodynamic phases, which determine retention (although not in size exclusion chromatography where everything occurs in a single phase), and the moving and static zones, which largely determine band spreading. The distinction between thermodynamic phases and kinetic zones is illustrated in Fig. 1.

It is particularly important to note that the mobile phase is now divided into two parts, the flowing part in the interparticle zone, and the stagnant part held within the pores of the packing material. The particles of packing material thus comprise a support structure (usually silica), a stationary phase (often chemically bonded to the support structure), and stagnant mobile phase. Generally the pores of the packing material for LC are very small, with dimensions from 10 to 50 nm, and its specific surface areas are from 50 to 300 m^2/g. The stationary

THERMODYNAMIC PHASES	SECTORS	KINETIC ZONES
Mobile Phase	Flowing Mobile Phase (in the interparticle space)	Moving Zone
	Stagnant Mobile Phase (within the pores of the particles of packing material)	Static Zone
Stationary Phase	Stationary Phase (on the surface of the support structure)	
Support Structure		

Fig. 1 Structure of the chromatographic column.

phase is often in the form of a monolayer, for example octadecyl-bonded to a silica support, and is just a few nanometers. This "thickness" is of the same order as the "thickness" of a typical liquid-liquid phase boundary.

The mobile phase or eluant is usually a mixture (e.g., acetonitrile/water), and one of its components will inevitably be preferentially adsorbed by the bonded layer. The net result is that the bonded layer is swollen by selective sorption of eluant. It now becomes difficult to distinguish clearly between the mobile and stationary phases, since the bonded layer contains adsorbed mobile phase. Indeed, if one were to start from the surface of the support and go though the stationary phase perpendicular to the support surface, one would observe a gradual change in composition from something approaching that of the original bonded stationary phase, through bonded phase selectively swollen by eluant, to pure eluant. This "local difficulty" mainly affects the thermodynamic interpretation of retention, and makes it difficult to be clear about the meaning of quantities like $d \ln k'/dT$, which for a true two-phase system would equal $\Delta H/RT^2$, ΔH being the heat of transfer from the mobile to the stationary phase. Fortunately, this complication does not greatly affect the understanding and interpretation of band spreading.

Because of the distinction between thermodynamic phases and kinetic zones one has to recognize two ways in which the linear velocity may be defined. The linear velocity averaged over the mobile phase is denoted by u_m, while the linear velocity averaged over the moving zone is denoted by u_z. Clearly u_z is greater than u_m. We also have to define concentrations in two ways. Following Giddings, concentrations will be defined per unit volume of the column as a whole. The total concentration will be denoted by c, the concentrations in the mobile and stationary phases by c_m and c_s, and the concentrations in the static and moving zones by c_{mz} and c_{sz}. Equilibrium concentrations, that is concentrations which would be achieved if the flow were stopped and equilibration allowed to occur between the phases or zones (but with no axial diffusion), will be identified by an asterisk. We can thus write

Phases	$c = c_m + c_s$		$= c_m^* + c_s^*$	(1a)
Zones	$c = c_{mz} + c_{sz}$		$= c_{mz}^* + c_{sz}^*$	(1b)
	concentrations in general		equilibrium concentrations	

II. EARLY THEORIES

Early theories of band spreading, in the 1940s, were somewhat mathematical and generally did not address the molecular phenomena which were at the bottom of the chromatographic process [6–8]. The mass balance approach resulted in various general equations, but practicing chromatographers took little account of them and on the whole did not understand them.

Martin and Synge [9] took a more practical approach to the theory of chromatography, and their remarkable paper in 1941, giving the first description of liquid-liquid partition chromatography, laid the foundations of the theory we use today. They also predicted the development of both gas chromatography (GC) and high-performance liquid chromatography (HPLC):

> Very refined separations of volatile substances should therefore be possible in a column in which permanent gas is made to flow over gel impregnated with a non-volatile solvent in which the substances to be separated approximately obey Raoult's Law.
>
> Thus the smallest H.E.T.P. should be obtainable by using very small particles and a high pressure difference across the length of the column.

They likened the chromatographic column to a series of equilibration stages, in the same way that a real distillation column is likened to a series of idealized stages or plates, hence their use of the term HETP meaning the height equivalent to a theoretical plate. Large distillation columns do indeed consist of just such a series of plates. Vapor enters any given plate from the plate below through a bubble cap and (ideally) condenses into the plate without change of composition. The vapor in equilibrium with the liquid in that plate goes on to the plate above, while excess liquid in any plate overflows to the plate below. When no distillate is taken from the top of the column, each plate contains liquid and vapor in equilibrium. The liquid in each plate has the same composition (ideally) as the vapor in the plate below and under conditions of total reflux the flow of vapor upward is exactly balanced by the flow of liquid downward. Although Martin and Synge borrowed the plate concept from distillation, the way in which chromatography works is quite different from that of distillation.

Each stage or plate of the chromatographic column, as envisaged by Martin and Synge, is identical, and contains defined quantities of two immiscible phases called the mobile phase and the stationary phase. Chromatography takes place by injecting a small quantity of sample mixture into the first plate and then sequentially adding small portions of mobile phase to the column to "elute" the sample. Chromatography differs fundamentally from distillation in that (1) the fate of the sample is all-important, not what happens to the mobile and stationary phases whose composition is the same throughout the whole column, and (2) product is not taken continuously from one end of the column as in distillation; the sample moves along the column and eventually emerges as one or more bands from the end of the column. Any comparison of the numbers of plates in a distillation column (where 100 would be considered large) with the number in a chromatographic column (where one needs at least 1000 to get any reasonable separation) is meaningless.

In more detail, Martin and Synge imagined that a minute quantity of analyte sample was initially placed on the first plate and then allowed to distribute itself according to its equilibrium distribution ratio between the two phases. A "plate volume" of mobile phase was then added to the first plate so that the mobile phase in every plate was displaced instantaneously to the next plate. The plates were then allowed to reequilibrate. The process of adding one plate volume of mobile phase followed by equilibration was repeated numerous times. The packet of analyte molecules then moved gradually down the column and finally emerged from the outlet end. Solutes, with different partition ratios between the phases, would move at different rates and so become separated. But the process of equilibration and transfer inevitably spreads out any band of solute. The original process defined by Martin and Synge may be refined and made more realistic by adding the mobile phase quasi-continuously in infinitesimally small portions. The two main predictions made by the plate theory were as follows.

1. The peak maximum for any solute, moves at a rate determined by its equilibrium partition ratio between the mobile and stationary phases, namely,

Rate of movement of solute/rate of movement of mobile phase, R

= fraction of solute in mobile phase at equilibrium

= fraction of total time that average solute molecule spends in mobile phase (2)

The quantity R is called the retention ratio and is equivalent to the R_F-value of thin-layer or paper chromatography (apart from minor corrections for demixing at the solvent front). It is directly related to the capacity ratio k' more commonly used in GC and HPLC. By definition,

$$R = \frac{c_m^*}{c_m^* + c_s^*} \quad \text{and} \quad k' = \frac{c_s^*}{c_m^*} \tag{3}$$

whence

$$R = \frac{1}{1+k'} \quad \text{and} \quad k' = \frac{1-R}{R} \tag{4}$$

Most chromatographers now use the k'-notation, but Giddings always used the R-notation.

2. After a band of solute has moved through a reasonable number of plates (say over 100) its concentration profile approximates to Gaussian. With quasi-continuous elution, the standard deviation of the Gaussian profile according to the analysis of Martin and Synge is given very simply:

Standard deviation (measured in plates)
= (number of plates through which band has migrated)$^{1/2}$

This leads to the well-known equations for calculation of the plate height, H, from the bandwidth within the column, σ_z, and the distance migrated, z:

$$H = \frac{\sigma_z^2}{z} \quad \text{or} \quad \sigma_z = (Hz)^{1/2} \tag{5a, 5b}$$

Martin and Synge recognized, of course, that a real chromatographic column did not consist of plates. However, the idea of a series of plates operating as Martin and Synge first proposed was actually embodied in the "Craig machine," invented by Lyman Craig [10]. This machine consisted of a series or rack of ingeniously designed glass vessels, machines with up to 1000 vessels were constructed. Each vessel contained two immiscible liquids, the quantities being the same for each vessel. The rack of vessels could either be shaken to mix the contents, or it could be tipped to transfer the upper immiscible layer (or most of it) to the next vessel. To operate the machine, a small quantity of analyte mixture was added to the first vessel. The solutions in the vessels were shaken to equilibrate them; then, by tipping the rack, the upper liquid layer from each vessel was transferred to the next vessel. The vessels were then reequilibrated. The sequence of equilibration and transfer was repeated a large number of times so that the solutes moved gradually down the train of vessels. Substances with different partition ratios could then be separated. The Craig machine exactly reproduces the plate model of Martin and Synge (operated in the discontinuous mode): each vessel is equivalent to one of Martin and Synge's theoretical plates.

The analysis by Martin and Synge recognized that the main reason for the plate having a finite "height" was that perfect equilibrium could never exist between the mobile and stationary phases when one of the phases was flowing over the other. Equilibration takes a finite time, and while it is occurring the mobile phase moves on. As Martin and Synge state in their definition of the theoretical plate: "The H.E.T.P. is defined as the thickness of the layer such that the solution issuing from it is in equilibrium with the mean concentration of solute in the non-mobile phase throughout the layer." To put it another way, the concentration profile in the mobile phase runs very slightly ahead of the concentration profile in the stationary phase. This simple idea is the basis of Giddings' nonequilibrium theory.

Nevertheless, as pointed out by Giddings, Martin and Synge's theory had serious inadequacies. In particular, it did not allow for the effects of axial molecular diffusion, nor for any sources of dispersion other than nonequilibrium between the mobile and stationary phases. Yet, since diffusional spreading obeys the same laws (Fick's laws) as spreading due to slow equilibration, Eq. (5), relating the peak width to the plate height, is still correct even when additional sources of band spreading are added. The recognition of the importance of these other dispersion mechanisms came with the seminal work of van Deemter, Zuiderweg, and Klinkenberg [3] in 1956. They used chemical engineering principles to derive their famous equation for the plate height in gas chromatography. They identified two sources of band spreading in addition to the nonequilibrium identified by Martin and Synge. Their original equation, publicized by Keulemans [11] in his book (1959), took the form

$$H = 2\lambda d_p + \frac{2\gamma D_m}{u} + \left(\frac{8}{\pi^2}\right)\frac{R(1-R)d_f^2}{D_s} \tag{6a}$$

$$= A + \frac{B}{u} + Cu \tag{6b}$$

The three contributions to H arose from eddy diffusion, axial diffusion, and slow mass transfer in the stationary phase. Later [12] the original equation was modified, and a further contribution added for slow mass transfer in the mobile phase to give

$$H = 2\lambda d_p + \frac{2\gamma D_m}{u} + \left(\frac{1}{100}\right)\frac{(1-R)^2 d_p^2 u}{D_m} + \left(\frac{2}{3}\right)\frac{R(1-R)d_f^2 u}{D_s} \tag{7a}$$

$$= A + \frac{B}{u} + C_m u + C_s u \tag{7b}$$

These equations still form the basis of our interpretation of plate height velocity data, although some important modifications have been made as discussed below. The numerical values of the constants in the equations, particularly for the C-term contributions, have not generally been taken quantitatively. Most chromatographers have simply determined the values of A, B, and C empirically. The most important practical conclusion arising from the van Deemter equation is that H had a minimum value which occurs at an optimum linear flow rate:

$$H_{min} = A + 2(BC)^{1/2}, \quad u_{opt} = \left(\frac{B}{C}\right)^{1/2} \tag{8}$$

It was emphasized by Keulemans and many others that one should operate any gas chromatograph close to these "optimum conditions." The same applies for liquid chromatography, although until the 1970s it was difficult to achieve this, since separations of reasonable efficiency in a reasonable time required, in the words of Martin and Synge, "very small particles and a high pressure difference across the length of the column." The practical consequences, as shown by Knox and Saleem [13], were that pressures had to be in the region of hundreds of bar, and particles in the micron size range (e.g., 200 bar and 2 μm diameter). Over the last 20 years particle sizes in HPLC have been reduced from 30 to 3 μm, and, according to a very recent advertisement, 2-μm particles are now finally available [14].

In 1958 at the second International Symposium on Chromatography organized by the Gas Chromatography Discussion Group in Amsterdam, a remarkable and, indeed, totally unexpected development occurred. Marcel Golay, a mathematician working for Perkin-Elmer, was asked if he could model gas chromatography more precisely than van Deemter et al. had done. Golay decided to model the simplest geometrical column configuration, the open tube. The result was his famous equation [15].

$$H = \frac{2D_m}{u} + \frac{1}{96}\{11 - 16R + 6R^2\}\frac{d^2 u}{D_m} + \frac{2}{3}\{R(1-R)\}\frac{d_f^2 u}{D_s} \quad (9a)$$

$$= \frac{2D_m}{u} + \frac{1}{96}\frac{1 + 6k' + 11k'^2}{(1+k')^2}\frac{d^2 u}{D_m} + \frac{2}{3}\left\{\frac{k'}{(1+k')^2}\right\}\frac{d_f^2 u}{D_s} \quad (9b)$$

$$= \frac{B}{u} + C_m u + C_s u \quad (9c)$$

For open tubes there is, of course, no A-term, since the stream lines now run straight and parallel to the axis of the tube. But we now see, for the first time, an exact and rigorous calculation of the mobile- and stationary-phase mass transfer terms for a well-defined system. Unfortunately no such equation is derivable for the complex geometry of the packed bed without fairly massive assumptions. Because of this, one has to be somewhat less ambitions in the derivation and interpretation of plate height equations for packed columns. Empirical measurements have to be used, and there is no point in overcomplicating the analysis. It is here that Giddings, in *Dynamics of Chromatography* [1], has been particularly successful treading a path between the oversimplified and the overcomplex.

The Golay equation is in fact a development of earlier equations. Taylor [16] was the first to calculate the dispersion of a packet of solute caused by flow in an open tube, and obtained the well-known result for the plate height of an unretained solute:

$$H = \frac{d^2 u}{96 D_m} \quad (10)$$

This is the limiting case of the Golay formulation with $k' = 0$. Aris [17] independently obtained an equation for dispersion in an open tube for a retained solute but with plug rather than parabolic flow.

$$H = \left\{\frac{k'}{1+k'}\right\}^2 \frac{d^2 u}{16 D_m} \tag{11}$$

The Aris equation is not correct for pressure-driven open tube, but turns out to be important in capillary electrophoresis.

Capillary columns were rapidly taken up in GC and have become the norm for high-efficiency separations and routine GC separations. No such development has taken place in LC. Knox and Gilbert [18] showed why. They examined the conditions for optimum performance in open tubular liquid chromatography following the original analysis for packed columns of Knox and Saleem [13]. They concluded that the bore of the tube required for optimum performance was extremely small, of the order of 1 µm or less. This seemed impracticable at the time, but recently the Dutch school [19,20] has demonstrated that high efficiencies can be achieved in LC using tubes as little as 5 µm in bore. Open tubes will probably never be used for routine LC, but they have come into their own for electrically driven separations where the flow profile is close to that of plug flow and the Aris equation applies. When $k' = 0$, that is when there is no surface adsorption, there is no term in the equation for H arising from slow mass transfer, and the tube bore at first sight appears to be unlimited. This is not, of course, true since self-heating arising from the electric current gives rise to convection and other dispersive effects. The maximum practicable bore lies in the range 100–200 µm.

III. FIRST MEETING WITH CALVIN GIDDINGS—REDUCED PARAMETERS

I first met Calvin Giddings at the International Gas Chromatography Symposium [21], organized by the redoubtable Al Zlatkis and held in Houston, Texas, in January 1963. This was to be the first of a long series of meetings which were subsequently called International Symposia on Advances in Chromatography. The present series of volumes, initiated by Calvin Giddings and Roy Keller in 1965, uses the same title, *Advances in Chromatography* [22]. Al Zlatkis invited me, on this my first trip to the United States, to present work which Lilian Mclaren and I were carrying out on spreading of air peaks in open and packed columns [23]. In our paper we defined what we called "reduced parameters." These reduced or dimensionless parameters were related to

H_{min} and u_{opt} and defined as $h = H/(BC)^{1/2}$ and $v = u/(B/C)^{1/2}$. This formulation enabled us to determine A of the van Deemter equation by curve fitting. However, our reduced parameters were not the same as those derived by Giddings [24]. His were much more fundamental, scaling the plate height to the particle diameter (or the tube diameter for a capillary), and the linear velocity, to the diffusion rate over a particle (the chemical engineer's Peclet number). What everyone now calls the reduced plate height and reduced velocity are defined as

$$\text{Reduced plate height} \quad h = \frac{H}{d_p} \tag{12}$$

$$\text{Reduced velocity, or Peclet number} \quad v = \frac{u d_p}{D_m} \tag{13}$$

By using Gidding's reduced parameters it is possible to compare different chromatographic systems on the same basis (for example, gas and liquid chromatography [25]) since the particle diameter and rate of diffusion in the mobile phase are factored out. Most of the plate height equations are greatly simplified by casting them into reduced forms. Taking the van Deemter and Golay equations as examples, Eqs. (7) and (9) take the forms:

$$h = 2\lambda + \frac{2\gamma}{v} + \frac{1}{100}(1-R)^2 v + \frac{2}{3}R(1-R)\left(\frac{d_f}{d_p}\right)^2 \left(\frac{D_m}{D_s}\right) v \tag{14a}$$

$$= A + \frac{B}{v} + C_m v + C_s v \tag{14b}$$

$$h = \frac{2}{v} + \frac{1}{96}(11 - 16R + 6R^2)v + \frac{2}{3}R(1-R)\left(\frac{d_f}{d}\right)^2 \left(\frac{D_m}{D_s}\right) v \tag{15a}$$

$$= \frac{B}{v} + C_m v + C_s v \tag{15b}$$

IV. OBSTRUCTED DIFFUSION AND THE B-TERM

Cal and I had many fascinating discussions at the first Houston Symposium, often in the company of Howard Purnell. I remember well a vigorous argument we had about pressure correction factors which should be applied to the various terms in the van Deemter equation. Everyone recognized that the C-term for the stationary phase had to be multiplied by the James-Martin pressure correction factor, $(3/2)(\pi^2-1)/(\pi^3-1)$ (where π = inlet over outlet pressure), but Howard and I could not understand why the gas-phase terms had to be corrected by the factor $(9/8)(\pi^4-1)(\pi^2-1)/(\pi^3-1)^2$. We argued for many hours before Cal convinced us that he was right.

This discussion focused my attention on a unique feature of GC in relation to band spreading. It may be noted from Eqs. (7), (9), (14), or (15) that the plate height contributions are divided into those which arise in the gas phase (A, B, and C_m terms), and the single contribution which arises in the liquid or stationary phase (C_s term). These contributions have different dependences upon the pressure since the diffusion coefficient D_m is inversely proportional to pressure while D_s is independent of pressure. Because of this feature it is possible to separate the mobile-phase and stationary-phase contributions to H, and establish their velocity dependences. This can be done by carrying out parallel experiments with the same column and solute(s), but with two different carrier gases or two different pressure regimes. Perrett and Purnell [26] had used the first method as early as 1962. We used both methods in 1972, and showed that by far the largest contribution to H in GC arose from mobile-phase processes [27]. Regrettably this method is not available to LC, and it is more difficult to distinguish between mobile- and stationary-phase contributions to H.

Later in the same conference Cal and I had a key discussion in a taxi. We came up with the elegant idea of measuring B-term dispersion by stopping the flow of gas, allowing the band to disperse within the column, and then restarting the flow. We argued that since the axial diffusion process occurred by molecular diffusion, the rate of diffusion would be unaffected by whether or not the mobile was actually moving. Indeed the velocities of the random movement of molecules between collisions, which is the basis of diffusion, are many thousand times faster than the linear flow rates in gas chromatography. The molecules would therefore be "unaware" of any slow drift while they were diffusing. In the packed column they would of course find their diffusion obstructed by the packing. Typically any molecule experiencing random collisions with other molecules would ultimately encounter the packing material which would bar its random movement in that direction. It would have to "back off," and this would lead to slower overall diffusion than in the absence of the packing.

The effective diffusion coefficient could readily be obtained by plotting the band variance against the time for which the flow was arrested. The gradient of this plot would be γD_m of the van Deemter equation. The arrested elution method promised to be a simple and unequivocal method of determining the obstructive factor γ for packed columns. It was likely to be very much better than trying to measure H at very low flow rates with the attendant problem of maintaining a steady baseline. Following the symposium, Lilian McLaren and I set about making measurements of γD_m in a variety of packed columns, using the arrested elution method. Ethylene was the test solute and nitrogen the carrier gas. We determined D_m by making equivalent measurements in an open tube. The value of γ was about 0.6 for columns packed with glass bead, about 0.5 for columns packed with firebrick, and about 0.7 for columns packed with Celite, Celite being the forerunner of the modern packings based upon calcined diatomaceous earths. This work was presented at the Second International Symposium on Advances in Gas Chromatography [28].

In gas chromatography diffusion in the stationary phase is so slow in comparison with diffusion in the mobile phase (it is some 10^5 times slower) that it can be ignored, and the B-term is that given in the van Deemter equation. This is not true of liquid chromatography where the diffusion rates in both phases are comparable. Diffusion occurs independently in the two phases and the individual dispersions must be summed. We thus obtain for the overall B-term dispersion:

$$\sigma_z^2 = \sigma_z^2 \text{ (mobile phase)} + \sigma_z^2 \text{ (stationary phase)} \tag{16a}$$
$$= 2\gamma_m D_m t_m + 2\gamma_s D_s t_s \tag{16b}$$
$$= t_m(2\gamma_m D_m + 2\gamma_s D_s k') \tag{16c}$$

An effective or average diffusion coefficient can be defined as

$$D_{\text{eff}} = \frac{\sigma_z^2}{2(t_m + t_s)} = \frac{\gamma_m D_m + \gamma_s D_s k'}{1 + k'} \tag{17a}$$

whence

$$\frac{(1+k')D_{\text{eff}}}{D_m} = \gamma_m + \frac{\gamma_s D_s k'}{D_m} \tag{17b}$$

The contribution of axial diffusion to H from Eq. (5) then becomes

$$H_B = \frac{\sigma_z^2}{z} = \frac{2\gamma_m D_m + 2\gamma_s D_s k'}{u_m} = \frac{2D_{\text{eff}}(1+k')}{u_m} \tag{18}$$

In these equations both u_m and k' refer to the two phases. The tortuosity factor for the mobile phase, γ_m, applies to the mobile phase as a whole and with porous particles will be closer to unity than for glass beads. Knox and Scott [29] used the arrested elution method for the determination of the apparent diffusion rates of solutes in HPLC columns. They measured D_{eff}/D_m as a function of k' for oxygenated benzene derivatives and for some polynuclear aromatics chromatographed on ODS Hypersil. They showed, Eq. (17b), that $(1+k')D_{\text{eff}}/D_m$ rose gently with k'. The intercepts of the plots at $k' = 0$ were about 0.9, indicating that the tortuosity factor in the mobile phase was much higher than the 0.6 obtained by Knox and McLaren for glass beads. It appeared to be almost unobstructed by the relatively porous packing (the particle porosity is around 60% for Hypersil). The gradients gave $\gamma_s D_s$, which were about half $\gamma_m D_m$. Contrary to expectation, diffusion in the bonded ODS phase was actually quite fast. This is an area where more data is certainly desirable, and could readily be obtained with modern HPLC equipment.

V. EDDY DIFFUSION, COUPLING, RADIAL DISPERSION, AND THE INFINITE DIAMETER EFFECT

Our discussions at the 1963 Houston Symposium led to an invitation from Cal to spend a sabbatical year in his group at Salt Lake City as a NSF Senior Visiting Research Scientist Fellow. This was an invitation I eagerly accepted. My wife and I were particularly attracted to the idea of working in Salt Lake City because of the excellent skiing which Cal assured me was waiting for us, and which we subsequently greatly enjoyed. It was a particular privilege to work with Cal and his colleagues at this exciting time from January to September 1964. Cal was deeply engaged in finalizing *Dynamics of Chromatography—Part 1*. Part 2 was intended to be a more practical guide to the practice of chromatography but was never published. Perhaps this was for the best, for while the theory of chromatography has remained more or less untarnished by the passage of time, its practice has changed dramatically in 30 years. Part 1 remains a permanent intellectual masterpiece. We all attended fascinating and highly illuminating lectures by Cal on the various sections of the book. They demonstrated Cal's remarkable insight into the chromatographic process.

At this time Cal had just propounded his "coupling theory" of eddy diffusion [30]. While most of us thought that the A-term of the van Deemter equation arose from the dispersive effects of the column packing which made the stream and diffusion paths through the bed tortuous, Cal took a different view. If one starts from the open tube, the stream lines are straight and, if molecules kept strictly to these stream lines, the dispersion would be more or less infinite. Excessive dispersion is prevented only by diffusion across the stream lines, so that over a period of time any given molecule samples all the stream lines in a random fashion. Thus, we obtain the gas-phase C_m-term in the Golay equation (Eq. 9). Another way to avoid the potentially catastrophic dispersion in an open tube, according to Cal, was to mix up the stream lines—for example, by introducing turbulence (an idea later studied and exploited by Pretorius [31]) or to pack the column so that the flow velocity in any stream line would vary rapidly. Thus, far from the packing producing dispersion, it actually provided an efficient means of reducing it. This alternative view allowed Cal to see that in a packed column there were actually two independent methods whereby the catastrophic dispersion, which would otherwise arise from the parallel stream lines of the open tube, could be prevented. The two mechanisms *cooperated to reduce* dispersion rather than *adding to increase* it. Accordingly the plate height contributions should be combined harmonically, not simply added as in the modified van Deemter equation (Eq. 7); that is;

$$H_A = \left\{ \frac{1}{A} + \frac{1}{C_m u} \right\}^{-1} \tag{19}$$

Cal quickly recognized that this equation was oversimplified, and he soon replaced it by a sum of five contributions, each contribution requiring different values of A and C_m depending upon the range and extent of the velocity inequalities which had to be accommodated. This formulation also turned out to be an oversimplification.

For my period at Utah we decided that I should examine the validity of Cal's coupling theory by studying band spreading of unsorbed solutes in columns packed with glass beads. So that I could cover a wide range of velocity above that giving minimum H, I decided to work with a liquid eluant. I judged that it would be impossible to get to high enough u-values with a gaseous eluant. Either I would have to work with very large particles or at very high pressures with extremely fast recording equipment. Neither alternative seemed attractive. Furthermore with a gaseous eluant, one would soon get into the turbulent regime, which would introduce yet another mechanism for containing band spreading. Thus my first liquid chromatograph was built [32]. It was in fact the first LC system to be used in Cal's laboratory. Glass columns were packed with glass beads of various diameters. The eluant, 10% aqueous potassium nitrate, was driven through the column by a head of mercury via a restricting capillary. The detector was made with a tungsten filament bulb equidistant from two photocells. One photocell viewed the lamp through a dummy cell, and the other looked through the eluate from the column. The solute was potassium permanganate. The work showed clearly that h rose gradually as velocity increased with a velocity exponent of about 1/3 declining as the velocity rose. Curves were parallel and independent of the column to particle diameter ratio, ρ. From this it was concluded that transcolumn inhomogeneities were the main cause of A-term broadening. At very high reduced velocities (above about 4000) h actually declined, probably due to the onset of turbulence (the Reynolds number being about 1/1000 of the reduced velocity). The study showed once and for all that the original velocity-independent A-term of the van Deemter equation was incorrect and should be replaced by a term which was weakly dependent upon velocity. The best representation of the data was given by an integral form replacing the five-term summation.

This work, with glass beads of different diameters, showed a strong dependence of h upon the column to particle-diameter ratio, ρ. Later Parcher [33], who used particles of the same diameter in columns of different diameters, found less dependence. A step change in efficiency occurred at ρ around 8. Columns with $\rho > 8$ showed h-values about twice those of columns with $\rho < 8$. The log h versus log ν plots were still parallel, showing a velocity exponent of around 1/3. The magnitude of the A-term clearly depended upon how well the column was packed, although its velocity dependence did not. Knox and Parcher found a simple representation of the data, over a range of ν from 10 to 3000, namely

$$h_A = \left(\frac{1}{A} + \frac{1}{C\nu^{1/3}} \right)^{-1} \quad (20)$$

B-term dispersion was negligible under these conditions. This equation resembles that previously proposed by Huber and Hulsman [4] on the basis of chemical engineering data, except that their exponent was 1/2. For columns with $\rho > 8$, and over the narrower range of reduced velocity 30 to 300, the data were well fitted by

$$h_A = 0.33 v^{0.35} \tag{21}$$

Knox and Parcher also showed that with a very wide column, which was sampled only from the core region, the plate height was much less dependent upon velocity, with an exponent around 0.15. They called this an "infinite diameter column." The comparison with a walled column is shown in Fig. 2.

For the infinite diameter column, a 100-fold increase in velocity was required to double h. This result confirmed the original contention of Knox [32] that the main contribution to the A-term in regular columns must come from transcolumn inhomogeneities in the packing. Most probably these arise from different packing structures near the wall, which lead to substantial variations in local flow rates. With no wall influence such transcolumn inhomogeneities are absent. The term "infinite diameter column" is now generally applied to any column which is sufficiently wide that a centrally injected sample will never reach the walls.

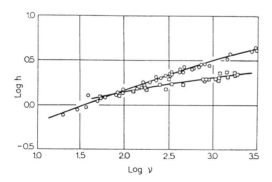

Fig. 2 Comparison of (h, v) plots for walled (o) and "infinite diameter" columns (□) from Knox and Parcher [33]. Walled column $\rho > 8$; infinite diameter column 40-mm diameter with central take-off. Both packed with 480-μm glass beads. (Reproduced with permission of authors and publisher.)

Horne, Knox, and McLaren [34] examined the literature on radial dispersion in packed beds and concluded that this could be characterized by a radial plate height equation (in reduced form) given by

$$h_r = 0.15 + \frac{1.2}{\nu} \tag{22a}$$

$$= A_r + \frac{B}{\nu} \tag{22b}$$

The B-term was the same as for normal axial dispersion and the underlying process was the same, namely obstructed molecular diffusion. The radial A-term, however, was much smaller and had no velocity dependence; under practical operating conditions in HPLC, it was about 10 times smaller than the axial A-term. They could explain the radial A-term by a simple stream-splitting model which predicted a value between 0.1 and 0.2 in agreement with the literature data.

Subsequently Knox, Laird, and Raven [35] measured radial dispersion directly using a 12-mm-bore glass tube packed with 64-µm glass beads. The column terminated in a full-width frit below which was mounted a dual-electrode electrochemical detector. One of the electrodes was fixed, and the other could be moved by means of a micrometer so as to sample different radial positions. This experiment showed that the radial A-term in a well-packed chromatographic column was even smaller than had been anticipated, but confirmed that the radial B-term was normal. They obtained

$$h_r = 0.06 + \frac{1.4}{\nu} \tag{23}$$

The radial A-term was considerably smaller than that derived by Knox, Horne, and McLaren [34] from the chemical engineering literature.

Knox, Laird, and Raven [35] also showed convincingly that, with conventional equipment, wall effects have a significant effect on band spreading. With *central injection* into columns dry-packed with 20-µm particles, performance deteriorated seriously whenever the flow rate was low enough that solute diffused into the wall region. They showed that in most practical HPLC columns (say 5-mm-bore, 100-mm-long packed with 3- or 5-µm particles) there is insufficient radial dispersion to achieve transcolumn mixing, and elements of mobile phase which starts near the core of the column will never disperse sufficiently to reach the walls, and vice versa. They gave the condition for a column to behave as if of infinite diameter as

$$\frac{d_c^2}{Ld_p} > 1 + \frac{32}{v} \qquad (24)$$

It has often been misrepresented that columns for which Eq. (24) holds behave as if of infinite diameter. This is regrettably incorrect. A column will only behave in the infinite diameter mode if none of the solute which passes to the the detector has arrived there via the wall region. When Eq. (24) holds, this is achieved only when injection or sampling of eluate (or better, both) is effected by a curtain flow system. These conditions are never met in commercial HPLC systems, where one injects via a valve and takes all the eluate to the detector.

Typically in HPLC we are operating at v-values in the region of 5 to 15 (e.g., $u = 1$ mm/s, $d_p = 3$ to 5 µm, $D_m = 0.5$ to 1.0×10^{-9} m²/s). The factor $1 + 32/v$ is then in the range 3 to 7. If we take a typical 100×4.7 mm column packed with 3-µm particles then d_c^2/Ld_p, is about 75, substantially greater than required for "infinite diameter" operation. There is thus little exchange between the core and wall regions in standard HPLC columns, and they could well be operated in the infinite diameter mode with modified injection and/or detection systems. Higher efficiencies would then be obtained.

The overall result of the studies of A-term broadening was the emergence of the so-called Knox equation for the plate height. In reduced form this is normally written

$$h = \frac{B}{v} + Av^{1/3} + Cv \qquad (25)$$

An example of a plot of log h against log v taken from the paper by Knox, Laird, and Raven [35] is shown in Fig. 3. The values of A, B, and C giving the best fit to the data were 0.4, 3, and 0.1–0.2, depending upon k'.

A and B seemed to be independent of k' although C increased with k'. Later Bristow and Knox [36] recommended that the "standard" values of A, B, and C should be taken as 1, 2, and 0.1. This has sometimes been misrepresented as implying that there is a theoretical minimum value of h at just over 2. There is, in fact, no theoretical justification for this. These values of A, B, and C were simply recommended as a yardstick against which column performance could be compared and gross differences from the norm identified. The best columns have A in the

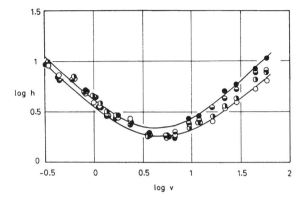

Fig. 3 (h, v) plot for 21.5-mm Spherisorb Alumina over a wide velocity range from Knox, Laird, and Raven [35]. Column: 82 × 5 mm. Solutes: aromatic hydrocarbons, eluent: hexane. (Reproduced with permission of authors and publisher).

region of 0.4, while C according to the theory should be about 0.02 or less. There is, however, no way of avoiding the B-term: B cannot reasonably be less than about 2. With these optimal values, the theoretical minimum for h would be in the region of 1.25 at a reduced velocity of about 5. There are, however, ways of improving on this. In particular, the use of electrodrive instead of pressure drive can reduce A still further, giving the possibility of h-values in the region of 1 and maybe even less. This is discussed later.

A cautionary note must be added: the eddy diffusion contribution to H is, in reality, only well established for nonporous nonretaining particles (glass beads). It is much more difficult to determine how the A-term depends upon velocity for porous particles, since the A- and C-terms both have velocity dependence. The lower exponent of the A-term dependence suggests that the best way to eliminate it and determine C on its own will be to work at very high reduced velocities. This was examined by Knox and Scott [29] and is discussed below.

VI. NONEQUILIBRIUM BETWEEN THE STATIONARY AND MOVING ZONES OR PHASES

Giddings' major contribution to the theory of chromatography was undoubtedly his nonequilibrium theory [37], which enabled him to evaluate mass transfer contributions to H for a wide variety of configurations and velocity profiles in the stationary and mobile phases. Thus, it can deal with variations of velocity within a phase, as in the derivation of the mobile-phase contribution to H in the Golay and Aris equations, it can deal with variations in velocity between phases (e.g., the mobile and stationary phases in gas chromatography) or between zones (e.g., the mobile zone and the stationary zone in liquid chromatography), and it can deal with slow chemical reaction between different forms of a solute undergoing chromatography.

Giddings explains his theory in Chapters 3 and 4 of *Dynamics of Chromatography*. Qualitatively the basic situation is illustrated in Fig. 4 for the case of a moving and a static zone between which the equili-

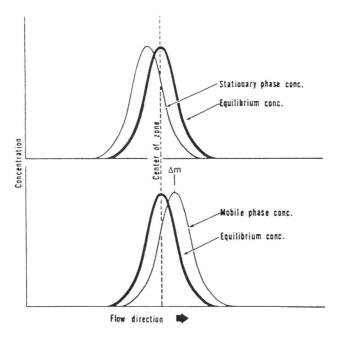

Fig. 4 Taken from *Dynamics of Chromatography* [1]. Upper diagram shows actual concentration profile for stationary zone lagging behind the concentration profile which would exist at equilibrium (in bold); lower diagram shows concentration profile for the mobile zone leading the equilibrium profile (in bold). The equilibrium concentration profiles are Gaussian and are identical apart from scale. (Reproduced with permission of publisher.)

bration process takes place at a finite rate. The concentration profiles in the two zones are Gaussian.

The solute band profile in the mobile zone runs ahead of the profile in the stationary zone. These two profiles lie on opposite sides of the profiles which would be reached if the flow were stopped and full equilibrium achieved, the equilibrium profiles for the two zones being superimposable (apart from differences in concentration). Giddings presents his theory in four stages on pages 102–107 of *Dynamics of Chromatography* I have always found this presentation difficult to grasp, and I think the reason is that the order of presentation which suits my cast of mind is the reverse of that given by Giddings. Perhaps other readers have experienced similar difficulty. In the hope that others may find this reversal of order easier to follow, I give the following presentation of the nonequilibrium theory. The mathematics is unaltered.

I take the case of a solute which exchanges between a static and a mobile zone, where mass transfer in the stationary zone is the rate-limiting process. The theory applies equally well to mass transfer between a moving and a static phase. Giddings defines two sets of concentrations: the actual concentrations, c_m and c_s, present during chromatography and the concentrations, c_m^* and c_s^*, which would be present if flow were stopped and equilibration allowed to occur. As noted, all concentrations are defined as amounts per unit volume of the column. Thus, the total concentration, c, is the sum of the concentrations in the zones (the subscript z is omitted for convenience); that is,

$$c = c_m + c_s = c_m^* + c_s^* \tag{1}$$

The mobile zone moves down the column at a linear velocity, u_m. The stationary zone has a linear velocity of zero, and the solute band as a whole moves down the column at a rate given by

$$u_{band} = R u_m = u_m \left\{ \frac{c_m^*}{c_m^* + c_s^*} \right\} = u_m \left\{ \frac{c_m^*}{c} \right\} \tag{26}$$

Away from the band maximum, the concentrations deviate from their equilibrium values, and the local segment velocity differs slightly from the overall band velocity. It is given by

$$u_{local} = R_{local} u_m = \left\{ \frac{c_m}{c_m + c_s} \right\} = \frac{u_m c_m}{c} \tag{27}$$

From Fig. 4 it is seen that, on the upstream side of the band, R_{local} is less than R so this segment of the band moves more slowly than the band overall, whereas on the downstream side of the band the opposite applies and any segment moves faster than the band overall. This is a clear recipe for band spreading. The band spreads because of the lack of equilibrium. From the point of view of someone sitting on the band maximum, this is seen as an outward flux. Giddings' particular insight was to note that the outward flux is related to the fractional excesses (positive or negative) of the concentrations in the two phases relative to the equilibrium concentrations.

Giddings defines the fractional excesses, ε, by writing

$$c_m = c_m^*(1+\varepsilon_m) \quad \text{and} \quad c_s = c_s^*(1+\varepsilon_s) \tag{28a}$$

$$\varepsilon_m = \frac{c_m}{c_m^*} - 1 \quad \text{and} \quad \varepsilon_s = \frac{c_s}{c_s^*} - 1 \tag{28b}$$

Since $c_m + c_s = c_m^* + c_s^* = c$, we obtain

$$c_m^* \varepsilon_m + c_s^* \varepsilon_s = 0 \quad \text{or} \quad c_m^* \varepsilon_m = -c_s^* \varepsilon_s \tag{29}$$

In other words the excess concentrations in the two phases are equal but opposite.

The next step is to consider how the outward flux resulting from nonequilibrium can be related to the fractional excesses, ε_m (or ε_s). A flux in general is defined as the quantity of material passing through a surface of unit area in unit time, the surface being perpendicular to the direction of flux. The total flux in the present case is given by

J = sum of (velocity × concentration) for each zone

$$= u_z c_m + u_s c_s \tag{30}$$

Since u_s, the velocity of the stationary zone, is zero, we obtain

$$J = u_m c_m = u_m c_m^*(1+\varepsilon_m) = J^* + \Delta J \tag{31}$$

with

$$J^* = u_m c_m^* \quad \text{and} \quad \Delta J = u_m c_m^* \varepsilon_m \tag{32a}$$

The contribution J^* is the downstream flux at equilibrium due to the flow, so it is only ΔJ which contributes to the spreading. This outward spreading flux is then

$$\Delta J = u_m c_m^* \varepsilon_m = u_{band} c \varepsilon_m \tag{32b}$$

The final equality arises from Eq. (26). It is noted that ε_m is zero at the peak maximum, positive downstream of the maximum and negative upstream of the maximum. The flux for any diffusion-like process is given by Fick's first law, namely

$$J_{diff} = -D_{eff}(dc/dz) \tag{33}$$

where dc/dz is the concentration gradient in the direction of the flux. If J_{diff} is set equal to ΔJ, we obtain

$$D_{eff} = \frac{-u_{band} c \varepsilon_m}{dc/dz} = \frac{-u_{band} \varepsilon_m}{d(\ln c)/dz} \tag{34}$$

For any process of diffusion the profile, starting from a very narrow band, and after a period of time, t, takes the form of a Gaussian. The standard deviation of the Gaussian profile, σ_z, is given by

$$\sigma_z^2 = 2Dt = 2D_{eff}(t_m + t_s) \tag{35}$$

It is noted that the time for diffusion is the combined time the solute spends in the two phases; i.e., $t = t_m + t_s$. The plate height is given in terms of the variance of the profile, $H = \sigma^2/z$, by Eq. (5); that is,

$$H = \frac{2D_{eff}(t_m + t_s)}{z} = \frac{2D_{eff}}{u_{band}} \tag{36}$$

the ratio $z/(t_m + t_s)$ being the band velocity. Inserting D_{eff} from Eq. (34) gives finally

$$H = \frac{2\varepsilon_m}{d(\ln c)/dz} \tag{37}$$

This is a very simple equation and shows that the fractional excess concentration, ε_m, is the key to the determination of H arising from nonequilibrium. What is not obvious from this equation is that the calculation of ε_m for any first-order rate process gives an expression which contains the factor $d(\ln c)/dz$, so that this always cancels out in the final equation for H. Virtually all processes which are slightly disturbed from equilibrium relax by first-order processes.

The implications of Eq. (37) may be clarified by taking two Gaussians, one slightly displaced from the other (as in Fig. 4). Both represent concentration profiles for the mobile zone or phase: the first represents the equilibrium profile, with its maximum at z_0, and the second represents the actual concentration profile, which is displaced by an amount Δz to the right (downstream) with its maximum at $z_0 + \Delta z$. Compared to the peak width; Δz is very small:

$$c_m^* = \exp\left[\frac{-(z-z_0)^2}{2\sigma_z^2}\right] \quad \text{and} \quad c_m = \exp\left[\frac{-(z-\Delta z-z_0)^2}{2\sigma_z^2}\right] \qquad (38)$$

It is readily shown, by taking logs and differentiating, that, for the first curve,

$$\frac{d(\ln c_m^*)}{dz} = \frac{-(z-z_0)}{\sigma_z^2} \qquad (39)$$

The fractional excess is given, Eq. (28b), by

$$\varepsilon_m = \frac{c_m}{c_m^*} - 1 \qquad (40)$$

If the two Gaussians are expanded as a series and Δz is assumed much less than $z-z_0$, and if furthermore only the first-order terms are considered we obtain

$$\varepsilon_m = \frac{(z-z_0)\Delta z}{\sigma_z^2} \qquad (41)$$

Then, using Eq. (39), and since c_m is a constant fraction of c,

$$\varepsilon_m = -\Delta z \, d\ln c_m^* / dz = -\Delta z \, d\ln c / dz \qquad (42)$$

Finally by Eq. (37),

$$H = 2\,\Delta z \quad \text{and} \quad \Delta z = H/2 \qquad (43)$$

This is a neat, elegant, and eminently sensible result: the separation between the true and equilibrium concentration profiles in the mobile zone is equal to the half the plate height contribution arising from the nonequilibrium.

In order to get a useful equation for the mass transfer contribution to H, we need to get an analytical expression for the fractional excess concentration ε_m (or if more convenient ε_s). Two aspects of the process have to be considered. The first arises from the flow, since it is obvious that ε_m is going to be larger the larger the flow rate; the second arises from the rate and mechanism of mass transport within the stationary zone since it is obvious that ε_m is going to be larger the slower the mass transport. One has to set up two equations: one relating to the flow, and the other relating to the mechanism of mass transfer. The connecting link for these two equations is the rate of mass transfer from the stationary to the mobile phase, s_m.

The equation relating to flow is obtained from mass balance considerations. For any short segment of the column, say on the upstream side of the peak maximum, flowing eluent is carrying solute into the segment. At the same time solute is being desorbed from the stationary zone into the segment to augment the concentration in the mobile zone. The eluent flowing out of the segment is thus carrying a little more solute than it did when it entered the segment. Giddings [38] shows by a simple proof that the rate of mass transfer from the stationary to the mobile phase is given by

$$s_m = (1-R)u_m(dc_m^*/dz) \tag{44}$$

or alternatively from the mobile into the stationary phase by

$$s_s = -Ru_m(dc_s^*/dz) \tag{45}$$

s_m and s_s must, of course, be equal and opposite. This is confirmed by noting that $(1-R)\,c_m^* = Rc_s^*$ by Eq. (3). This stage in the derivation requires the assumption that the process is occurring close to equilibrium (in other words, that the deviations from equilibrium are small throughout the band) and that second-order effects may be ignored. This is a crucial assumption of the nonequilibrium theory. By a corollary it can be argued that if a chromatographic peak spreads in a Gaussian fashion then the process must be occurring very close to equilibrium. This is a crucial general principle for "well-behaved" chromatography. The near-equilibrium hypothesis establishes that, when peaks are symmetrical, one cannot ascribe selectivity of solute retention to differences in diffusion rates or to differences in any other kinetic parameters which determine their equilibration. One can ascribe differences in selectivity only to thermodynamic differences between solutes which govern their partition between the mobile and stationary zones or phases. An example of a theory which was demonstrably incorrect for this reason was that of Ackers [39] for gel permeation chromatography. This theory proposed that selectivity in GPC arose from the different times it took for different solutes to diffuse into the gel particles. Large molecules supposedly took longer to diffuse and so did not have time to explore the whole of the gel space. This theory is demonstrated to be untenable by the Giddings analysis. Selectivity in GPC must arise primarily from steric effects.

Relating ε_m to the mechanism of mass transfer is generally quite complex. However, the case of one-site adsorption is straightforward and is used here to illustrate the method. The rates of the forward and reverse processes can be expressed as

$$\text{Rate of adsorption} = k_a c_m \tag{46a}$$

$$\text{Rate of desorption} = k_d c_s \tag{46b}$$

Under equilibrium conditions these rates must be equal, thus

$$k_a c_m^* = k_d c_s^* \tag{47}$$

The net rate of mass transfer into the mobile zone is

$$\begin{aligned} s_m &= k_d c_s - k_a c_m \\ &= k_d c_s^* (1 + \varepsilon_s) - k_a c_m^* (1 + \varepsilon_m) \\ &= k_d c_s^* \varepsilon_s - k_a c_m^* \varepsilon_m \\ &= -(k_d + k_a) c_m^* \varepsilon_m \end{aligned} \tag{48}$$

The final equality uses Eq. (29). We now have two equations for s_m which can be used to evaluate ε_m, namely:

$$s_m = (1-R)u(dc_m^*/dz), \text{ and } s_m = -(k_d + k_a)c_m^* \varepsilon_m \tag{44,48}$$

whence

$$\varepsilon_m = \frac{(1-R)u(dc_m^*/dz)}{(k_d + k_a)c_m^*} = \frac{(1-R)u(d\ln c_m^*/dz)}{(k_d + k_a)} \tag{49}$$

Since c_m^* is a constant fraction of c, $d\ln c_m^*/dz = d\ln c/dz$. Thus using Eq. (37) we obtain

$$H = \frac{2\varepsilon_m}{d(\ln c)/dz} = \frac{2(1-R)u}{k_d + k_a} \tag{50}$$

Now k_a and k_d are related to R, for $k_a c_m^* = k_d c_s^*$, and $R = c_m^*/(c_m^*+c_s^*)$. Thus $1/(k_d + k_a) = R/k_d$, and we finally obtain

$$H = \frac{2R(1-R)u}{k_d} = \frac{2\left\{\dfrac{k'}{(1+k')^2}\right\}u}{k_d} \tag{51}$$

This result is given on page 108 of *Dynamics of Chromatography*.

When diffusion in a stationary liquid phase is involved, the calculation of the fractional excess becomes more complex. The two important cases are that of a uniform thin layer of stationary phase (relevant to GC), and of a spherical body making up the stationary zone (relevant to LC). The reader is referred to *Dynamics of Chromatography* for the details. The first three steps are as given above, but the mathematics of the final stage—that is, the calculation of ε_m (or ε_s)—is complex. In these two cases one has to divide the stationary phase or zone into infinitesimally thin slices or spherical shells and then compute the mass transfer between each slice and the next. The fractional excess for each slice is averaged over the layer to give the mean excess corresponding to ε_s. Two boundary conditions have to be accommodated. The first is that there is no mass transport through the solid interface at the bottom of the layer of stationary phase (in the thin-layer case) or through the center of the sphere (in the spherical particle case); the second is that the local excess at the surface is zero: in other words there is perfect equilibrium at the boundary between the stationary and mobile zones (or phases). Then ε_m, required in Eq. (37), can be found from ε_s by using Eq. (29).

The result for a thin layer of stationary phase is

$$H_c = \frac{2}{3} \cdot \frac{R(1-R)u_m d_f^2}{D_s} = \frac{2}{3}\left\{\frac{k'}{(1+k')^2}\right\}\frac{u_m d_f^2}{D_s} \tag{52}$$

where D_s is the diffusion coefficient of the solute in the stationary phase.

The result for a spherical porous particle is

$$H_c = \frac{1}{30} \cdot R(1-R) u_z \frac{d_p^2}{D_{sz}} = \frac{1}{30} \left\{ \frac{k''}{(1+k'')^2} \right\} \frac{u_z d_p^2}{D_{sz}} \tag{53}$$

In Eq. (53), I have used the notation k'' rather than k', and u_z rather than u_m, to indicate that the mass transfer in question is between kinetic zones rather than between thermodynamic phases. In practice this means that the correct capacity ratio is obtained from the excluded solute peak, not from the fully permeating solute peak. The D_{sz} term has to be interpreted as the effective diffusion coefficient of the solute in the stationary zone (that is, within the porous particle). Generally in LC the particle, as explained above, will comprise an impermeable structure such as silica, whose internal surface carries a bonded monolayer (say ODS) and whose pores are filled with stagnant mobile phase. Then D_{sz} has to be a weighted fraction of the diffusion coefficients in the stagnant mobile phase and in the stationary phase as described above for evaluation of the B-term contribution to H. Knox and Scott [29] considered this matter in detail and arrived at the formula

$$D_{sz} = \frac{\gamma_{sm} \phi D_m + [k''(1-\phi) - \phi] \gamma_s D_s}{k''(1-\phi)} \tag{54}$$

where ϕ is the fraction of mobile phase which is stagnant within the particles, and k'' is the zone capacity ratio. This value for D_{sz} can then be put into the equation for H_c to give

$$H_c = \frac{1}{30} \cdot \left\{ \frac{k''}{(1+k'')} \right\}^2 \frac{(1-\phi) u_z d_p^2}{\gamma_{sm} \phi D_m + [k''(1-\phi) - \phi] \gamma_s D_s} \tag{55}$$

This equation looks complicated, but its significance can be seen by looking at limiting values. If diffusion occurs at the same rate in the mobile and stationary phases (i.e., if $\gamma_{sm} D_m = \gamma_s D_s$), we obtain

$$H_c = \frac{1}{30} \left\{ \frac{k''}{(1+k'')^2} \right\} u_z \frac{d_p^2}{\gamma_s D_s} \tag{56}$$

This is the normal equation obtained from mass transfer into a uniform spherical particle.

If the diffusion rate in the stationary phase is zero ($D_s = 0$), we obtain

$$H_c = \frac{1}{30}\left\{\frac{k''}{(1+k'')}\right\}^2\left\{\frac{(1-\phi)}{\phi}\right\}\frac{u_z d_p^2}{\gamma_{sm} D_m} \qquad (57)$$

The k'' dependence in this equation is typical for mass transfer occurring only in the mobile phase.

These limiting equations show that the dependence of H_c upon the capacity ratio k'' depends upon the ratio D_s/D_m. When $D_s \approx D_m$, Eq. (56) applies, and we have the normal dependence for a uniform stationary zone, with a maximum when $k'' = 1$. When D_s is much less than D_m, Eq. (57) applies, and H_c rises gradually to a maximum when k'' is large. In retentive chromatography the unretained solute still has a k'' value which is finite, being given by $k'' = \phi/(1-\phi)$. Only in exclusion chromatography do we find k'' values starting at zero. Typically ϕ is about 0.4 for a modern silica gel, giving a lowest accessible value of k'' in retentive chromatography of about 0.66.

VII. SEPARATION OF THE A- AND C-TERMS

In gas chromatography dispersion occurring in the gas phase can be separated from dispersion occurring in the stationary (liquid) phase by manipulating the inlet/outlet pressure ratio or by changing the carrier gas. This was done by Knox and Saleem [25], following earlier work by Perrett and Purnell [26]. In GC since diffusion in the mobile phase is so much faster than in the stationary phase one might expect the major contribution to H to come from resistance to mass transfer in the stationary phase. Surprisingly the results of Knox and Saleem, using 6% squalane-loaded Chromosorb, showed that the main source of dispersion arose from mobile-phase effects. The plate height contributions from mobile-phase dispersion arise from B-term dispersion, A-term dispersion, and slow diffusion in and out of the particles of packing, i.e., C_m-term dispersion. Figure 5, taken from Knox and Saleem, shows that the log-log plots of the mobile phase contribution to the reduced plate height against reduced velocity had gradients of 0.4 to 0.5, the gradients increasing somewhat with k'. These are somewhat higher than the gradi-

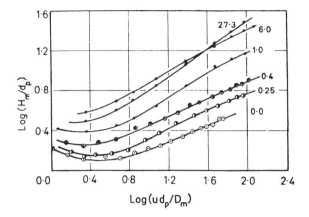

Fig. 5 Plot of mobile phase contribution to reduced plate height against reduced velocity in GC from Knox and Saleem [27]. Packing: 235-μm Chromosorb G. Lower three curves, bare Chromosorb G; upper three curves, Chromosorb G loaded with 6% squalane. Solutes: ethylene, butane, cyclohexane. k'-values given on individual lines. (Reproduced with permission of authors and publisher.)

ent of 0.35 found for unretained solutes with glass bead columns by Knox and Parcher.

The question arises as to how much of this contribution arises from C-term dispersion, that is from slow diffusion within the particles of the packing material. Since D_s in GC is effectively zero, this contribution would be given by Eq. (57):

$$H_c = \frac{1}{30}\left\{\frac{k''}{(1+k'')}\right\}^2 \left\{\frac{(1-\phi)}{\phi}\right\} \frac{u_z d_p^2}{\gamma_{sm} D_m} \tag{57a}$$

or

$$h_c = \frac{1}{30}\left\{\frac{k''}{(1+k'')}\right\}^2 \left\{\frac{(1-\phi)}{\phi}\right\} \frac{\nu}{\gamma_{sm}} \tag{57b}$$

To obtain an approximate value for h_c we assume $\phi = 0.5$, $k'' = 2$, and $\gamma_{sm} = 0.8$. The reduced plate height from this effect cannot be larger than about 0.02v. Appling this to the 6% squalane-loaded column which gives significant retention, C_m-term dispersion accounts for only about 10% of the observed plate height at the highest reduced velocities used by Knox and Saleem (100); at reduced velocities near to those for minimum H, the contribution is only about 2%. The data for the solutes retained on the 6% squalane-loaded column can be represented roughly by

$$h = \frac{2}{v + 1.0v^{0.5}} \tag{58}$$

We conclude that in GC the major contribution to the plate height at the usual operating velocities in GC (around those giving minimum H) arises from A-term dispersion. For retained solute this is somewhat larger than given by Knox and Parcher [33] for unretained solutes on glass beads. Poor column packing is probably the reason.

Knox and Scott [29] carried out experiments to separate the A- and C-terms under LC conditions. They noted that although the velocity dependence of the A-term for unretained solutes in glass bead columns was known, there was little information for retained solutes on porous particles. Both the velocity dependence and the k''-dependence of the A-term were uncertain. It could, however, be assumed that the velocity dependence of the A-term was likely to be much less than that of the C-term, and therefore that one could reduce the impact of the A-term by working at high reduced velocities. To do this it was necessary to use large particles. Knox and Scott used 50-μm and 540-μm particles, which enabled them to reach reduced velocities or 4000 and 15,000 respectively. Figure 6 gives their results for a solute with $k'' = 1.86$.

A key deduction from this graph is that A cannot be independent of velocity, since one would then have to draw a straight line through the points (the broken line in Fig. 6), which would give $A = 21$. This cannot be correct since the minimum value of H is well known to be in the region of 2 for any reasonably well packed HPLC column. The A-term must be velocity dependent. Knox and Scott fitted their data to the Eq. (59), assuming an exponent of 1/3:

$$h = Av^{1/3} + Cv \tag{59}$$

The best fit, given by the full line in Fig. 6, had $A = 2.5$ and $C = 0.0125$. The rather high value of A suggests that the exponent 1/3 may be too low. If, for example, the exponent were increased to 0.45, the data

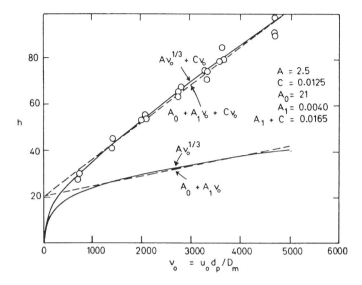

Fig. 6 (h, ν) plot for phenol eluted from 50-mm ODS silica gel from Knox and Scott [29]. Eluant: methanol/water (60/40). k″ = 1.86. Full lines give fit of Knox equation to experimental data points; broken lines give fit of original van Deemter equation. Lower curves gives estimates of A-term dispersion. (Reproduced with permission of authors and publisher.)

would be fitted with $A = 0.9$ without significant change in C. This would be more in keeping with (h, ν) data obtained at low reduced velocities and agrees with the data of Knox and Saleem [27].

With any reasonable formulation of the A-term, Fig. 6 shows clearly that, at low reduced velocities, the C-term contribution to H is going to be very small. Nearly all the dispersion arises from the A-term. The actual value of the C-term, 0.0125, found by Knox and Scott [29] for $k″$ of 1.86, is much smaller than generally assumed, and is well below 0.1 recommended by Bristow and Knox [36]. It is, however, of the right order of magnitude from the theoretical point of view taking into account the leading factor of 1/30 in the theoretical expression for C and the fractional value of the $k″$-factor. It has always been something of an anomaly that values of C found empirically using the van Deemter equation should be so high.

Assuming that their value of the A-term applied to other solutes,

Knox and Scott next examined the dependence of C upon k''. They found that C did indeed rise as k'' increased from 0.6 (its lowest value for a fully permeating unretained solute) to a maximum at k'' around 2 and then fell gradually as k'' increased further. This was a clear indication that the true dependence of C upon k'' lay somewhere between the dependences given by Eqs. (56) and (57). They were able to derive an approximate value of $\gamma_s D_s/D_m$ of about 0.45. This was in excellent agreement with the value of about 0.55 derived from measurement of the same ratio using the arrested diffusion method. The agreement lends strong support to the analysis as a whole.

The studies by Knox and Parcher [33]. Knox and Saleem [25], and Knox and Scott [29] effectively settle the question of the relationship between the A- and C-terms in the classical plate height equations. The major conclusions which can be derived are as follows:

1. The velocity dependence of the A-term is convincingly established. The exponent lies between 0.33 and 0.5.
2. The magnitude of the A-term is strongly dependent on how well the column is packed.
3. The major contribution to band broadening in chromatography columns whether GC or HPLC columns, operated at low reduced velocities (i.e., those giving near optimum performance and minimum H) is A-term dispersion. This amounts to over 90% of the observed dispersion at low reduced velocities (say, $v < 100$).
4. The theoretical equations for the C-term are sound and give an accurate value for the contribution to H from slow mass transfer in the stationary zone.
5. With bonded packing materials in HPLC, diffusion in the stationary phase occurs at about half its rate in the mobile phase.

These important conclusions are relevant to improving the performance of current HPLC columns, and they explain some of the apparent anomalies in plate height velocity plots which have been obtained for packed columns. The A-term dispersion as we have seen [32,33,35] is highly dependent upon the goodness of packing, and upon the influence of wall effects. It also depends upon the method of packing the column, on the precise column geometry, and on the method of injection, takeoff, and so on. Since A-term dispersion is now seen to be the major cause of dispersion, it is not surprising that there is wide variation in the reproducibility of plate height curves. There is no evidence that C-term dispersion, now seen to be a small proportion of the total, is affected by column packing methods. It is not surprising that the assumption, widely made, that C-term dispersion is the major contributor to the plate height has led to difficulties in explaining both its magnitude and its variability.

If we are to learn more about the complex dependence of the A- and C-term dispersion on relative diffusion rates and upon retention, further careful experiments will be required using large- and small-diameter particles and a very wide range of v values.

VIII. EXCLUSION CHROMATOGRAPHY

One of the major problems which worried practitioners of exclusion chromatography in the early days was the effect of kinetic band spreading (or chromatographic band broadening as they called it) on the calculation of polymer molecular weight distribution curves. Kinetic band spreading widens the chromatogram of any polymer and thus leads to an overestimate of the width of the distribution. To generate the true "unbroadened" chromatogram from the actual chromatogram requires a complex deconvolution procedure such as the one developed by Tung [40]. However, any deconvolution procedure requires knowledge of the band-broadening parameter (i.e., the plate height), which depends upon the molecular weight and the degree of exclusion of the polymer. The major problem was therefore not so much the unfolding of the original chromatogram, but the value to be given to the broadening parameter. How can one determine it without monodisperse polymers? One way is to use the reverse elution method of Tung [41], whereby a band of narrow MWD polymer is eluted some distance along a column and then back again to the beginning at the same velocity. The dispersion of the band due to its polydispersity is thereby reversed and eliminated, while the kinetic dispersions for the forward and backward elutions are combined. Regrettably this elegant method does not seem to have been widely used, and unsatisfactory assumptions have generally had to be made about the dispersiveness of the column.

Knox and McLennan [42,43] developed a different method for tackling the problem. Band dispersion for a narrow MWD polymer (e.g., a polystyrene standard) arises partly from its intrinsic polydispersity and partly from kinetic band spreading. The bandwidth arising from polydispersity increases linearly with migration distance, while the bandwidth arising from the kinetic effects increases as the square root of the migration distance. Thus, by using columns of different lengths and plotting the apparent H against column length the kinetic dispersion is obtained as an intercept and the polydispersity from the gradient. This method of determing the polydispersities of standards appeared to be new, as was the concept of determining H-values for high polymers by such a method. Knox and McLennan also determined the velocity dependence of H_{app}. Since the polydispersity contribution is unaffected by velocity, any dependence arose purely from kinetic effects, namely the A- and C-term broadening (B-term broadening being unimportant at the flow velocities used). Knox and McLennan (K & M) [42,43] noted that the reduced plate plots for a range of polystyrenes (MW's 2000 to 33,000, k'' 0.74 to 0.15, v-values up to about 300) were superimposable within experimental error. Dawkins and Yeadon (D & Y) [44,45] later used a similar method to determine H-values for polymer standards as did van Kreveld and van den Hoed (VK & VDH) [46] except that they used extremely high velocities with large particles (v-values up to 120,000).

Knox and McLennan and D & Y assumed that the more or less linear dependence of h upon v meant that C-term dispersion dominated. This enabled them to calculate the effective diffusion coefficients of the polymers in the stationary zone (i.e., within the pores of the particles). In contrast, the H/u curves of VK & VDY for their highest MW polymers were severely curved toward the velocity axis. They attributed this to flow through the pores of the packing material assisting diffusion or possibly to unspecified turbulence effects. The effective diffusion coefficients within the particles were calculated by curve fitting, which was equivalent to extrapolation of the results at the low-velocity ends of the curves. The combined results of three sets of authors are listed in Table 1. All three groups found that D_m/D_{eff} fell as permeation increased. The numerical results of K & M and of D & Y agreed well, with D_m/D_{eff} falling from about 30 for nearly excluded polymers to about 10 for well-permeating polymers. The values of VK & VDH were about half. The main difference between the studies of K & M, D & Y, and VK & VDH is that K & M, D & Y, worked at relatively low reduced velocities (up to 300), while K & Y worked at extremely high reduced velocities (10,000 to 100,000) At these high reduced velocities, following the work of Knox and Scott, nearly all the band broadening would have arisen from C-term effects, whereas under the conditions used by K & M and D & Y, a considerable proportion of the band broadening would have come from A-term effects.

Table 1 Effective Diffusion Rates of Polystyrenes Partially Exluded from Silica Gel

Authors[a]	MW	k''	D_m/D_s [b]
K&M	33,000	0.12	22.2(16.9)
	20,000	0.22	14.2(10.7)
	10,000	0.32	11.8(7.7)
	4,000	0.52	9.6(6.0)
	2,000	0.58	10.7(6.5)
	Toluene 78	0.79	
D&Y	11,000	0.10	27.7
	35,000	0.37	11.2
	9,800	0.66	7.3
	3,600	0.82	6.6
	600	0.94	11.0
VK&VDH	160,000	0.105	8.4
	97,200	0.19	7.4
	51,000	0.29	4.7
	20,400	0.60	3.2
	Toluene 78	0.96	1.0

[a]K&L = Knox and McLennan [43], D&Y = Dawkins and Yeadon [45], VK&VDH = van Kreveld and van den Hoed [46].
[b]For explanation of values in parentheses see text.

In the experiments of Knox and Scott with a low MW retained solute, something like 80% of the band broadening would arise from A-term effects at $v \approx 300$. However, with such solutes, diffusion within the particles was only slightly obstructed. Had it been significantly obstructed, the contribution from C-term broadening would have been greater. If the values of VK & VDH are taken as accurate, then it seems that about half of the H-value observed in the experiments of K & M and of D & Y would have arisen from A-term broadening. This is not an unreasonable proportion. Correction for A-term broadening was indeed attempted by K & L using the standard A-term, $1.00v^{1/3}$. Those results are given in parentheses and are lower than the results without A-term correction. A larger A-term, for example that proposed by Knox and Scott [29], would reduce D_m/D_{eff} further and could bring K & L's values into line with those of VK & VDH.

Another puzzling feature of exclusion chromatograms is the sharpness of the peaks of excluded solutes when compared to those of permeating solutes. Most chromatographers explain this by saying that the peak of a totally excluded polymer is not broadened by polydispersity. While this is correct, the peak widths of excluded polymers are still narrow on the basis of purely kinetic dispersion. For example, VK & VDH found that H for a totally excluded polymer (MW 411,000) was comparable to that for toluene, a fully permeating solute. They found $h = 30$ for PS 411,000 at $\nu = 14,000$. Knox and McLennan made a similar observation. They found $h = 5$ for PS 200,000 at $\nu = 600$. The "standard" A-term, $1.0\nu^{1/3}$, would give $h = 24$ at $\nu = 14,000$ and $h = 8$ at $\nu = 600$, values that are not inconsistent with the observations and are close to what one might expect for a completely excluded solute in a GPC column which would, in fact, behave much like a column of nonporous spheres.

Comparison of the results of K & L, D & Y, and VK & VDH along with those of Knox and Scott suggests that A-term broadening is significant in exclusion chromatography and can account for about half of the observed broadening under normal operating conditions. The results confirm that A-term broadening is probably larger for partially permeating solutes than for excluded solutes.

There are unquestionably aspects of A-term broadening in GPC, and indeed LC generally, which are still not fully understood and further careful experiments will be required. Ideally a study like that of Knox and Scott [29] is require for partly excluded polymers using relatively large particles, say around 50 μm.

IX. SLOW CHEMICAL EQUILIBRATION AND WIDE PEAKS

Giddings, in *Dynamics of Chromatography* (p.130), treats the case of slow reversible chemical reaction leading to band broadening. A simple slow equilibration scheme is given in Fig. 7, where A_1 to A_4 are different forms of a single solute. The original scheme of Giddings envisages slow chemical reaction within the phases and possibly slow mass transfer from one phase to the next, which could involve further slow chemical reaction. This reaction scheme produces a somewhat complex expression for H_c which involves the rate constants of the various steps and the overall retention ratio, R, of the analyte A. If, however, it is assumed that the phase transition steps are fast while only the chemical steps within the phases are slow, the original expression is greatly simplified to

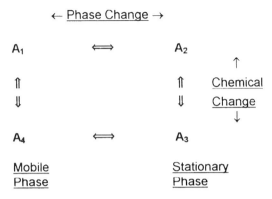

Fig. 7 Reaction scheme for slow chemical reaction between various forms of analyte, A_1, A_2, A_3, and A_4, according to Giddings *Dynamics of Chromatography* [1].

$$H_c = \frac{2(X_1 X_3 - X_2 X_4)^2}{k_M X_1 + k_s X_2} \frac{u}{R} \qquad (60)$$

where the X_i's are the fractions of A in each of the different forms, and k_M, k_S are the rate constants for the forward reactions in the mobile and stationary phases. The ratios of the X_i are related to the equilibrium constants for the reactions and to the k'-values of the chemical forms. Further rationalization of the equation is achieved by replacing the X_i's by equilibrium and chromatographic parameters.

When Dr. Shibukawa was working in my group studying the HPLC of some metal-EDTA complexes, he noticed that among the normal sharp peaks of test compounds and reactants some very wide peaks appeared [47]. An example is shown in Fig. 8. The H-values for the wide peaks were large (of the order of a few millimeters), and their dependence on flow rate was strictly linear as seen from Fig. 8. The plots provided perfect examples of C-type dispersion. A ligand exchange reaction, known to be slow, was identified as the cause of the excessive peak broadening:

$$[Cr(EDTA)]^- + CH_3COO^- \underset{k_b}{\overset{k_f}{\rightleftharpoons}} [Cr(EDTA)(CH_3COO)]^{2-} \qquad (61)$$

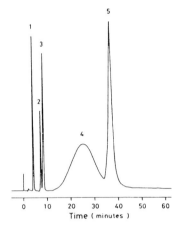
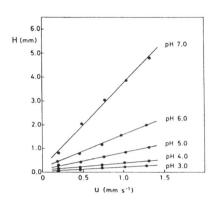

Fig. 8 (Left) Chromatogram of metal(II,III)-EDTA compexes on 5-μm Hypersil C8, showing very wide peak arising from slow chemical reaction (from Knox and Shibukawa [47]). Eluant: 0.1 mM cetrimide in 0.1 M posphate buffer pH 7/acetonitrile (90/10). Peak identification: 1 Co(III)-EDTA; 2 = Co(II)-EDTA; 3 = Cu(II)-EDTA; 4 = Cr(III)-EDTA; 5 = Fe(III)-EDTA. (Right) Plot of H against u for peak 4 with eluants of various pH, showing excellent linear relationship for C-term dispersion. (Reproduced with permission of authors and publisher.)

By carrying out experiments at different temperatures it was possible to evaluate the activation energies of the forward reaction as 71 kJ/mol, a value which agreed with the literature.

X. BAND SPREADING IN ELECTROSEPARATION SYSTEMS

Modern capillary electrophoresis was initiated by Jorgenson and Lukacs in 1981 [48]. Jorgenson appreciated that, because electroosmotic flow in an open tube originates in the electrical double layer at the wall, there would be no dispersion arising from transcolumn velocity variations. This meant that with pure electroosmotic flow the only dispersion mechanism for an unretained solute would be axial diffusion (B-term dispersion), giving rise to exceptionally high plate efficiencies. A further consequence of the plug nature of electroosmotic flow was that the tube bore could be quite wide, much wider than would be appropriate for pressure-driven capillary LC. Indeed the only real limitation to the tube bore was likely to be set by heating effects, which initiate convection and increase analyte diffusion coefficients, both of which lead to band broadening. The practical limit is somewhere in the region of 100–200 µm. At the Avignon Symposium in 1981 [49] those attending the last lecture of the meeting were astonished by the remarkable separations which Jorgenson presented and which resembled those obtainable by capillary gas chromatography. The success of CE (capillary electrophoresis) depended upon the absence of wall retention. As soon as any wall retention occurs, the Aris equation, Eq. (11), operates, and C-term dispersion seriously increases H from its value of $2D_m/u$. Thus, even with electrodrive, the column bore for retentive capillary LC is still impractically small, and it cannot be carried out in the open tubes currently used for CE.

The effective channel diameter can, however, be reduced by packing the tube as in conventional LC, thereby allowing retention to occur without catastrophic loss of efficiency. Packed-column capillary LC driven by an electric field is called electrochromatography and is a relatively new liquid separation technique. It was invented by Pretorius [50] in 1974, but was forgotten until revived by Jorgenson and Lukacs in 1981. Around 1986, Ian Grant and I decided to investigate the method from the point of view of its performance and potential as a future analytical method. Our results were presented in three papers [51–53]. Because electrochromatography is driven by electroosmosis, and since this itself depends upon the existence of a thin electrical double layer (typically about 1 nm thick) surrounding the particles of packing, the flow rate in a bed packed with particles of say silica gel is independent of the particle size until this drops to about 100 electrical double-layer thicknesses. In practice this means that the particle size in electrochromatography can be reduced to below 1 micrometer without loss of electroosmotic flow. This has important benefits for chromatography because band spreading is greatly reduced while chromatographic retention is unaffected. With reasonable linear flow rates, it should be possible to reduce A- and C-term band broadening to a negligible proportions, leaving axial diffusion as the only important bandbroadening process. As in normal CE, the column bore is still limited by self-heating to somewhere in the region of 100–200 µm. The technique, now called "capillary electrochromatgraphy" (CEC), is carried out in typical CE equipment.

Knox and Grant confirmed that the flow rate was unaffected by particle size down to at least 1.5 µm. They also showed that the efficiency of electrochromatography was higher than that of pressure-driven LC using the same column. This was attributed to lower A-term dispersion.

In a pressurized system, the local flow rate varies substantially across the column since the flow channels are of different diameters and are oriented at all angles to the axis of the column. With electroflow some of these dispersive effects are removed or at least reduced. Since the flow is generated at the surface of the particles in the thin double layer, the channel diameter becomes unimportant and the flow along all channels with the same orientation to the axis should be very similar irrespective of their dimensions. On the other hand, there will still be dispersion due to the variety of orientations of the channels. The improved efficiency using electrodrive is shown in Fig. 9, taken from Knox and Grant's paper of 1991 [53]. The best reduced plate height achieved using 5-μm particles and a sample of polynuclear hydrocarbons was 1 particle diameter, the 0.5-m-long column giving an efficiency of 100,000 plates.

Another remarkable feature of the chromatograms achieved by electrodrive in capillaries is the extreme symmetry of peaks. This is demonstrated in Fig. 10, where the differential chromatogram is shown

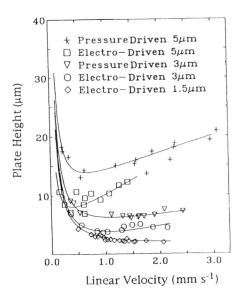

Fig. 9 Comparison of plate heights for pressure- and electrically driven chromatography on 5-, 3-, and 1.5-mm ODS materials from Knox and Grant [53]. (Reproduced with permission of authors and publisher.)

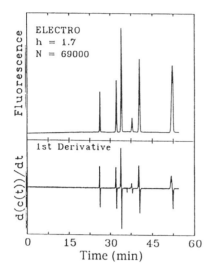

Fig. 10 Chromatogram of simple aromatics on 5-μm ODS Hypersil in a drawn-packed capillary from Knox and Grant [53]. Lower chromatogram is first derivative of upper chromatogram. Column: 600 μm × 60 μm. (Reproduced with permission of authors and publisher.)

below the normal chromatogram. It is noted that the positive and negative excursions from the baseline in the differential chromatogram are of equal height, apart from those for the last peak which is slightly asymmetric. Figure 10 is encouraging for those of us who have argued that the asymmetric peaks so often seen in published chromatograms arise not from deficiencies in the chromatographic process itself but from deficiencies in the design of chromatographic equipment, in particular from badly swept volumes in injectors, column terminators, and detectors. The excellent symmetry in CEC is achieved since injection is made directly onto the column, while detection is also on-column immediately after the frit which terminates the packing.

The development of capillary electrochromatography is now occurring very rapidly, and it promises to become a truly high performance LC system with efficiency comparable to that of capillary gas chromatography. If columns can eventually be packed successfully with submicron particles, efficiencies in excess of 500,000 plates should be attainable. Within the next 10 years CEC is likely to supplant conventional HPLC as the major liquid separation method.

XI. CONCLUDING REMARKS

The insight of Calvin Giddings has illuminated the whole field of chromatographic peak broadening and led to a huge improvement in our understanding of the topic. In particular it has enabled us to think much more clearly about the nature of diffusional band spreading in a packed bed, the role of the various complex processes which occur in the mobile zone, and the importance of slow equilibration between the mobile and stationary zones or phases. In spite of the advances in technology, for example the introduction of electroseparation methods, there is still much experimental work to be done to clarify the magnitudes of the different contributions to bandbroadening in the different forms of chromatography and electrophoresis. Chromatography has advanced greatly since 1965 when *Dynamics of Chromatography* was published. Indeed *Dynamics* was published before HPLC had been invented. Remarkably Cal Giddings' book is still the best book on the subject. Chromatography is now one of the most powerful methods of determining the magnitudes of dispersive factors in packed beds, and its results are directly applicable in other areas, particularly chemical engineering. It is now time to merge the insights of chromatographers and chemical engineers once again.

SYMBOLS

Subscripts

app	apparent as in H_{app}
A, B, C	representing A-, B-, and C- contributions to the plate height as in H_A, H_B, H_C
c	column: as in column diameter, d_c
band	band: as in band velocity, U_{band}
diff	diffusion: arising from as in J_{diff}
eff	effective: as in D_{eff} the effective diffusion coefficient
f	film: as in film thickness d_f
m	mobile phase

min	minimum: minimum value as in H_{min}
mz	mobile zone
opt	optimum: as in optimum linear velocity, U_{opt}
p	particle: as in particle diameter, d_p
r	radial: as in A_r
s	stationary phase
sm	stagnant mobile phase
sz	stationary zone
z	distance in the column: as in σ_z
z	zone: as in u_z

Roman Symbols

A	eddy diffusion constant in plate height equation, or in reduced plate height equation
A_r	radial eddy diffusion constant in reduced plate height equation
B	axial diffusion constant in plate height equation, or in reduced plate height equation
c	total concentration per unit volume of column
c_m, c_s, c_{mz}, c_{sz}	concentrations in phases or zones per unit volume of column
c^*, etc	equilibrium concentrations.
C, C_m, C_s	mass transfer constant in plate height equation or in reduced plate height equation
$D, D_m, D_s, D_{sz}, D_{sm}$	diffusion coefficients
d, d_p, d_c	open tube diameter, particle diameter, column diameter.
d_f	film thickness of stationary phase
$H, H_{app}, H_r, H_A, H_B, H_C$	plate height
h, h_r, h_A, h_B, h_C	reduced plate height, $h = H/d_p$
$J, J^*, J_{diff}, \Delta J$	flux down column, flux at equilibrium, flux arising from diffusion, excess flux $\Delta J = J - J^*$
k'	phase capacity ratio
k''	zone capacity ratio
k_a, k_d	rate constants for adsorption and desorption
k_M, k_S	rate constants for chemical reaction in mobile and stationary phases
L	column length

R	retention ratio (for zones or phases)
s_m, s_s	rate of mass transport into mobile and stationary phases
t_m, t_s	time solute spends in mobile and stationary phases
$u, u_m, u_z, u_{opt}, u_{band}, u_s$	linear velocity in column
X_1, X_2, X_3, X_4	fractions of solute in various forms in chemical reaction scheme
$z, z_0, \Delta z$	distance along column measured from inlet end, z_0 position of peak maximum, Δz = increment in z_0

Greek Symbols

$\varepsilon_m, \varepsilon_s$	fractional excess concentrations in nonequilibrium theory
ϕ	fraction of mobile phase which is stagnant within particles
$\gamma, \gamma_{sm}, \gamma_s$	obstructive factor in plate height equations
λ	tortuosity factor in plate height equation
ν	reduced velocity $\nu = u d_p / D_m$
π	constant = 3.141
	ratio of inlet to outlet pressure in GC
ρ	column to particle diameter ratio, $\rho = d_c / d_p$
σ_z	standard deviation of Gaussian concentration profile measured as a distance

REFERENCES

1. J. C. Giddings, *Dynamics of Chromatography—Part 1*, Marcel Dekker, New York, 1965.
2. J. C. Giddings, *J. Chem. Phys.*, 26:1755 (1957).
3. J. J. van Deemter, F. J. Zuiderweg, and A. Klinkenberg, *Chem. Eng. Sci.*, 5:271 (1956).
4. J. F. K. Huber and J. A. R. J. Hulsman, *Anal. Chim. Acta*, 38:305 (1967).
5. J. J. Kirkland, *J. Chromatogr. Sci.*, 7:7 (1969).
6. J. N. Wilson, *J. Am. Chem. Soc.*, 62:1583 (1940).
7. D. De Vault, *J. Am. Chem. Soc.*, 65:532 (1943).

8. E. Glueckauf, *Disc. Faraday Soc.*, 7:12 (1949); *Trans Faraday Soc.*, 51:34, 1540 (1955).
9. A. J. P. Martin and R. L. M. Synge, *Biochem. J.*, 35: 1358 (1941).
10. L. G. Craig, *Analy. Chem.* 22:1346 (1950); (see also B. L. Karger, L. R. Snyder, and C. Horvath, *An Introduction to Separation Science*, Wiley-Interscience, New York, 1973, p. 110).
11. A. I. M. Keulemans, *Gas Chromatography*, 2nd. ed., Reinhold, Baltimore, 1959.
12. J. J. van Deemter, 2nd Informal Symposium of Gas Chromatography Discussion Group, Cambridge Sept. 1957 (quoted in *Gas Chromatography*, J. H. Purnell, Wiley, London, 1962).
13. J. H. Knox and M. Saleem, *J. Chromatogr. Sci.*, 7:614 (1969).
14. *Supelco Reporter 15(6)*:6 (1996).
15. M. Golay, in *Gas Chromatography 1958*, Desty, Ed. Butterworth, London, 1958, p.35.
16. Sir G. I. Taylor, *Proc. Roy. Soc. (London)*, A219:186 (1953).
17. R. Aris, *Proc. Roy. Soc. (London)*, A235:67 (1956).
18. J. H. Knox and M. T. Gilbert, *J. Chromatogr.*, 186:405 (1979).
19. O. van Berkel-Geldof, J. C. Kraak, and H. Poppe, *J. Chromatogr.*, 499:345 (1990).
20. S. Eguchi, J. G. Kloosterboer, C. P. G. Zegers, P. J. Schoenmakers, P. P. H. Tock, J. C. Kraak, and H. Poppe, *J. Chromatogr.*, 516:301 (1990).
21. Proceedings published as *Analy. Chem. 35(4)*:425–549 (1963).
22. *Advances in Chromatography*, Vol. 1, Giddings and Keller, Eds., Marcel Dekker, New York, 1965.
23. J. H. Knox and L. McLaren, *Analy. Chem.*, 35:449 (1963).
24. J. C. Giddings, *Analy. Chem.*, 35 1338 (1963); see also Ref. 1, pp. 57, 58.
25. J. H. Knox and M. Saleem, *J. Chromatogr. Sci.*, 7:745 (1969).
26. R. H. Perrett and J. H. Purnell, *Analy. Chem.*, 34:1336 (1962).
27. J. H. Knox and M. Saleem, *J. Chromatogr. Sci.*, 10:80 (1972).
28. J. H. Knox and L. McLaren, *Analy. Chem.*, 36:1477 (1964).
29. J. H. Knox and H. P. Scott, *J. Chromatogr.*, 282:297 (1983).
30. J. C. Gidding, *Analy. Chem.*, 34:1186 (1962); 35:1338 (1963); see also Ref. 1, p. 54.
31. V. Pretorius and T. W. Smuts, *Analy. Chem.*, 38:274 (1966).
32. J. H. Knox, *Analy. Chem.*, 38:253 (1966).
33. J. H. Knox and J. F. Parcher, *Analy. Chem.*, 41:1599 (1969).

34. D. S. Horne, J. H. Knox, and L. McLaren, *Separation Science*, *1*:531 (1966).
35. J. H. Knox, G. R. Laird, and P. A. Raven, *J. Chromatogr.*, *122*:129 (1976).
36. P. A. Bristow and J. H. Knox, *Chromatographia*, *10*:279 (1977).
37. J. C. Giddings, *Analy. Chem. 35*:439 (1963): see also Ref. 1, p 95.
38. See Ref. 1, pp. 105, 106.
39. G. K. Ackers, *Biochemistry*, *3*:723 (1964); (see also W. W. Yau, J. J. Kirkland, and D. D. Bly, *Modern Size-Exclusion Liquid Chromatography*, Wiley-Interscience, New York, 1979. p. 46).
40. L. H. Tung, *J. Appl. Pol. Sci. 10*:375 (1966).
41. L. H. Tung and J. R. Runyon, *J. Appl. Pol. Sci.*, *13*:2397 (1960).
42. J. H. Knox and F. McLennan, *Chromatographia*, *10*:75 (1977).
43. J. H. Knox and F. McLennan, *J. Chromatogr. 185*:289 (1979).
44. J. V. Dawkins and G. Yeadon, *J. Chromatogr.*, *188*:333 (1980).
45. J. V. Dawkins and G. Yeadon, *J. Chromatogr.*, *206*:215 (1981).
46. M. E. van Kreveld and N. van den Hoed, *J. Chromatogr.*, *149*:71 (1978).
47. J. H. Knox and M. Shibukawa, *J. Chromatogr.*, *545*:123 (1991).
48. J. W. Jorgenson and K. D. Lukacs, *Analyt Chem.*, *53*:1298 (1981).
49. J. W. Jorgenson and K. D. Lukacs, *J. Chromatogr.*, *218*:209 (1981).
50. V. Pretorius, B. J. Hopkins, and J. D. Sheike, *J. Chromatogr.*, *99*:23 (1974).
51. J. H. Knox and I. H. Grant, *Chromatographia*, *24*:135 (1987).
52. J. H. Knox, *Chromatographia*, *26*:329 (1988).
53. J. H. Knox and I. H. Grant, *Chromatographia*, *32*:317 (1991).

2
The Stochastic Theory of Chromatography

Francesco Dondi, Alberto Cavazzini, and Maurizio Remelli
University of Ferrara, Ferrara, Italy

I.	INTRODUCTION	52
II.	THEORETICAL APPROACHES IN CHROMATOGRAPHY	52
III.	LEVEL ONE THEORY: STRUCTURAL STOCHASTIC CONCEPTS	55
IV.	LEVEL TWO THEORY: FUNDAMENTALS AND BASIC ACHIEVEMENTS OF THE STOCHASTIC THEORY OF CHROMATOGRAPHY	56
V.	LEVEL THREE THEORY: SOME APPLICATIONS OF THE STOCHASTIC THEORY TO EXPERIMENTAL PEAK SHAPE ANALYSIS	63
VI.	THE CHARACTERISTIC FUNCTION METHOD IN THE STOCHASTIC THEORY OF CHROMATOGRAPHY	64
	REFERENCES	73

I. INTRODUCTION

The paper entitled "Molecular Dynamic Theory of Chromatography" was the first, and most significant, contribution to the theory of chromatography made by J. Calvin Giddings, in conjunction with Eyring [1]. It was further developed into a second paper entitled "Stochastic Consideration on Chromatographic Dispersion" [2]. A new and original theory of chromatography came forth from these two papers: the stochastic theory of chromatography. In this theory the process of chromatography, considered from a molecular point of view as a chain of sorption-desorption processes, is ruled by probabilistic laws [3,4]. Reading these historical papers of the young Cal (the first paper made in cooperation with Eyring was sent to the *Journal of Physical Chemistry* on November 1, 1954, when he was 24 years old) one can perceive the fascination of a mind carefully studying physical processes and building a model. Then, in a nearly pedagogical fashion, he accompanies the reader to see and touch first the most significant aspects, and then the details, in the proper and necessary order. One steadfast characteristic of Giddings was his ability to adopt the degree of theoretic complexity that the object in question required, producing as simple as possible a description which could be understood by everyone.

II. THEORETICAL APPROACHES IN CHROMATOGRAPHY

The structure of theoretical investigation in chromatography was deeply focused by Giddings in the opening lecture at the Brighton Fifth Symposium on Gas Chromatography in 1964 [5]. He singled out three levels of chromatography theory.

The first level involves structural concepts of the chemistry and physics of both the chromatographic process and system: the spatial structure of both flow and stationary phase, flow hydrodynamics, thermodynamics of the phase exchange processes, kinetics of the mass transfer processes. Several theoretical fields can be involved such as solution theory, surface adsorption theory, transport processes, classical thermodynamics, reaction rate theory, and porous media science. At this level, chromatography has not yet been recognized.

The second level is the core of true chromatographic theory: zone evolution is considered and a link is established with the flow, the kinetics, and the equilibrium parameters handled with at the first level.

The stochastic theory of chromatography belongs to level two; i.e., it is truly a chromatographic theory. However, stochastic concepts widely appear in the first level of theoretical description. For example the stochastic aspects of flow in porous media, the space configuration of the stationary liquid phase, the space configuration of the sorption sites for adsorption processes, all have stochastic features which are described by frequency functions, $f(x)$, with the pertinent statistical attributes such as the mean (m), variance (σ^2) and higher moments about the origin (μ_j) and central moments (μ'_j) [6]:

$$m = \int x f(x)\, dx \tag{1}$$

$$\sigma = \int (x-m)^2 f(x)\, dx \tag{2}$$

$$\mu_j = \int (x-m)^j f(x)\, dx \tag{3}$$

$$\mu'_j = \int x^j f(x)\, dx \tag{4}$$

Most important, the phase exchange and transport processes occurring in the column are stochastic processes the statistical parameters of which are a function of time, t, as in the basic Einstein diffusion equation [7]:

$$\sigma_l^2 = 2Dt \tag{5}$$

where σ_l is the spatial thickness of the spreading band and D the diffusion coefficient. Thus, stochastic aspects are fundamental in many aspects of chromatography and they appear in *all* types of chromatographic theory.

The third level contains theory applied to the laboratory and practical problem solving (e.g., elution programming, optimization, preparative scale columns, data handling), and lives out the achievements attained in level two. For example, one of the achievements of the stochastic theory is to provide a general description of peak shape properties and their evolution along the column. This is done through the Edgeworth-Cramér expansion [8], where the parameters can, in turn, be related to the first-level theory parameters. The first-level parameters, such as diffusion coefficient and other column parameter data, can thus be determined from a chromatographic experiment [9].

The method employed for developing theory is of primary importance for success. However, it is not to be confused with the theory itself. Here a brief mention of the three different methodological tools of the stochastic theory of chromatography can be significant. There is the intuitive approach aiming to provide a conveniently simple picture of the problem; in his scientific production, Giddings mastered all aspects of this approach, most prominently the stochastic description of the chromatographic process [3,4,10]. Obviously, there are rigorous mathematical methods for solving the pertinent model [1,8,11–13]; there are, in addition, computational approaches of two types: numerical simulation methods [14] and mathematical-symbolical methods very useful now in exploiting complex model cases [15].

The intuitive approach, based on stochastic considerations, has played important historical role in chromatography theory. This is the case of the derivation of the key **A** term of the van Deemter equation [3,4], also called the eddy diffusion term:

$$H_A = 2\lambda d_p \tag{6}$$

where λ is a constant and d_p is the packing particle diameter.

Mathematical methods employed in chromatography theory are generally highly complex. One can look, for example, at the classical mass balance approach papers of Walter [16], Thomas [17], papers on linear chromatography, or Golay's highly successful papers [18] establishing the use of capillary columns, and, in general, all those papers dealing with nonlinear chromatography [19]. The technical difficulties and the general level of mathematics involved in the stochastic description of chromatography were highly complex, and this difficulty hindered full development [1,2]. The use of powerful methods based on the characteristic function (CF, see below) [20] made it possible to fully develop the stochastic theory of chromatography, thus solving the major problems of linear chromatography [8,15,21]. Hence, the role of the limiting theories, the rate of convergence to them, the peak shape in the most general cases, and the kinetic tailing problem were better focused as one aspect of the central limit theorem of probability theory [9,21]. However, it must be emphasized that the basic idea of the stochastic theory of chromatography belongs to Giddings and Eyring, whereas the subsequent achievements are, instead, due to a most proper choice of the mathematical method together with the proper use of the main results of modern probability theory [20]. In practice only the nonlinear case has escaped rigorous handling in the stochastic description context. Most likely the point can be handled using a numerical simulation approach [22].

III. LEVEL ONE THEORY: STRUCTURAL STOCHASTIC CONCEPTS

Migration of an individual molecule through the chromatographic medium is a highly complex, random process. It can be generally described as a complex chain of random processes of different nature: ordinary diffusion, flow pattern effects (eddy diffusion), and sorption-desorption kinetics. However, it is the sorption-desorption process which determines differential chromatographic migration.

Figure 1 describes, from a stochastic point of view, the behavior of a molecule sorbing from the mobile (gas) phase on a single sorption site. This may be an adsorption site of a surface or a liquid phase. It is a basic physicochemical property that the time spent by a single molecule on a single site, τ_s, is a random variable [23]. In fact, a molecule passing near a sorption site is "captured" by the sorption site when the sorption energy, E_a, overcomes the molecule kinetic energy, E_k.

The reverse process—that is the desorption step—will occur when the kinetic energy of the sorbed molecule once more overcomes the sorption energy ($E_k > E_a$). The probability, per time unit, of such an

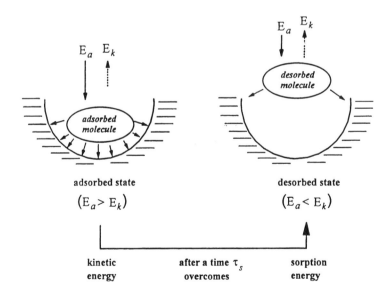

Fig. 1 Kinetics model for the adsorption-desorption process.

event is essentially related to the continuous energy exchange between the sorbed molecule and sorption site and to the fluctuations in energy. According to this simple picture the total sorption time is an example of a "waiting time" stochastic process [24]. The simplest reported expression for the frequency function of τ_s is

$$f(\tau_s) = \mathbf{k}_d \exp(-\tau_s \mathbf{k}_d) \tag{7}$$

where $\mathbf{k}_d$ is the process time constant related to the average time spent by the molecule on the sorption site:

$$\bar{\tau}_s = \frac{1}{\mathbf{k}_d} \tag{8}$$

This quantity is not easy to measure experimentally [25]. Its theoretical evaluation can be obtained from the Frenkel equation [23,26]:

$$\bar{\tau}_s = \tau_0 \exp\left(\frac{E_a}{RT}\right) \tag{9}$$

where τ_0 is approximately 1.6×10^{-13}s, $\mathbf{R}$ is the gas constant, and T is the absolute temperature. For example, in physical adsorption processes, E_a ranges from 0.1 to 20 kcal/mol and the mean sorption time varies from 1.2×10^{-13} to 100 s at 298 K [23].

In a more general way the phase exchange process can be considered as a stepwise kinetic process where the theory of reaction rates must be applied. The problem is to find the appropriate activation parameters [5]. It seems that this way has not been fully exploited; rather, a simplified equilibrium approach was preferred. The point would deserve major consideration, especially in the case of complex separations involving complex stepwise kinetics.

IV. LEVEL TWO THEORY: FUNDAMENTALS AND BASIC ACHIEVEMENTS OF THE STOCHASTIC THEORY OF CHROMATOGRAPHY

At this point, a detailed description of the sorption-desorption kinetics is not required. Instead, it is worth nothing that, due to the random nature of the basic chromatographic step, no single molecule repeats the chromatographic process in the same way over equivalent sites, nor do different molecules perform it identically on the same site [4,8,27].

Figure 2 describes the general features of the simplest stochastic model of chromatography: all the molecules are assumed to move in the mobile phase at a constant velocity, $\mathbf{v}_m$ (constant mobile phase stochastic model). In such a model, every molecule spends the same time, t_M, in the mobile phase. The progression of an individual molecule through the column can be described by a trajectory in the coordinate plane time-column length (t, l), respectively the time elapsed from the injection and the column distance covered. The process ends when the solute molecule has covered a column distance equal to the column length L. The retention time of a single solute molecule corresponds to the intersection of its trajectory with a horizontal axis located at $l = L$. The chromatographic elution peak is the probability density function of this trajectory cross section. Moreover, the cross with a vertical axis located at a time t gives the band profile inside the chromatographic medium at a given time t. In the first case (elution mode) the random quantity to be considered is time, while in the second case (development mode) it is distance. With reference to Fig. 2, the elution peak of an unretained compound is a Dirac function located at $t = \mathbf{t}_M$:

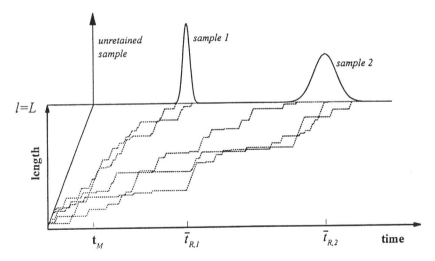

Fig. 2 Example of random trajectories (t, l) of solute molecules moving forward in the chromatographic medium.

$$\mathbf{t}_M = \frac{L}{\mathbf{v}_m} \tag{10}$$

A horizontal segment of a molecule trajectory represents the sorption time, τ_s, in a sorption site. This quantity is random and independent since the molecule has no memory of its past history. All the slanting segments refer to the molecule in the mobile phase and they have the same slope, $\mathbf{v}_m$. The column distance covered between two successive sorption steps is a random quantity since the time elapsed between a desorption step and the subsequent sorption step, τ_m, is random. Moreover, the total number of entries in the stationary phase (n) is a random quantity of integer type. The total time spent by an individual molecule inside the column is, thus, a sum of random quantities, but the number of terms in this sum is not the same for each molecule. The mathematics of the simplest stochastic model of chromatography reduces to the problem of finding a solution for the sum of a random number of random independent quantities [24]. A specific chromatographic system is then characterized by its pertinent distributions of n and τ_s. The CF method is able to find the general solution for this problem, provided some very general conditions are respected. However, in order to avoid plunging the reader directly into a complex mathematical procedure, an intuitive approach has been developed. This proves most convenient in singling out the basic aspects of the stochastic approach.

The basic parameter in chromatography is the retention ratio:

$$R = \frac{\bar{t}_M}{\bar{t}_R} \tag{11}$$

or the capacity factor:

$$k' = \frac{\bar{t}_R - \bar{t}_M}{\bar{t}_M} \tag{12}$$

which is known to be related to the phase ratio β:

$$\beta = \frac{V_s}{V_m} \tag{13}$$

and to the distribution coefficient, K_d:

$$k' = \frac{K_d}{\beta} \qquad (14)$$

In Eq. (13), V_s and V_m are the stationary-phase and the mobile-phase volumes, respectively. In Eq. (12) $\bar{t}_R - \bar{t}_M$ represents the mean total time spent by a molecule in the stationary phase, $\bar{t}_s$. If $\bar{n}$ is the mean number of sorption-desorption steps performed by a molecule,

$$\bar{t}_s = \bar{n}\,\bar{\tau}_s \qquad (15)$$

$$\bar{t}_M = \bar{n}\,\bar{\tau}_m \qquad (16)$$

where $\bar{\tau}_s$ and $\bar{\tau}_m$ are the mean times spent in the stationary and mobile phases for a single step. If the means performed in the two phases with respect to the space or to time are equal [4]; i.e., if the ergodic hypothesis holds, one has

$$\bar{\tau}_s \propto c_s V_s \qquad (17)$$

$$\bar{\tau}_M \propto c_m V_m \qquad (18)$$

where c_s and c_m are respectively the stationary and mobile equilibrium concentrations, and by combining Eqs. (15)–(18) one has Eq. (14).

Even if derived by an intuitive approach, Eq. (14) makes reference to a basic ergodic hypothesis. However, it can be derived by using more rigorous mathematical methods, and, indeed it has been demonstrated that it holds not for the peak maximum but for the peak mass center. In the case of tailing effects, this aspect can be significant. By using the CF method in stochastic theory, it was possible to derive a similar expression for two-site case [15] as well, and this can be logically extended to any complex multiphase system; i.e.,

$$k' = \sum_i \frac{K_{d,i}}{\beta_i} \qquad (19)$$

where the subscript i refers to different stationary sites.

The HETP derivation is the other goal of any theory of chromatography. The fundamental **C** term of the van Deemter expression is

$$H_C = 2R(1-R)\mathbf{v}_m \bar{\tau}_s \qquad (20)$$

It is interesting that this equation, or an equivalent form of this equation, has been derived many times, even by using the intuitive random walk model approach [3,4]. Equation (20) was also obtained and generalized by using the CF method [8]. The general expression is indeed well structured:

$$H_C = \left[\left(\frac{\sigma_s}{\bar{\tau}_s}\right)^2 + \frac{\sigma_n^2}{\bar{n}}\right] R(1-R) \mathbf{v}_m \bar{\tau}_s \qquad (21)$$

The term in brackets (corresponding to factor 2 in Eq. 20) contains two dispersion contribution factors: the first coming from stationary-phase kinetics ($\sigma_s/\bar{\tau}_s$) and the second from mobile-phase kinetics ($\sigma_n/\sqrt{\bar{n}}$), σ_s being the standard deviation of the site sorption time and σ_n the standard deviation of the number of sorption-desorption steps. If both the stationary- and mobile-phase entry processes follow a first-order kinetics, both the dispersion terms in Eq. (21) are equal to 1 and their sum is just the factor 2 appearing in the simplified expression. Hence, the slope value in the **C** term has a clear stochastic explanation and can be a guide for interpreting in a general way the separate influence of the mobile-phase and stationary-phase processes over column efficiency.

As a second example, if a first-order kinetics for the stationary-phase entry process is hypothesized, while the sorbing molecule behaves in a bifunctional mode, e.g., performing either a strong H-bonding interaction with $\tau_{s,1}$ or a weak lipophylic interaction with $\tau_{s,2}$, respectively with weighting factors p and $1-p$, the H_C term becomes [8]

$$H_C = R(1-R) \frac{2p\bar{\tau}_{s,1}^2 + 2(1-p)\bar{\tau}_{s,2}^2}{p\bar{\tau}_{s,1} + (1-p)\bar{\tau}_{s,2}} \mathbf{v}_m \qquad (22)$$

In this expression the site sorption-time-dependent part has a greater dispersivity value than in homogeneous models under equal conditions of R and $\mathbf{v}_m$ values. This explains the significant efficiency loss in these cases [8]. It is relevant that the same equation holds for a column made up of two types of sites by simply interpreting the weighting factors as the effective abundance of each of the two sites [15]. This result has been derived by means of CF applied to stochastic theory: that site heterogeneity and complexity of the stationary phase interaction thus have equivalent effects on H. This can be interpreted as a sort of theorem in stochastic theory of chromatography, in the sense of equivalence between the sorbing molecule and the sorption site.

Another significant result attained with the CF approach is the simultaneous derivation of the **B** and **C** terms by handling a stochastic model of chromatography with nonconstant mobile-phase velocity [21]:

$$H = H_B + H_C \qquad (23)$$

which is an example of a more general property of the independence and of the additivity of the different H terms.

The k' and H values define respectively the band position and its spreading. If the band is Gaussian, these two quantities alone define the its profile:

$$Z(c) = \frac{1}{\sqrt{2\pi}} \exp\left(-\frac{c^2}{2}\right) \qquad (24)$$

where c is the standardized variable:

$$c = \frac{t - \bar{t}_R}{\sigma} \qquad (25)$$

and σ is the standard deviation of the peak, related to column length, L, H, and $\bar{t}_R$ by

$$\sigma = \bar{t}_R \sqrt{\frac{H}{L}} \qquad (26)$$

This representation is thus possible if H and $\bar{t}_R$ are given.

In general, one searches for the exact peak shape function. Giddings [3] was able to obtain the only "exact" solution available to date:

$$f(t_s) = \left(\frac{\mathbf{k}_a \mathbf{k}_d t_M}{t_s}\right)^{1/2} \exp(-\mathbf{k}_a t_M - \mathbf{k}_d t) \sum_{r=0}^{\infty} \frac{\left(\sqrt{\mathbf{k}_a \mathbf{k}_d t_M \tau_s}\right)^{2r+1}}{r!(r+1)!} \qquad (27)$$

where $\mathbf{k}_a$ and $\mathbf{k}_d$ are, respectively, the adsorption and the desorption rate constants and r is the running index. Its attempt to solve, in a similar manner, the two-site case did not provide tractable expressions [2]. It is obvious that the unsuccessful result in handling even simplified models discouraged any further attempts to "exactly" solve more complex cases. These appeared workable only under the "long-time" or "long-column" or "ideal" approximations; in these conditions, the peak can be well approximated by a Gaussian [5,19].

The long-time approximation is a key question in chromatography.

The basic fact giving rise to the Gaussian function relies on the intimate structure of the chromatographic process. As described above, it consists of a great number of elementary incremental processes (similar to diffusion or Brownian movement) and related to the central limit theorem in probability theory [28]. The rate of convergence to the Gaussian law [8] is obviously the underlying question in the problem of long-time approximation of chromatography and, connected with it, there are the following problems:

1. Why and how does the peak depart from the Gaussian shape?
2. How can we gain better peak shape approximation?
3. How can we express a "distance" between the actual shape and the Gaussian one?
4. How does this distance go to zero when the long-time condition is reached?

The elucidation of these points in linear chromatography was a significant achievement of the subsequent developments of stochastic theory of chromatography [8,21,29] and can be referred to the theory of the addition of random variables. This theory was clearly presented in the classical booklet by Cramér [20], where the theory of the Edgeworth-Cramér expansion is also reported. The Edgeworth-Cramér expansion, being intimately related to the central limit theorem, is practically the general answer to the above-reported questions. It must, therefore, be considered not as a simple fitting tool, but as the fundamental peak shape function in linear chromatography, since the Gaussian function is only the zeroth-order term and the limiting expression. The Edgeworth-Cramér expansion of kth-order development for a distribution function $F(c)$ is expressed as [20]

$$F(c) = P(c) + \sum_{j=1}^{k} Q_j(-P) + R_k(c) \tag{28}$$

where

$$P(c) = \int_{-\infty}^{c} Z(u)\, du \tag{29}$$

is the Gaussian distribution function, the $Q_j(-P)$ terms are linear aggregates as a function of $P(c)$-derivatives of maximum order $3j$ and of cumulant coefficients, and $R_k(c)$ is the remainder term expressing the "distance" between the actual function and the approximating Edgeworth-Cramér expansion [29,30]. The most relevant fact is that under wide conditions the peak shape can be computed at any degree of approximation provided that the cumulant coefficients are available. How to compute the cumulant coefficients for a chromatographic model will be specified in the next section. Another relevant fact is that the remainder term also has a well-defined structure, and its dependence on the number of sorption-desorption steps—equivalent to either column length or inverse mobile-phase velocity—is given by

$$|\ln R_k(c)| \lesssim M_k - \frac{k+1}{2}\ln\overline{n} \qquad (30)$$

where M_k is an undefined constant depending on k. In addition, for $k = 0$, i.e., for the Gaussian approximation, the constant M_0 is dependent on the absolute third moment of the sorption site [8]. The remainder term was approximated by the "Levy distance" [8,29]. For example, it can be shown that in bifunctional interaction the absolute third moment is greater than in homogeneous interaction, and its value accounts fairly well not only for the Levy distance from the Gaussian shape but also for the rate of convergence to the Gaussian law upon increasing column length (see Eq. 30) [8]. In principle, all these points exhaust the entire content of the limiting theories and of their approximation degrees and thus of the problem of the peak shape associated with slow kinetics and surface heterogeneity [31] under linear conditions.

V. LEVEL THREE THEORY: SOME APPLICATIONS OF THE STOCHASTIC THEORY TO EXPERIMENTAL PEAK SHAPE ANALYSIS

Applied chromatographic theories have the most general validity when they are based on the fundamental concepts as developed in levels one or two [5]. One of the most important functions of these theories is to describe the peak profiles as measured in a real chromatographic system through the detector response. It is well known that the total peak variance, σ^2_{tot}, is the sum of that produced into the column, σ^2_{col}, and that produced in different parts of the chromatographic systems, called extra-column contributions, $\sigma^2_{i,extra}$ [32]:

$$\sigma_{\text{tot}}^2 = \sigma_{\text{col}}^2 + \sum_i \sigma_{i,\text{extra}}^2 \tag{31}$$

This relationship holds true for higher cumulants of order j [6]:

$$\kappa_{j,\text{tot}} = \kappa_{j,\text{col}} + \sum_i \kappa_{j,i,\text{extra}} \tag{32}$$

(The first three cumulants are equal to the central moments (see below), whereas the subsequent ones are a linear combination of them [30].) The Edgeworth-Cramér expansion, which is a function of the cumulants, even holds for the peak shapes resulting from the column and extra-column processes and can thus properly represent the real chromatographic shape. The Edgeworth-Cramér series was thus the base for a nonlinear fitting of the chromatographic peak shape in order to evaluate both column and extra-column processes [9].

VI. THE CHARACTERISTIC FUNCTION METHOD IN THE STOCHASTIC THEORY OF CHROMATOGRAPHY

In this section the basic properties of the CF are recalled and their usefulness for theoretical chromatographic modeling is focused [6,8,20,21,24]. The CF, $\Phi(\xi)$, is the expectation of a particular function $g(X)$:

$$g(X) = \exp(i\xi X) \tag{33}$$
$$\Phi(\xi) = \int \exp(i\xi x) f(x) dx \tag{34}$$

where $\mathbf{i}$ is the imaginary unit, ξ is an auxiliary variable, and $f(x)$ is the density function of the *continuous random variable* X (of argumental values x). In the case of the *discontinuous random variable*, the CF is defined as

$$\Phi_{\text{disc}}(\xi) = \sum_k q_k \exp(i\xi x_k) \tag{35}$$

where q_k is the probability of x_k.

CF transformation of distributions has two basic properties. (1) Since the random independent variable addition results in a convolution product of the respective distributions; i.e.,

$$f_{\text{tot}}(y) = \int f_A(y-x) f_B(x) dx = f_A * f_B \tag{36}$$

(where the asterisk denotes convolution and A and B denote the random variables), the corresponding CF transformation will simply be the product of the respective CFs (Φ_A and Φ_B):

$$\Phi_{tot}(\xi) = \Phi_A(\xi)\Phi_B(\xi) \tag{37}$$

(2) In the combination of mutually exclusive events, it exhibits distributive properties:

$$\Phi_{tot}(\xi) = q_A\Phi_A(\xi) + q_B\Phi_B(\xi) \tag{38}$$

where q_A and q_B are the weighting factors for the random events A and B.

The original stochastic model proposed by Giddings and Eyring [1] can be expressed by independent event addition (see above) and hence can be formulated as a CF product (thus avoiding convolution integrals which are difficult to handle mathematically).

The name "characteristic function" derives from the fact that it fully defines the pertinent probability function since all the moments can be obtained from it, according to the following expression [6]:

$$\Phi^{(j)}(0) = \left[\frac{d^j\Phi(\xi)}{d\xi^j}\right]_{\xi=0} = \mathbf{i}^j\mu'_j \tag{39}$$

where $\Phi^{(j)}(0)$ represents the jth CF derivatives in $\xi = 0$.

Since the chromatographic peak can be considered as the probability density function of the column sojourn time, its CF fully defines it. $\Phi(\xi)$ is also called the first CF (1stCF). For many applications, the logarithm of CF (called the second CF, 2ndCF) is considered [6,20]:

$$\Psi(\xi) = \ln[\Phi(\xi)] \tag{40}$$

In the 2ndCF the jth derivative (in $\xi = 0$) defines the cumulant of order j, κ_j:

$$\Psi^{(j)}(0) = \left[\frac{d^j\Psi(\xi)}{d\xi^j}\right]_{\xi=0} = \mathbf{i}^j\kappa_j \tag{41}$$

The first two cumulants are respectively the mean and the variance:

$$\kappa_1 = \mu'_1 = m \tag{42}$$
$$\kappa_2 = \sigma^2 \tag{43}$$

If the cumulants are normalized with respect to σ, the cumulant coefficients are defined:

$$\gamma_{r-2} = \frac{\kappa_r}{\sigma^r} \quad r = 3, 4, \ldots \qquad (44)$$

The first two cumulant coefficients are called, respectively, skewness (S) and excess (E):

$$S = \frac{\kappa_3}{\sigma^3} \qquad (45)$$

$$E = \frac{\kappa_4}{\sigma^4} \qquad (46)$$

S and E are commonly referred to as the frequency function shape properties, i.e., respectively, tailing and extent of flatness with reference to the Gaussian peak shape for which S and E are both zero.

By means of moments and cumulants, it is also possible to express the most important chromatographic quantities [21]. For example, plate height, H, is

$$H = \left(\frac{\sigma_R}{t_R}\right)^2 L = \frac{\kappa_2}{\kappa_1^2} L \qquad (47)$$

where σ_R^2 is the standard deviation referred to total retention process.

If the aim is only to derive the **C** term of the van Deemter equation, i.e., investigating only the mobile to stationary-phase kinetics, a useful reformulation of the H expression is

$$H = R(1-R)\frac{\sigma_R^2}{\bar{t}_s}\bar{v}_m \qquad (48)$$

Under the hypothesis that all sorption sites are equal (homogenous phase model), the total time a molecule spends in the stationary phase, when it performs exactly n entries, is the sum of n equally distributed independent random variables, τ_s. The resulting sorption time frequency function is the n-fold convolution of the same frequency function, $f_s(\tau_s)$ with itself. Therefore, symbolically

$$f_{s,\text{tot}}|n = f_s^{n*} \qquad (49)$$

where the symbol |n emphasizes the fact that it is a conditional distribution, given n.

Since different n values can produce the same time spent in the stationary phase, the total frequency function, $f_{s,\text{tot}}$, will be a "weighted mixture" of f^{n*}, with weighting factors, q_n, the n entry process probabilities:

$$f_{s,\text{tot}} = \Sigma_n q_n f_s^{n*} \tag{50}$$

The corresponding 1stCF (for the CF properties 1 and 2) will be

$$\Phi_{s,\text{tot}}(\xi) = \Sigma_n q_n \Phi_s^n(\xi) \tag{51}$$

The sum is to be extended to all possible n values, and q_n has the properties of a frequency function; that is,

$$\sum_n q_n = 1 \tag{52}$$

and

$$\bar{n} = \sum_n n q_n \tag{53}$$

By using the identity

$$x^r = \exp[r \ln(x)] \tag{54}$$

Equation (51) can be written as

$$\Phi_{s,\text{tot}}(\xi) = \sum_n q_n \exp\left[\frac{i n \ln \Phi(\xi)}{i}\right] \tag{55}$$

Since the right-hand term of Eq. (55) corresponds to the CF definition of the integer random variable n (see Eq. 35), with auxiliary variable $\ln \Phi_s(\xi)/i$, one has

$$\Phi_{s,\text{tot}}(\xi) = \Phi_{n,\text{tot}}\left[\frac{\ln \Phi_s(\xi)}{i}\right] \tag{56}$$

The 2ndCF is obtained by combining Eqs. (40) and (56):

$$\Psi_{s,\text{tot}}(\xi) = \Psi_{n,\text{tot}}\left[\frac{\Psi_s(\xi)}{i}\right] \tag{57}$$

Equations (56) and (57) represent the solution for the most general chromatographic model under the constant-mobile-phase-velocity condition. Here, the CF of the *total sorption time* t_s is expressed as CF of number of entry processes in stationary phase, supported by the CF of a single sorption process. Taking into account the constant mobile-phase time, by using the CF shifting properties [6,20], the 2ndCF of the *retention time* t_R is obtained:

$$\Psi_{R,\text{tot}}(\xi) = \Psi_{n,\text{tot}}\left[\frac{\Psi_s(\xi)}{i}\right] + i\xi t_M \tag{58}$$

It is worth noting that Eqs.(56)–(58) are absolutely general, holding for all distribution types of the random variables "number of entries" and "time spent at the site," since no hypothesis has yet been advanced for these variables.

If one considers the relevant case of a Poisson distribution [30] of n:

$$q(n=r) = \frac{\bar{n}^r \exp(-\bar{n})}{r!} \tag{59}$$

whose 2ndCF is

$$\Psi(\xi) = \bar{n}[\exp(i\xi) - 1] \tag{60}$$

Equation (57) becomes

$$\Psi_{R,\text{tot}}(\xi) = \bar{n}[\Phi_s(\xi) - 1] + i\xi t_M \tag{61}$$

Moreover by using Eqs. (10) and (16), Eq. (61) can be expressed as

$$\Psi_{R,\text{tot}}(\xi) = \frac{L}{v_m\left\{\left(\frac{1}{\tau_m}\right)[\Phi_s(\xi) - 1] + i\xi\right\}} \tag{62}$$

where the linear dependence on L or $1/v_m$ is singled out. This property allows one to classify the chromatographic process described above (Eqs. 59—62) as a *stochastic process with stationary and independent increments*, which can be correctly represented by the Edgeworth-Cramér series expansion [20]. In technical terms it is established that the Edgeworth-Cramér series are asymptotically convergent approximations for the general linear chromatographic process. In this way a general solution to the peak shape problem under linear condi-

Table 1 Stochastic Model of Chromatography Investigated by CF Method (constant mobile-phase velocity)

	Hypotheses on surface type	Hypotheses on entry distribution in stationary phase	Hypotheses on single site sorption time distribution	Equivalence	References
Model I	Homogeneous	None	None		8
Model II	Homogeneous	Poisson	None		8
Model III	Homogeneous	None	Exponential		8
Model IV	Homogeneous	Poisson	Exponential	Giddings-Eyring McQuarrie	1,8,12
Model V	Homogeneous	Constant	Exponential	Martin-Synge	8,35
Model VI	Homogeneous (bifunctional molecule on a sorption site)	Poisson	Mixed exponential	Model VIII	8,15
Model VII	Heterogeneous (two different sorption sites)	Poisson	Sites 1 and 2 Different exponentials	Model VII	8,15

tions is attained. In practice, this condition holds under all cases where the column is homogeneous in length, i.e., the most general condition in chromatography. It must be emphasized that only the Edgeworth-Cramér expansion is suitable for representing the chromatographic peak shape. For example, the Gram-Charlier series expansion, proposed by other authors [12,33], does not exhibit the same fundamental properties, since its terms are not ordered according to their magnitude [34].

A series of chromatographic models has been investigated by using the CF method, and they are summarized in Table 1. In addition, the above treatment has also been extended, in order to consider nonconstant mobile-phase velocity [21].

At the end of the 1980s, during a stay at the Department of Chemistry of the University of Utah, where one of the authors discussed with Cal the recent advances in the stochastic theory of chromatography mentioned above, he focused on how many things are still open to research in chromatography: most notably the problem of the complete structure of the flow pattern in chromatographic beds and the problem of kinetic tailing. Both problems still require a solution, but can be dealt with in the context of stochastic theory. Along these lines, a number of questions still calling for investigation are the main lines and the key questions clearly set out by J.C.G.

SYMBOLS

A, B, C	van Deemter equation terms
A, B	random variables
c	standardized random variable
c_s, c_m	stationary and mobile phase equilibrium concentrations
CF	characteristics function
1stCF	first characteristic function
2ndCF	second characteristic function
d_p	packing particle diameter
D	diffusion coefficient
$f, f_A, f_B, f_{s,\text{tot}}$	probability density functions respectively undefined, for the random variables A and B, and referred to a total stationary phase process
E	excess
E_a, E_k	adsorption and kinetic energy respectively

F	distribution function
H	plate height
H_A, H_B, H_C	van Deemter equation terms (respectively the *eddy* diffusion, or **A**, term, the longitudinal diffusion, or **B**, term, and the kinetic, **C**, term)
i	imaginary unit
$\mathbf{k}_a, \mathbf{k}_d$	adsorption and desorption process time constants
k'	capacity factor
K_d	distribution coefficient
$K_{d,i}$	site **i** distribution coefficient
l	distance from the column inlet
L	column length
m	mean or mass center of distribution
M_k	constant in the expression of the Edgeworth-Cramér series remainder term
n	number of stationary phase entries (or number of equally distributed random variables)
p	statistical factor of the sorption state or heterogeneous site abundance
P	normal distribution function
q_j	probability or weight referred to j
$Q_j(-P)$	jth term of the Edgeworth-Cramér series expansion of a distribution function
R	retention ratio
R	gas constant
$R_k(c)$	remainder term in Edgeworth-Cramér series development
S	skewness
t	time variable
t_s, t_M, t_R	random time, respectively unspecified, spent by a molecule in the stationary phase, in the mobile phase, in the column (retention time)
$\mathbf{t}_M$	constant mobile-phase time
T	temperature (K)
v_m	mobile-phase velocity
$\mathbf{v}_m$	constant mobile-phase velocity
V_s, V_m	stationary- and mobile-phase volumes
x	value of a random variable
X	random variable
Z	normal density function

β	phase ratio
β_i	site i phase ratio
γ_j	cumulant coefficient of order j
κ_j	cumulant of order j
μ_j	central moment of order j
μ'_j	moment about the origin of order j
σ	standard deviation
σ_l	spatial thickness of the spreading band
$\sigma_R, \sigma_s, \sigma_n$	standard deviation referred respectively to total retention process, to single site stay, and to total stationary entry process
τ_s, τ_m	single stepwise time spent by the solute molecule respectively in the sorption and mobile phase
$\tau_{s,1}, \tau_{s,2}$	single step time spent by a bifunctional molecule over a two-energy (1 and 2) sorption site. Or single step time spent by a molecule over a heterogenous surface with two different site types (1 and 2)
Φ	first characteristic function, 1stCF
Ψ	second characteristic function, 2ndCF
ξ	auxiliary variable

Superscripts

(j)	derivative of order j
j	j power
$-$	mean quantity
$n*$	n-fold convolution

Subscripts

a	adsorption process
col	column contribution
d	desorption process
extra	extra-column contribution
j	order (of the moment about the origin, of the cumulant coefficient) or type
m or M	mobile-phase process
n	number of stationary-phase entry process
R	retention quantity
s	stationary-phase process
tot	a total quantity, in general to the whole column process (mobile phase plus stationary phase)

Symbols

* symbol for convolution integral
| conditioning

REFERENCES

1. J. C. Giddings and H. Eyring, *J. Phys. Chem.*, *59*: 416 (1955).
2. J. C. Giddings, *J. Chem. Phys.*, *26*: 169 (1957).
3. J. C. Giddings, *Dynamics of Chromatography*, Marcel Dekker, New York, 1965.
4. J. C. Giddings, *Unified Separation Science*, Wiley, New York, 1991.
5. J. C. Giddings, in *Gas Chromatography*, A. Goldup, Ed., The Institute of Petroleum, London, 1965, p. 3.
6. H. Cramér, *Mathematical Method of Statistics*, Princeton University Press, Princeton, NJ, 1974.
7. A. Einstein, *Investigations on the Theory of the Brownian Movement*, Dover, New York, 1956.
8. F. Dondi and M. Remelli, *J. Phys. Chem.*, *90*: 1885 (1986).
9. M. Remelli, G. Blo, F. Dondi, M. C. Vidal-Majar, and G. Guiochon, *Anal. Chem.*, *61*: 1489 (1989).
10. J. C. Giddings, *J. Chem. Educ.*, *35*: 588 (1958).
11. N. G. van Kampen, *Physica*, *96A*: 435 (1978).
12. D. A. McQuarrie, *J. Chem. Phys.*, *38*: 437 (1963).
13. C. P. Woodbury, *J. Chromatogr. Sci.*, *32*: 339 (1994).
14. J. B. Phillips, N. A. Wright, and M. F. Burke, *Sep. Sci. Technol.*, *16*: 861 (1981).
15. A. Cavazzini, M. Remelli, and F. Dondi, *J. Micro. Sep.*, in press.
16. J. E. Walter, *J. Chem. Phys.*, *13*: 229 (1945).
17. H. C. Thomas, *Ann. N. Y. Acad. Sci.*, *49*: 161 (1948).
18. M. J. E. Golay, in *Gas Chromatography*, D. H. Desty, Ed., Academic Press, New York, 1958, p. 3.
19. G. Guiochon, S. Golshan Shirazi, and A. M. Katti, *Fundamentals of Preparative and Nonlinear Chromatography*, Academic Press, Boston, 1994.
20. H. Cramér, *Random Variables and Probability Distributions*, Cambridge University Press, Cambridge, UK, 1961.
21. F. Dondi, G. Blo, M. Remelli, and P. Reschiglian, in *Theoretical Advancement in Chromatography and Related Separation Tech-*

niques, NATO ASI Series C- Vol. 383, F. Dondi and G. Guiochon, Eds., Kluwer, Dordrecht, 1992, p. 173.
22. F. Dondi, unpublished results.
23. J. H. de Boer, *The Dynamical Character of Adsorption*, 2nd ed., Clarendon Press, Oxford, UK, 1968.
24. W. Feller, *An Introduction to Probability Theory and Its Applications*, Vol. II, Wiley, New York, 1966.
25. W. A. Steele, *The Interaction of Gases with Solid Surfaces*, Pergamon Press, Oxford, UK, 1974.
26. J. Frenkel, *Kinetic Theory of Liquids*, Dover, New York, 1955.
27. S. Weber, *Anal. Chem.*, *56*: 2104 (1984).
28. S. Chandrasekhar, *Rev. Mod. Phys.*, *15*: 1 (1943).
29. F. Dondi, *Anal. Chem.*, *54*: 473 (1982).
30. M. Abramowitz and I. Segun, *Handbook of Mathematical Functions with Formulas, Graphs and Mathematical Tables*, Dover, New York, 1965.
31. J. C. Giddings, *Anal. Chem.*, *35*: 1999 (1963).
32. J. C. Sternberg, in *Advances in Chromatography*, J. C. Giddings and R. A. Keller, Eds., Marcel Dekker, New York, 1956, Vol. 2, p. 205.
33. O. Grubner, in *Advances in Chromatography*, J. C. Giddings and R. A. Keller, Eds., Marcel Dekker, New York, 1968, Vol. 6, p. 173.
34. F. Dondi, A. Betti, G. Blo, and C. Bighi, *Anal. Chem.*, *53*: 496 (1981).
35. A. J. P. Martin and R. L. Synge, *Biochem. J.*, *35*: 1358 (1941)

3
Solvating Gas Chromatography Using Packed Capillary Columns

Yufeng Shen and Milton L. Lee *Brigham Young University, Provo, Utah*

I.	INTRODUCTION	75
II.	PACKED-COLUMN HPGC, SGC, SFC, AND LC	77
III.	FLOW TRANSPORT IN SGC	79
IV.	EFFICIENCY IN SGC	83
V.	SEPARATION SPEED IN SGC	91
VI.	POLARITY AND SOLVATING POWER IN SGC	101
	REFERENCES	111

I. INTRODUCTION

While the similarities of gas and liquid chromatography are much more prominent than the differences, they can be stated much more succinctly; zone migration and separation occur by virtue of the same kinds of thermodynamic, flow, kinetic, and diffusion processes, and are thus subject to the same theoretical laws.

The differences between gas and liquid chromatography are

naturally centered around the differences in gases and liquids. As far as chromatography is concerned, the most significant differences are related to some of the physical properties and their immense variation from liquids to gases.

—J. C. Giddings [1]

The different forms of chromatography are determined by the differences in physical state of the mobile phase. These include gas chromatography (GC), supercritical fluid chromatography (SFC), and liquid chromatography (LC). Adjusting the temperature and pressure can change the mobile phase from liquid to supercritical fluid to gas, with concomitant changes in their physical properties.

The physical properties of the mobile phase affect the flow characteristics, column efficiency (kinetics), and retention (thermodynamics) in the chromatographic process. The physical properties of the mobile phase involved in chromatography are mainly its viscosity, diffusivity, and solubility [1,2]. The diffusivities of solutes in the mobile phase determine the selection of column dimensions to obtain the desired column efficiency in reasonable time. Highly diffusive gases (diffusion coefficient $D_m \sim (0.01–1.0) \times 10^{-4}$ cm^2 s^{-1} [3]) allow the use of relatively large-bore (200–300 μm i.d.) open tubular columns in GC. The high column permeability allows the use of long columns (e.g., 100 m) to obtain high column efficiency [4]. The lower diffusivities in supercritical fluids ($D_m \sim (0.5–3.3) \times 10^{-4}$ cm^2 s^{-1} [3]) and liquids ($D_m \sim (5–20) \times 10^{-4}$ cm^2 s^{-1} [3]) require the use of narrow-bore open tubular columns, e.g., 50 μm i.d. for SFC and less than 10 μm i.d. for LC. When using packed columns, the column parameters that influence diffusion and separation efficiency are the size and surface structures of the packed particles. Small particles (e.g., 1.5–15 μm) are required to obtain high efficiency in SFC and LC, and large particles (e.g., 60–200 mesh, 75~250 μm diameter) are typically used in packed-column GC. The use of large particles in packed column GC is mainly based on considerations of experimental convenience such as low column inlet pressure and easy sample introduction. The use of microparticles should be an effective method to produce high column efficiency per unit column length in all three chromatographies (GC, SFC, and LC).

The viscosity of the mobile phase is related to mobile-phase flow transport. A high liquid viscosity $((30–240) \times 10^{-4}$ poise [3]) requires a high pressure drop to maintain the desired linear velocity. However, this pressure drop little affects the column performance in LC because the pressure has little effect on the properties of the mobile phase in the commonly used pressure range of LC. Although supercritical fluids and gases have relatively low viscosities (($2.0–9.9$) $\times 10^{-4}$ and ($0.5–3.5$) $\times 10^{-4}$ poise, respectively [3]), column pressure drop in GC or SFC can lead to complex problems because of the compressibility of the mobile phase. In GC, the main concern of a pressure drop is its influence on column efficiency, which primarily results from the nonuniformity in mobile-phase linear velocity [5]. In SFC, the influence of column pressure drop on both column efficiency and retention must be considered because the solvating power of the supercritical mobile phase, and the resultant solute retention, are associated with its pressure or density [6,7].

The solubility of the mobile phase is also related to solute retention in SFC and LC. The solvating power of the mobile phase allows the elution of low-volatility molecules from the column. In SFC and LC, a specific mobile phase can be selected to adjust solute retention to a desired value and, furthermore, a certain degree of selectivity [8,9].

All of the physical properties of substances mentioned above are affected by temperature. By increasing the temperature, solute diffusivities in liquids and supercritical fluids can be enhanced, liquid and supercritical fluid viscosities can be decreased, and solute solubilities in liquids and supercritical fluids can be improved.

An ideal mobile phase in chromatography should provide a diffusivity and viscosity like those of gases to provide good mobile-phase flow transport, solute mass transfer, and a certain degree of solubility for solutes which can extend the polarity and molecular weight ranges of solutes that can be analyzed. Solvating gas chromatography (SGC) has been developed for this reason in our laboratory. In this Chapter we describe the characteristics of this technology and illustrate some resultant experimental results.

II. PACKED-COLUMN HPGC, SGC, SFC, AND LC

Figure 1 illustrates the chromatographic terminology which describes the nature of the mobile phase when using packed columns. In high-pressure GC (HPGC), high pressure gases such as He and N_2 are used. Because of the low boiling points (T_b) of these light gases, gaseous conditions exist throughout the column, although high pressure is imposed on the column inlet end. In this situation, the solubility of the mobile phase is very limited and can be neglected. In LC, a single liquid state exists in the column. Solute solubility in the mobile phase is the

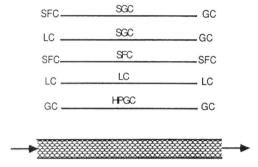

Fig. 1 Representation of chromatographic conditions at the column ends for different chromatographic modes using packed columns.

major advantage of LC. However, high mobile-phase viscosity and low solute diffusivity negatively affect mobile-phase flow and solute mass transfer. In order to improve these properties, high-temperature LC (HTLC) and CO_2-modified LC (enhanced fluidity LC, EFLC) have been developed [10–13]. In order to keep the mobile phase in a uniform liquid state at high temperature (higher than the boiling point), a restrictor or back pressure must be connected to the column outlet. However, this restrictor or back pressure limits the possible maximum mobile-phase linear velocity that a specific column inlet pressure can provide. In SFC, a restrictor or back pressure must be imposed on the column outlet in order to preserve the mobile phase as a supercritical fluid. The same limitation in mobile-phase linear velocity as is characteristic of HTLC and EFLC is a consideration also in SFC. In SGC, two conditions exist in the column and a transformation occurs from one end of the column to the other. In one case, a supercritical mobile phase is introduced into the column and a gas exits because the column outlet end is open to ambient (1 atm). The temperature is an important factor which largely determines the elution of solutes from the column. The solvating power of the mobile phase assists in this elution, and can lower the required temperature. In this situation, the mobile phase has a viscosity and diffusivity between those of gas and supercritical fluid, and a carrier gas having solvating power is used. In the other SGC case, a liquid is introduced into the column which is maintained at a temperature that is higher than the liquid boiling point, and the column outlet is open to ambient. In this situation, LC with a superheated liquid mobile phase occurs near the column inlet and a vapor carrier gas exists near the outlet. The superheated vapor mobile phase can provide low viscosity and high diffusivity while maintaining the solvating power of the mobile phase.

Because both dense phase (superheated liquid or supercritical fluid) and gas phase exist, a large difference in the physical properties of the mobile phase occurs between the column inlet and outlet. The capability of SGC to obtain high column efficiency, high speed, and separation of polar compounds has been studied. In this chapter these results are summarized and specific SGC applications are illustrated.

III. FLOW TRANSPORT IN SGC

Because of the elimination of restrictor or back pressure at the column outlet in SGC, a greater mobile-phase linear velocity can be obtained compared with SFC, HTLC, and EFLC. In pressure driven chromatography, the mobile-phase flow resistance along the column results from the friction of the mobile-phase flow along the column. The mobile-phase characteristic that accounts for this friction is viscosity, and the column parameters affecting friction include the packed bed structure, packed particle physical properties, and column length. When average values are used, the following equation can be used to describe the relationship between mobile-phase linear velocity (u), particle diameter (d_p), column length (L), mobile-phase viscosity (η), and column pressure drop (ΔP) [2]:

$$\Delta P = \frac{\phi \eta u L}{d_p^2} \tag{1}$$

In SGC, the column outlet pressure (P_o) remains at 1 atm, and the column inlet pressure (P_i) is much higher than 1 atm. Therefore, ΔP approximates P_i.

The column resistance factor (ϕ) is determined by the packing structure and other physical properties of packed particles. A previous study showed that nonporous particle packed columns provide a smaller column resistance or better column permeability than porous particle packed columns in GC, SFC, and LC [14]. Figure 2 shows the

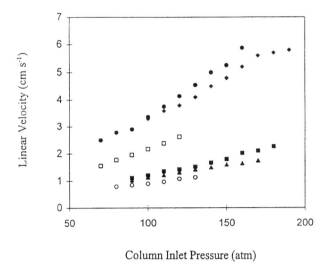

Fig. 2 Relationship between mobile-phase linear velocity and column inlet pressure in SGC. Conditions: CO_2; 130°C FID; methane used as unretained marker; (●) 20 cm × 250 μm i.d. capillary packed with 3-μm nonporous ODS particles; (◆) 10 cm × 250 μm i.d. capillary packed with 3-μm porous (80-Å pores) ODS particles; (□) 30 cm × 250 μm i.d. capillary packed with 3-μm nonporous ODS particles; (■) 20 cm × 250 μm i.d. capillary packed with 3-μm porous ODS particles; (▲) 57 cm × 250 μm i.d. capillary packed with 3-μm nonporous ODS particles; and (○) 20 cm × 250 μm i.d. capillary packed with 1.5-μm nonporous ODS particles. (From Ref. 15.)

relationship between P_i and u under SGC (SFC-GC) conditions [15]. Columns packed with 3-μm nonporous particles produced a larger mobile-phase linear velocity than those packed with the same size porous (80-Å pores) particles under SGC (SFC-GC) conditions. For example, a linear velocity of 4.17 cm s^{-1} was obtained for a 20-cm-long column containing 3-μm nonporous particles, while 1.52 cm s^{-1} was obtained for the same length column containing 3-μm porous (80-Å pores) particles at a column inlet pressure of 130 atm [15]. Because of the better column permeability, relatively long columns can be used when using nonporous particles as the packing material. A 57-cm-long column packed with 3-μm nonporous particles exhibited a similar relationship between column inlet pressure and mobile-phase linear velocity as a 20-cm-long column packed with the same size porous (80-Å pores) particles.

Since porous particle packed columns provide lower column permeability, reducing the column length produces a more significant effect on the change in mobile-phase linear velocity with pressure. For example, mobile phase linear velocities of 1.52 and 4.10 cm s^{-1} were obtained for 3-μm porous (80-Å pores) particle packed columns with column lengths of 20 and 10 cm, respectively, at 130°C and 130 atm [15]. When the column was reduced to half the length, the mobile-phase linear velocity increased 1.7 times. However, for 3-μm nonporous particle packed columns, reducing the column length from 57 to 30 cm increased the mobile-phase linear velocity from 1.32 to 2.62 cm s^{-1} at 130°C and 120 atm [15]. Reducing the column length increased the mobile-phase linear velocity by approximately the same factor.

For a specific column (set L, d_p, and ϕ), the column inlet pressure required to obtain a desired mobile-phase linear velocity is dependent on η. In SGC where the mobile phase at the inlet is a supercritical fluid, the average viscosity of the mobile phase in the column is between that of HPGC and SFC. Compared with HPGC, it can be expected that SGC would require an increased column inlet pressure to obtain a desired mobile-phase linear velocity. The properties of mobile-phase flow transport and solute mass transfer under these conditions are presently under investigation.

In SGC where the mobile phase at the inlet is a liquid, high temperatures can be used to decrease the mobile-phase viscosity. The relationship between mobile-phase linear velocity and temperature can be expressed as follows [2]:

$$u = \frac{20 P_i d_p^2}{\Phi L \eta_b} e^{-3T_b/T} \tag{2}$$

For a specific column (specific L, d_p, and ϕ) and column inlet pressure (P_i), u exponentially increases with increasing T. Figure 3 shows experimental relationships between u and $1/T$ when using acetonitrile as the mobile phase at column inlet pressures of 200 and 400 atm [16]. The experimental relationship between u and T is in agreement with that predicted by Eq. (2), even when the temperature is higher than the boiling point.

From Eq. (2), u is proportional to P_i at a specific temperature. However, both T_b and η_b are influenced by pressure. The relationship between pressure and T_b can be estimated using the Clausius-Clapey-

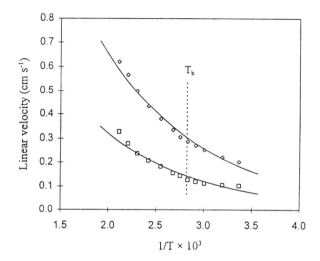

Fig. 3 Experimental relationship between mobile-phase linear velocity and temperature in RTLC, HTLC, and SGC. Conditions: 260 cm × 250 μm i.d. fused silica capillary column packed with 10-μm porous (300-Å pores) ODS particples; nitromethane used as unretained marker; acetonitrile used as mobile phase; UV detector (254 nm). ◇: 200 atm column inlet pressure, □: 400 atm column inlet pressure. (From Ref. 16.)

ron equation, however, the relationship between η_b and pressure is not clear. Experiments have shown that an excellent linear relationship (linear regression coefficient larger than 0.999) exists between u and P_i, as illustrated in Fig. 4 [16]. This suggests that the net influence of η_b and T_b on u can be neglected.

A decrease in viscosity improves the flow in SGC, which is especially significant when using long packed columns. For example, a 260-cm-long capillary column packed with 10-μm porous (300-Å pores) particles produced mobile-phase linear velocities of 0.20, 0.26, and 0.62 cm s^{-1} for RTLC (24°C), HTLC (60°C), and SGC (200°C), respectively, using a column inlet pressure of 400 atm (16).

It is noteworthy that the relationship between mobile-phase linear velocity and temperature in SGC is different from that in HTLC when the temperature is higher than the boiling point. In HTLC, a restrictor or back pressure controls the mobile-phase linear velocity.

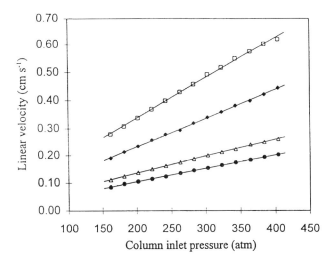

Fig. 4 Experimental relationship between mobile-phase linear velocity and column inlet pressure in SGC, HTLC, and RTLC. Conditions: □: SGC (200°C), ◆: SGC (140°C), △: HTLC (60°C), and ●: RTLC (24°C); other conditions are the same as in Fig. 3. (From Ref. 16.)

IV. EFFICIENCY IN SGC

The effect of mobile phase on column efficiency becomes more complex in the order LC, HPGC, SFC, and SGC. In LC, although high pressure may be imposed at the column inlet, the noncompressible liquid leads to little variation in physical properties of the mobile phase, such as viscosity, diffusivity, solubility, and so on, under typical conditions. The mobile phase can be considered as constant along the column. The column efficiency can be theoretically treated using Giddings' coupling theory [1], the van Deemter equation [17], or the Knox equation [18]. In HPGC, the main mobile-phase effect on column efficiency is the mobile-phase linear velocity gradient along the column [5]. In SFC, effects of both mobile-phase linear velocity and solute retention gradients on column efficiency must be considered [19,20]. In SGC, in addition to mobile-phase linear velocity and solute retention gradients, a change from liquid to gas, or supercritical fluid to gas occurs in the column, which makes the theoretical analysis of the mobile-phase effect on column efficiency extremely difficult.

For HPGC, Giddings [5] theoretically calculated the effect of the mobile-phase linear velocity gradient on column efficiency, and pointed out that the loss in column efficiency could be corrected using a compressibility factor (f_2). The maximum value of f_2 (P_I>>P_o) is 1.125, which means that the maximum loss in column efficiency would be 12.5%. Myers and Giddings packed a 4000-ft column containing 50–60 mesh particles [21]. They obtained a total column efficiency of approximately 10^6 plates, corresponding to a reduced plate height of 2, using a column inlet pressure of 2500 psi. This result experimentally illustrated that a high pressure drop and resultant large mobile-phase linear velocity gradient had limited effect on column efficiency in HPGC. However, an intolerably long analysis time was needed to obtain this high efficiency because of the use of large particles. Microparticle packed columns must be used in order to improve column efficiency per unit time or per column length.

Myers and Giddings packed a 2-m-long column with 13-µm particles [22]. A minimum plate height of 82 µm was obtained, corresponding to 24,000 plates per column and a reduced plate height of 6.3. Corcia et al. used 20–25 µm particles to pack a 21-cm-long column, and they obtained 17,000 plates per meter, corresponding to a plate height of 60 µm and a reduced plate height of 2.5–3.0 [23]. This is the lowest reduced plate height reported for packed column GC when using microparticles as packing materials, although this result is difficult to obtain in a repeatable manner. Lu et al. packed a 10-cm-long column using 7 ± 2 µm particles, and 45,000 plates per meter were obtained; corresponding to a reduced plate height of 3.1 [24]. This is the largest plate number per meter reported in packed column GC.

In packed-column GC, two apparent limitations have existed: (a) when using microparticles as packing materials, the reduced plate height has never reached as low as 2, which is commonly obtained in microparticle-packed-column LC, and (b) the total column efficiency and plate number per unit time were never as high as those obtained in open-tubular-column GC (~100,000 plates per column, 100–600 plates s^{-1}). This limited the practical use of packed columns in the separation of complex samples.

In packed-column SFC, both mobile-phase linear velocity and solute retention gradients exist. The effect of mobile-phase linear velocity gradient on column efficiency is similar to that observed in HPGC. However, There have been few publications which theoretically analyze the effect of solute retention gradient on column efficiency. In fact, it is difficult to isolate this effect from that of mobile-phase linear velocity gradient. The total effects of mobile phase on column efficiency were only experimentally illustrated for packed column SFC. Gere et al. showed that when 3-μm particles were used in packed column SFC, a reduced plate height of less than 2 could be obtained [25]. Berger and Wilson connected several LC columns together and found that the total column efficiency was proportional to the column length [26].

In SGC, the variation of mobile-phase physical properties becomes more significant compared to SFC. In this case, theoretical analysis becomes much more complex. In this chapter we only show experimental results to illustrate that little column efficiency loss is observed under SGC conditions.

Table 1 lists experimental values for total plate number (N) as a function of average mobile phase linear velocity (u) under SGC (supercritical fluid to gas) conditions [27]. Increasing the column inlet pressure and subsequently the mobile-phase linear velocity, the total column efficiency was increased for all columns packed with 10-μm porous (300-Å pores) particles within the experimental range of inlet

Table 1 Relationship between N (plates per column) and u (cm s^{-1}) in SGC (supercritical fluid to gas)[a]

Column 1		Column 2		Column 3		Column 4	
u	N	u	N	u	N	u	N
1.05	68,594	0.86	104,197	0.66	123,860	3.93	88,485
1.20	86,562	0.95	143,602	0.72	125,585	4.45	76,487
1.36	106,014	1.04	124,895	0.77	190,146	4.85	55,400
1.45	112,500	1.11	151,515	0.82	213,368	5.26	43,350
1.55	112,503	1.19	155,819	0.86	227,966	5.67	39,397
1.67	121,702	1.26	184,817	0.90	264,088	5.94	38,752
1.76	118,341	1.33	154,968	0.96	225,008	6.11	34,245

[a]Conditions: n-octane used as test solute, methane used as unretained marker, CO_2 used as mobile phase, 130°C, FID, Column 1: 180 cm × 250 μm i.d. capillary column packed with 10-μm (300-Å pores) ODS particles; column 2: 250 cm × 250 μm i.d. capillary column packed with 10-μm (300-Å pores) ODS particles; Column 3: 336 cm × 250 μm i.d. capillary column packed with 10-μm (300 Å pores) ODS particles; and Column 4: 225 cm × 250 μm i.d. capillary column packed with 15-μm (300-Å pores) ODS particles.
Source: Ref. 27.

pressure evaluated. All of these columns produced more than 100,000 total plates at most mobile-phase linear velocities studied. The highest plate number (264,000 plates) was obtained by using a 336 cm × 250 μm i.d. column packed with 10-μm porous (300-Å pores) particles at a mobile-phase linear velocity of 0.90 cm s^{-1}. This is the highest plate number per column reported to date in packed column chromatography using microparticles as packing materials. The column efficiency significantly decreased with increasing retention factor (k), and at a retention factor of 4, more than 118,000 plates were obtained, as illustrated in Fig. 5 [27]. These column efficiencies are comparable with those of typical open tubular column GC.

Table 2 lists data showing the relationship between reduced plate height and mobile-phase linear velocity under SGC (supercritical fluid to gas) conditions [27]. Minimum reduced plate heights of 1.48, 1.35, and 1.27 were obtained using 180-, 250-, and 336-cm-long columns packed with 10-μm particles, respectively. All of these 10-μm porous (300-Å pores) particle packed columns produced minimum reduced plate heights of less than 1.5. This suggests that repeatable and highly efficient capillaries for SGC can be prepared using the CO_2 slurry packing method. A minimum reduced plate height of 1.27 was obtained using a 336 cm × 250 μm i.d. column containing 10-μm porous (300-Å pores) particles at a mobile-phase linear velocity of 0.90 cm s^{-1}. This is the lowest reduced plate height reported in packed column chromatography for a column inner-diameter-to-particle-diameter ratio of larger than 10. It was found that the optimum mobile-phase linear velocity increased by reducing the column length in SGC [27].

Theoretical considerations indicate that a well-packed column can produce a minimum reduced plate height of 2, and when the ratio of column inner diameter to particle diameter is less than 8, this value decreases [18,28]. In LC, very narrow bore [25–50 μm i.d] packed capillaries have been prepared for which this ratio approached 8 [29], and a reduced plate height of ca. 0.8 was obtained. However, in SGC, a minimum reduced plate height of 1.27 was obtained even though the ratio of column inner diameter to particle diameter was 25.

A comparison between column efficiencies obtained in SGC and HPGC was recently made. Figure 6 shows representative separations of test solutes [27]. An amazing difference was observed. In HPGC using He as the carrier gas, naphthalene and 1-methylnaphthalene were not eluted under the experimental conditions. Lighter compo-

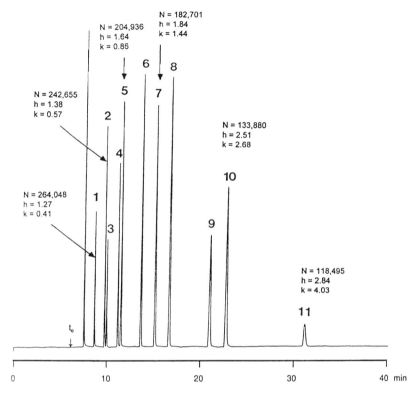

Fig. 5 SGC chromatogram of test solutes. Conditions: 336 cm × 250 mm i.d. fused silica capillary column packed with 10-μm porous (300-Å pores) ODS bonded particles; 130°C; 260-atm column inlet pressure; CO_2; FID. Peak identifications: (1) benzene, (2) toluene, (3) n-octane, (4) p-xylene, (5) n-nonane, (6) n-decane, (7) n-butylbenzene, (8) n-undecane, (9) n-dodecane, (10) naphthalene, (11) 1-methylnaphthalene. (From Ref. 27.)

nents were eluted, however, large retention and peak tailing were observed, as illustrated in Fig. 6A. At 180°C, all components were eluted, however, low column efficiency and peak tailing were also observed (Fig. 6B). When CO_2 was used as mobile phase, excellent results were obtained at the same conditions as those used for HPGC, as illustrated in Fig. 6C. The shorter retention times and im-

Table 2 Relationship between Reduced Plate Height (h) and u (cm s^{-1}) in SGC (supercritical fluid to gas)[a]

Column 1		Column 2		Column 3		Column 4	
u	h	u	h	u	h	u	h
1.05	2.26	0.86	2.40	0.66	2.70	3.93	1.70
1.20	2.08	0.95	1.71	0.72	2.28	4.45	1.97
1.36	1.70	1.04	2.00	0.77	1.77	4.85	2.74
1.45	1.60	1.11	1.65	0.82	1.57	5.26	3.51
1.55	1.60	1.19	1.40	0.86	1.48	5.67	3.86
1.67	1.48	1.26	1.35	0.90	1.27	5.94	3.92
1.76	1.52	1.33	1.61	0.96	1.49	6.11	4.44

[a]Conditions are the same as in Table 1.
Source: Ref. 27.

proved peak shapes resulted from the increase in the solvating power of the mobile phase and, probably, the decrease in interaction between the solutes and the stationary phase. A reduction in interactions between solutes and stationary phase could be due to adsorption of CO_2 on the stationary phase. When the same column, split tube, concentration of solutes (~2% total concentration), temperature, and inlet pressure were used, a much higher detection response was observed in SGC than in HPGC. Figure 6C was obtained by increasing the attenuation 10 times more than that in Fig. 6A, while all other experimental conditions were the same. In our experiments, high-efficiency HPGC was never obtained when using long, microparticle packed capillaries.

Temperature has a significant influence on the column efficiency in SGC, especially when using an organic liquid to vapor mobile phase. Figure 7 shows experimental relationships between reduced plate height and mobile-phase linear velocity for this type of SGC at 140 and 200°C, HTLC at 60°C, and RTLC at 24°C when using pure acetonitrile as the mobile phase [16]. Approximately the same minimum reduced plate height (h_{min}) of 1.8 was obtained for all of these processes. This suggests that the change in physical properties of the mobile phase, including large mobile-phase linear velocity, density, and solute retention gradients, has little effect on column efficiency, at least on the maximum column efficiency.

It was found that the optimum mobile phase linear velocity (u_{opt})

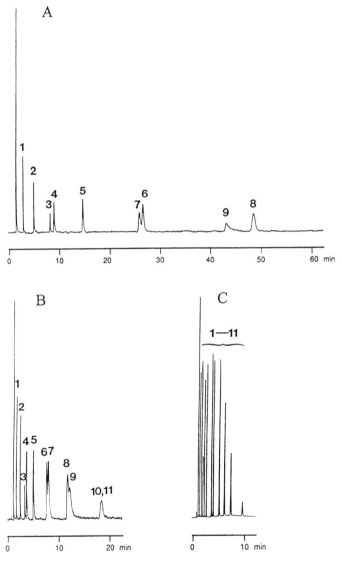

Fig. 6 HPGC and SGC chromatograms of test solutes. Conditions: 228 cm × 250 μm i.d. fused silica capillary column packed with 15-μm porous (300-Å) ODS bonded particles; 150-atm column inlet pressure; (A) 130°C, He carrier gas; (B) 180°C, He carrier gas; (C) 130°C, CO_2 mobile phase; flame ionization detector (FID). Peak identifications: (1) benzene, (2) toluene, (3) n-octane, (4) p-xylene, (5) n-nonane, (6) n-decane, (7) butylbenzene, (8) n-undecane, (9) naphthalene, (10) n-dodecane, (11) 1-methylnaphthalene. (From Ref. 27.)

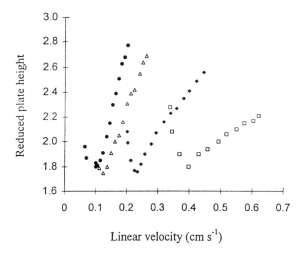

Fig. 7 Experimental relationship between reduced plate height and mobile-phase linear velocity in SGC, HTLC, and RTLC. Conditions: benzene used as test solute. □: SGC (200°C), ◆: SGC (140°C), △: HTLC (60°C), and ●: RTLC (24°C); other conditions are the same as in Fig. 3. (From Ref. 16.)

shifted toward a high value in SGC (liquid to vapor), and the relationship between u_{opt} and temperature (T) can be expressed as follows:

$$u_{opt} = \frac{2\alpha\gamma T}{bd_p} e^{-\frac{3T_b}{T}} \qquad (3)$$

where α, γ, and b are constants, d_p is the particle diameter, and T_b is the boiling point. Equation (3) shows that u_{opt} is a function of T multiplied by $e^{-1/T}$. The experimental u_{opt} values of 0.085, 0.124, 0.234, and 0.398 cm s^{-1} were obtained for RTLC (24°C), HTLC (60°C), and SGC (140 and 200°C), respectively, which confirmed the form of Eq. (3) [16].

The dependence of total column efficiency (N) on mobile phase linear velocity (u) became smaller in SGC than in RTLC and HTLC. This can be expected because an increase in T leads to an increase in D_m. Table 3 lists experimental data that show the relationship between N and u in SGC (liquid to gas), HTLC, and RTLC [16]. More than 100,000 plates per column were obtained in approximately 10 min under SGC conditions at 200°C when using 10-μm porous (300-Å pores) particles

Table 3 Relationship Between Column Efficiency (plates per column) and u (cm s^{-1}) in SGC (liquid to vapor)[a]

RTLC(24°C)		HTLC(60°C)		SGC(140°C)		SGC(200°C)	
u	N	u	N	u	N	u	N
0.066	132,384	0.111	145,599	0.201	125,170	0.338	114,035
0.070	139,037	0.124	148,451	0.204	130,653	0.345	125,001
0.085	144,454	0.137	144,442	0.213	140,540	0.369	136,842
0.100	144,444	0.150	136,180	0.225	146,893	0.398	144,447
0.105	143,093	0.162	129,730	0.234	146,291	0.429	136,803
0.110	142,077	0.175	126,828	0.245	142,657	0.457	134,024
0.114	140,541	0.187	122,510	0.257	136,842	0.492	130,041
0.123	136,126	0.201	112,613	0.278	131,501	0.517	126,214
0.135	127,451	0.212	108,782	0.293	125,466	0.549	123,810
0.145	120,901	0.223	107,552	0.319	120,370	0.574	120,933
0.154	113,043	0.240	101,891	0.340	116,592	0.602	119,816
0.164	108,963	0.250	98,489	0.359	114,537	0.619	117,647

[a]Conditions: 260 cm × 250 μm i.d. fused silica capillary column packed with 10-μm (300-Å pores) ODS particles, nitromethane used as unretained marker, acetonitrile used as mobile phase, UV detector (254 nm).
Source: Ref. 16.

as the packing material. Increasing the temperature led to a decrease in the time needed to obtain high total efficiency. It was found that when the temperature was increased from 24 to 200°C, the absolute value of $(\partial N/\partial u)_T$ decreased from 382,110 to 81,799 plates cm^{-1} s, or more than four times [16]. Of course, this change included the influence of temperature on solute retention.

V. SEPARATION SPEED IN SGC

Fast chromatography can be carried out using wall-coated open tubular columns, porous layer open tubular columns, and packed columns. The selection of column type should involve a consideration of both column efficiency and solute retention.

In considering column efficiency in fast separations, the solute mass transfer resistance in the mobile phase is a most important factor. According to the Golay [30] and van Deemter [15] equations, and for an open tubular column having an inner diameter (d_c) of 50 μm and a packed column containing particles having a diameter (d_p) of 10 µm, the ratio of the contributions from the solute mass transfer resistance in the mobile phase can be expressed as [31]:

$$\frac{H_{C,OT}}{H_{C,packed}} = 6.25 \frac{u_{OT}}{u_{packed}} \qquad (4)$$

where $H_{C,OT}$ or and $H_{C,packed}$ are the contributions to plate height from the solute mass transfer in the mobile phase for open tubular and packed columns, respectively. Equation (4) suggests that the same peak broadening from solute mass transfer in the mobile phase in the open tubular and packed columns is achieved when 6.25 times greater linear velocity is used for the packed column (u_{packed}) than for the open tubular column (u_{OT}). Therefore, in terms of mass transfer resistance in the mobile phase, microparticle packed columns favor high column efficiency per unit time.

Solute retention plays an important role in the separation. An estimation of the ratio of surface area provided by the wall of the open tubular column to the particles in the packed column is helpful for comparing the relative solute retention in open tubular and packed columns.

If the same film thickness is coated on the particle surface as on the wall of the open tubular column, the relationship between phase ratios in open tubular and packed columns can be expressed as [31]:

$$\frac{\frac{S_{packed}}{V_{packed}}}{\frac{S_{OT}}{V_{OT}}} = 4.5 \frac{d_c}{d_p} \qquad (5)$$

where S_{packed} and V_{packed} are the total particle surface area and void volume of the packed column, respectively, and S_{OT} and V_{OT} are the wall surface area and void volume of the open tubular column, respectively. If d_c = 50 µm and d_p = 10 µm, it can be estimated that approximately 22.5 times greater retention can be expected on the packed column than on the open tubular column.

The separating power of a column can be characterized by the number of peaks that can be separated in a specific time period. Figure 8 illustrates the dependence of separation on solute retention factor (k) and column efficiency (N) [31]. The situation in which the separation

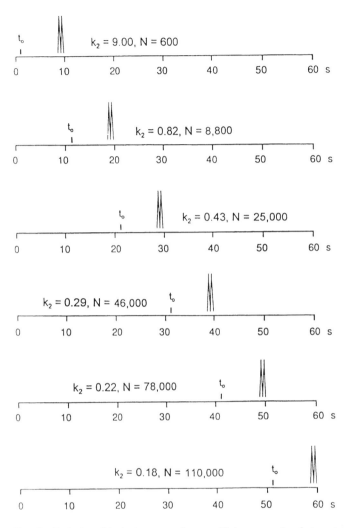

Fig. 8 Relationship between column efficiency and solute retention for a desired separation. (From Ref. 31.)

depends on large retention and low column efficiency is more suitable for high-speed separations.

It should be mentioned that when the analysis time period $\Delta t = t_n - t_1$ increases, the contribution of total column efficiency to peak capacity (n) becomes more significant [31,32]. Generally, short packed columns have characteristics of high solute retention and low total column efficiency, and they are suitable for high-speed separations; open tubular columns have properties of low retention and high total efficiency, and they are suitable for moderately fast separation of complex samples; porous layer open tubular columns are a compromise to carry out relatively fast separations of complex samples.

The peak capacity focuses on the resolution obtained in a time period of Δt. Low solute retention requires long dead time and high column efficiency to obtain the desired separation. Figure 8 is useful to compare the abilities of different column types to carry out fast separations as a function of efficiency and solute retention. If a particular column type (similar retention factor and selectivity) is selected to carry out a fast separation, the resolution per unit time (R_{st}) can be used for selection of column dimensions, including column length, column inner diameter for open tubular columns, and particle size and properties for packed columns. The resolution per unit time can be expressed as [33]:

$$R_{st} = \frac{\sqrt{N_t}}{4\sqrt{t_M}} \frac{\alpha - 1}{\alpha} \frac{k}{(1+k)^{1.5}} \tag{6}$$

This equation predicts that R_{st} depends on both column efficiency per unit time (N_t) and dead time (t_M) for constant solute retention factor (k) and selectivity (α). Reducing the dead time has the same significance as increasing the column efficiency per unit time in improvement of resolution per unit time.

In LC, extremely small particles (e.g., 1.5 µm) have been used for fast separations [34]. These small particles provide high column efficiency per unit column length, a large optimum mobile-phase linear velocity, and a small dependence of column efficiency on mobile-phase linear velocity. The practical problem with the use if these small particles in LC is the column pressure drop, which allows only the use of short columns (e.g., 3 cm). In SFC, the column pressure drop affects solute retention, and a compromise in particle size must be made to carry out fast separations. It has been found that 15-µm particles are suitable for fast SFC separations [33]. In SGC, solute elution depends on both the temperature and the solvating power of the mobile phase. Lower temperatures are required when greater solvating power is available.

Various length columns packed with various sizes of porous and nonporous particles have been investigated for fast SGC separations. Table 4 lists experimental data obtained (u versus N_t) for long columns packed with 10- and 15-µm porous (300-Å pores) particles in SGC [25]. Results (e.g., 100–800 plates s^{-1}) typical of open-tubular-column GC were obtained. For a 180-cm-long column packed with 10-µm porous (300-Å pores) particles, the plate number per second increased with increasing mobile-phase linear velocity until the linear velocity reached 1.8 cm s^{-1}. A maximum value of 780 plates s^{-1} was obtained for this column under SGC (supercritical fluid to gas) conditions. By increasing the column length to 250 and then to 336 cm, the rate of generation of plates increased with increasing mobile-phase linear velocity throughout the experimental range of linear velocities studied. By increasing the porous (300-Å pores) particle size from 10 to 15 µm, the plate number per second can be improved. For example, a 228-cm column packed with 15-µm particles easily produced 700–800 plates s^{-1} at high mobile-phase linear velocity, and a maximum value of 813 plates s^{-1} was obtained using this column [27].

For 10-µm porous (300-Å pores) particle packed columns, it was found that increasing the column length increases the plate number per second at a specific mobile phase linear velocity [27]. For example, at a linear velocity of 1.2 cm s^{-1}, a 180-cm-long column produced approxi-

Table 4 Relationship between Column Efficiency per Unit Time (plates s^{-1}) and u (cm s^{-1}) in SGC (supercritical fluid to gas)[a]

Column 1		Column 2		Column 3		Column 4	
u	N_t	u	N_t	u	N_t	u	N_t
1.05	187	0.86	180	0.66	131	3.93	736
1.20	289	0.95	297	0.72	156	4.45	813
1.36	426	1.04	305	0.77	271	4.85	693
1.45	525	1.11	421	0.82	340	5.26	628
1.55	598	1.19	483	0.86	401	5.67	631
1.67	732	1.26	633	0.90	504	5.94	702
1.76	780	1.39	669	0.96	464	6.38	738

[a]Conditions are the same as in Table 1.
Source: Ref. 27.

mately 350 plates s^{-1}, while a 250-cm-long column produced approximately 500 plates s^{-1}. However, reducing the column length can lead to an increase in mobile-phase linear velocity at a specific column inlet pressure, which favors high plates s^{-1} values. For example, at a column inlet pressure of 220 atm, 732 plates s^{-1} were obtained for a 180-cm-long column, while only 340 plates s^{-1} were obtained for a 336-cm-long column.

Although relatively high column efficiency per unit time (780 plates s^{-1} at a column inlet pressure of 260 atm) can be obtained using a 336-cm-long column packed with 10-μm porous (300-Å pores) particles, such a long column is not suitable to carry out fast separations using the currently available instrumentation. At a column inlet pressure of 260 atm and a temperature of 130°C, the dead time is more than 5 min, as illustrated in Fig. 6. However, by only increasing the particle size to 15 μm, the separation was completed in 3 min under the same experimental conditions, as illustrated in Fig. 9 [27]. Therefore, for fast separations using long packed columns (more than 2 m), large particle sizes (such as 15 μm) are required.

Using porous particles as packing materials, the typical column efficiency generation rate is less than 1000 plates s^{-1}. It has been found that nonporous particles can be used for improvement of column efficiency, especially at high mobile-phase linear velocity [14]. The use of small (sub-5-μm) nonporous particles has been investigated in fast SGC.

Figure 10 shows experimental relationships between plate number per second and mobile-phase linear velocity under SGC (supercritical fluid to gas) conditions [15]. When using small (1.5 and 3 μm) porous and nonporous particles as packing materials, increasing the mobile-phase linear velocity increased the plate number per second for all columns in the range of mobile-phase linear velocities studied. Reducing the column length from 57 to 30 to 20 cm decreased the maximum plate number from 1200 to 840 to 680 plates s^{-1} when using 3-μm nonporous particles as packing material. It should be mentioned that the achievement of the high plate numbers per second for short columns was limited by our experimental conditions. The 20-cm-long column packed with 3-μm nonporous particles produced a 0.25-s peak width at half-height using 100-atm column inlet pressure, at which a maximum of 680 plates s^{-1} was obtained. This peak width equals the response time constant of the system (0.25 s). An improvement in instrumentation is necessary to obtain the maximum efficiency that small nonporous-particle-packed columns can provide.

Columns packed with porous particles produced a lower plate

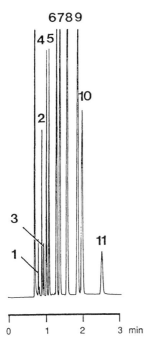

Fig. 9 Fast SGC chromatogram of test solutes. Conditions: 226 cm × 250 μm i.d. fused silica capillary column packed with 15-μm porous ODS bonded particles; CO_2 mobile phase, 130°C; 260-atm column inlet pressure; FID. Peak identifications: (1) benzene, (2) toluene, (3) n-octane, (4) p-xylene, (5) n-nonane, (6) n-decane, (7) n-butylbenzene, (8) n-undecane, (9) naphthalene, (10) n-dodecane, (11) 1-methylnaphthalene. (From Ref. 27.)

number per second than columns packed with the same size nonporous particles. Maximum values of 219 and 339 plates s^{-1} were obtained for 20- and 10-cm-long columns, respectively, when using 3-μm porous (80-Å pores) particles as packing material [15]. These low generation rates resulted from slow solute diffusion in the pores of the particles. Columns packed with porous particles have lower permeability than those packed with the same size nonporous particles, and higher column inlet pressures must be imposed on the column inlet end to obtain the desired mobile-phase linear velocity [14]. However, the large solute retention that porous-particle-packed columns can provide is ad-

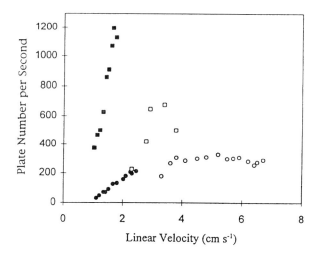

Fig. 10 Relationship between plate number per second and mobile-phase linear velocity in SGC. Conditions: (■) 57 cm × 250 μm i.d. capillary packed with 3-μm nonporous ODS particles; (□) 20 cm × 250 μm i.d. capillary packed with 3-μm nonporous ODS particles; (●) 20 cm × 250 μm i.d. capillary packed with 3-μm porous (80-Å pores) ODS particles; and (○) 10 cm × 250 μm i.d. capillary packed with 3-μm porous (80-Å pores) ODS particles; octane test solute; other conditions are the same as in Fig. 2. (From Ref. 15.)

vantageous for carrying out fast separations, as illustrated in Fig. 8. The low total column efficiency of short columns containing porous particles can be offset by their large solute retention. Figure 11 shows high-speed SGC chromatograms obtained using porous (80-Å pores) and nonporous particle packed columns [15]. Using a 20-cm-long column packed with 3-μm nonporous particles, the separation can be finished in 15 s with a column inlet pressure of 95 atm. Similar resolution was obtained by using a 10-cm-long column packed with 3-μm porous (80-Å pores) particles with a column inlet pressure of 250 atm. The effect of low column efficiency on resolution can be compensated for by larger solute retention. However, a much higher column inlet pressure is required when using the porous-particle-packed column because of low column permeability. Furthermore, with increasing analysis time, column efficiency becomes more important in obtaining high peak capacity [31], and nonporous-particle-packed columns become more applicable.

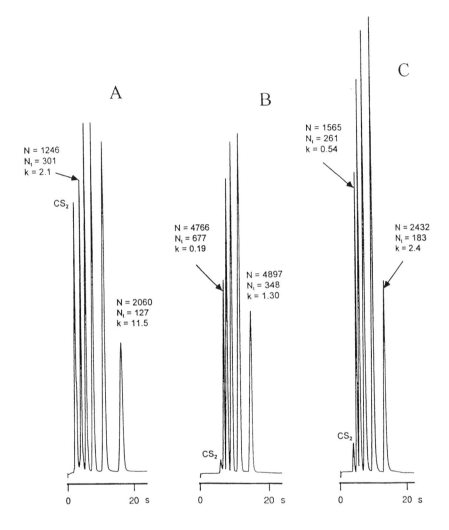

Fig. 11 High-speed SGC using porous and nonporous particles. Conditions: CO_2; 130°C; FID; n-hydrocarbons from octane to dodecane used as test solutes; (A) 10 cm × 250 μm i.d. capillary packed with 3-μm porous (80-Å pores) ODS particles using 250-atm column inlet pressure, (B) 20 cm × 250 μm i.d. capillary packed with 3-μm nonporous ODS particles using 95-atm column inlet pressure, and (C) 8.5 cm × 250 μm i.d. capillary packed with 1.5-μm NP-ODS particles using 100-atm column inlet pressure. (From Ref. 15.)

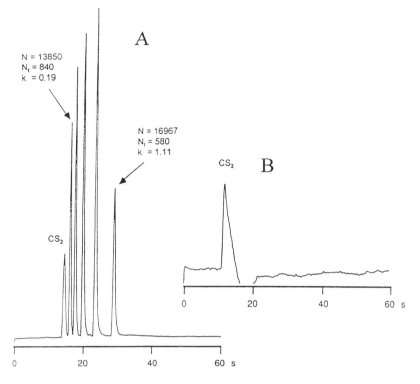

Fig. 12 Comparative chromatograms obtained using (A) SGC and (B) HPGC. Conditions: 30 cm × 250 μm i.d. fused silica capillary packed with 3-μm nonporous ODS; 130°C; 100-atm column inlet pressure; FID; n-hydrocarbons from n-octane to n-dodecane; (A) CO_2 and (B) He mobile phases. (From Ref. 15.)

Even nonporous particles as small as 1.5 μm have been used in SGC, and a very short column (8.5 cm) was used to obtain a similar resolution as that shown n Fig. 11, with a slight improvement in separation time [15].

In HPGC, the carrier gas has little effect on the retention of solutes. The choice of carrier gas is mainly determined by solute diffusion. Generally, lighter gases provide larger solute diffusion. However, the solvating power of the mobile phase in SGC produces mobile-phase interactions with solutes and with stationary phase, and these interactions significantly affect solute retention. The ability to carry out fast separations in SGC and HPGC are compared in Fig. 12, showing a significant difference [15]. In HPGC, only a poor CS_2 peak was obtained compared to an excellent SGC separation using 3 μm nonporous particles as packing material and 100-atm column inlet pressure. Even after increasing the oven temperature to 180°C, only CS_2 eluted in HPGC. This results from the severe adsorption of solutes on the stationary phase in HPGC, while the solvating power of the mobile phase in SGC greatly reduces this interaction.

In SGC (liquid to gas), fewer than 300 plates s^{-1} were obtained for a 260-cm-long, 10-μm porous (300-Å pores) particle packed column with pure acetonitrile as the mobile phase at 200°C and 400-atm column inlet pressure. The use of smaller porous and nonporous particles and more volatile liquid mobile phases are currently under study.

VI. POLARITY AND SOLVATING POWER IN SGC

Achieving a wide range of polarity in SGC is more difficult when using CO_2 as the mobile phase (supercritical fluid to gas) than when using liquids (liquid to vapor) which allows the selection of a variety of polar mobile phases. Currently, uniform microparticles with diameters ranging from 1.5 to 15 μm are available, which have been developed for LC. State-of-the-art microparticles used in LC provide excellent column efficiencies. However, the suitability of these particles in SGC for the separation of polar compounds was an area that needed to be investigated.

Carbon dioxide has been used as a mobile phase for the GC separation of polar compounds [35]. However, when bonded siliceous microparticles are used as column packing materials, only nonpolar and slightly polar compounds can be separated with good peak shapes. Much work has been done in our laboratory to deactivate various packing materials for packed column SFC [36–39]. These particles have allowed the successful separation of polar compounds, including free acids and amines using SFC. A higher temperature must be used in SGC in order to separate polar compounds.

The possibility of separating polar compounds using octadecyl-bonded silica (ODS) particle packed columns was investigated using supercritical CO_2 in packed-column SFC [36]. It was found that weakly polar compounds such as ketones, esters, and aldehydes could be separated using ODS particles as the packing material, while hydroxyl-containing compounds and amines either were not eluted or produced severe peak tailing.

Polymeric particles have also been studied in packed column SFC, and fatty acids can be separated at 150°C [40]. The suitability of these particles in SGC was also investigated [41]. Although these particles could be used for the separation of polar compounds such as anilines, low column efficiency was observed. Currently, modification of siliceous particles appears to be the best approach to obtain high efficiency separation of polar compounds.

Polymer-encapsulated particles have been found to produce the best results in packed capillary SFC. Polymer coatings on the deactivated silica particle surface can reduce the activity of residual silanol groups [37], and these polymer-encapsulated silica particles were found to be suitable for separation of medium polar compounds such as alcohols in SGC [41]. However, it was difficult to separate strongly polar acids and amines using these particles. Figure 13 shows an SGC chromatogram of *Peppermint oil* using SE-54 (polymethylphenylsiloxane) encapsulated particles as packing material [41]. The separation was completed within 8 min.

Using a specially deactivated diol-bonded and trimethylsilyl endcapped silica surface, hydrogen-bonding interactions exist between the residual silanol groups and the hydroxyl groups on the bonded phase [38]. These particles can be used for the separation of hydroxyl-containing compounds such as fatty acids in packed-column SFC. These particles were also found to be suitable for the separation of hydroxyl-containing compounds in SGC. Figure 14 shows a fast SGC separation of free fatty acids [41]. The separation was completed in 2 min. These particles are also suitable for the separation of weakly basic anilines and strongly basic tertiary alkylamines, as illustrated in Fig. 15 [41].

With a basic polymer encapsulated silica surface, the acid-base interaction between the basic centers of the coated polymers and the acidic residual silanol groups on the bonded silica surface serves to deactivate the residual silanol groups, and a basic particle surface is produced [39]. These particles were suitable for the separation of amines in packed-column SFC. They were also found suitable for the separation of basic amines in SGC [41]. Although relatively good peak shapes were obtained, the basic compounds have strong retention in columns

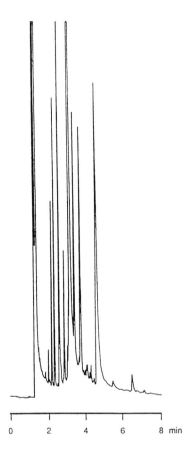

Fig. 13 Packed capillary SGC of *peppermint oil*. Conditions: 170 cm × 250 μm i.d fused silica capillary column (coated with 0.25 μm SE-54) packed with 10-μm SE-54 encapsulated particles; CO_2; FID; column inlet pressure programmed from 180 to 185 atm at 0.5 atm min^{-1}; temperature programmed from 160 to 185°C at 2.5°C min^{-1}. (From Ref. 41.)

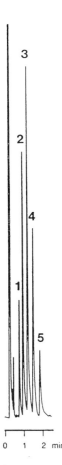

Fig. 14 Packed capillary SGC chromatogram of fatty acids. Conditions: 30 cm × 250 μm i.d fused silica capillary column (coated with 0.25 μm SE-54) packed with 10-μm diol bonded silica particles; CO_2; FID; 90-atm column inlet pressure; temperature programmed from 190 to 220°C at 8°C min^{-1}. Peak identifications: (1) caproic acid, (2) heptanoic acid, (3) caprylic acid, (4) nonanoic acid, and (5) capric acid. (From Ref. 41.)

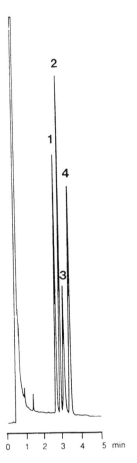

Fig. 15 Packed capillary SGC chromatogram of bases. Conditions: 55 cm × 250 μm i.d fused silica capillary column (coated with 0.25 μm SE-54) packed with 10-μm diol bonded silica particles; CO_2; FID; 110-atm column inlet pressure; 150°C. Peak identifications: (1) N,N-dimethylphenylethylamine, (2) N,N-dimethylaniline, (3) aniline, and (4) N-methylaniline. (From Ref. 41.)

packed with this type of packing material. Primary and secondary alkylamines eluted, but serious peak tailing was observed.

The suitability of HPGC and SGC for the separation of polar compounds has been investigated using diol-bonded particles. Much more serious peak tailing was observed in HPGC even for weakly polar 2-octanone, although a higher temperature and column inlet pressure was used. Figure 16 shows the elution of 2-octanone in SGC and HPGC [41]. In SGC, an excellent peak shape for 2-octanone (concentration of about 0.5% v/v in CH_2Cl_2) was obtained at 200°C and 110-atm inlet pressure. However, in HPGC, even after increasing the column inlet pressure to 150 atm and using a high concentration of 2-actanone (~20% v/v), a very poor peak shape was observed. A low concentration sample (e.g., 0.5% v/v) was not detected in HPGC.

As can be seen in Fig. 6, tailing peaks were obtained in HPGC, while excellent peak shapes were obtained in SGC even for nonpolar hydrocarbons. This reveals that the CO_2 mobile phase in SGC has solvating power and decreases the interaction between the solute and the stationary phase. The reduction of interactions between solutes and the stationary phase by CO_2 is probably due to adsorption of CO_2 on the stationary phase surface.

Many systems have been developed to describe interactions among solutes, mobile phases, and stationary phases in chromatography. Hildebrand (solubility) parameters are favored in SGC because they have been used in both GC [42,43] and LC [44,45], and hopefully can be used to link these chromatographic techniques together. Substances with similar Hildebrand parameters δ can be easily mixed ("like dissolves like") [46]. However, the δ-values are dependent on temperature. By increasing the temperature higher than the liquid boiling point, the δ-value decreases smoothly [46]. This means that one set of δ-values can be used to describe changes in solubility in both liquids and gases.

Because limited δ-data are available at a variety of temperatures, interactions in the system containing polycyclic aromatic hydrocarbons (PAHs), C18-bonded phase, and acetonitrile mobile phase were estimated by using δ-data for toluene, hexadecane (C16), and acetonitrile at various temperatures. Although this assumption is not accurate, it is helpful to investigate the changes in interactions among solute, stationary phase, and mobile phase under elevated-temperature SGC conditions.

The relationships between δ and temperature (T) for hexadecane,

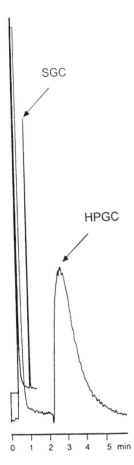

Fig. 16 Packed capillary SGC and HPGC chromatograms of 2-octane. Conditions: 55 cm × 250 μm i.d fused silica capillary column (coated with 0.25 μm SE-54) packed with 10-μm diol bonded silica particles; 200°C; CO_2 mobile phase for SGC (column inlet pressure of 110 atm); He carrier gas for HPGC (column inlet pressure of 150 atm); FID. (From Ref. 41.)

toluene, and acetonitrile are illustrated in Fig. 17. A smooth line exists even when the temperature crosses its boiling point (81°C). At elevated temperature, the Hildebrand polar parameter (δ_τ) for acetonitrile approaches the Hildebrand nonpolar parameter (δ_λ) for toluene (nonpolar compounds have only δ_λ). The difference in δ-values between nonpolar hexadecane and toluene increases at elevated temperature, which means that increasing temperature decreases the interaction between long-chain aliphatic hexadecane and aromatic toluene. The δ-values increase with increasing molecule size, and it can be expected that the total δ-value ($\delta^2 = \delta_\lambda^2 + \delta_\tau^2$) for acetonitrile becomes close to those of large PAHs at elevated temperature. This means that increasing temperature increases the solubility of PAHs in acetonitrile. Both the decrease in interaction between solute and stationary phase and increase in interaction between solute and mobile phase decrease the solute retention factor (k) in high-temperature SGC. Figure 18 shows the experimental relationships between k and T for 9,10-diphenylanthracene and 5,6,11,12-tetraphenylnaphthacene [16]. An approximately

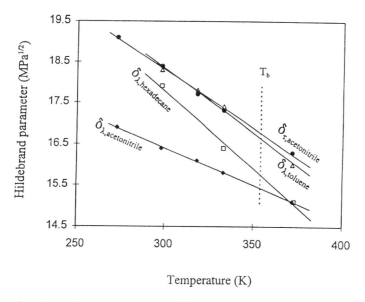

Fig. 17 Relationship between Hildebrand solubility parameters and temperature for acetonitrile, toluene, and hexadecane.

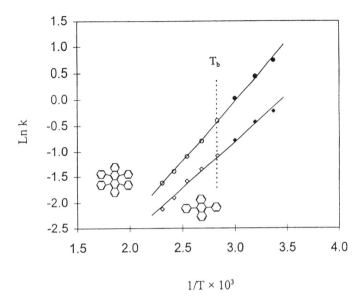

Fig. 18 Experimental relationship between solute retention factor and temperature in SGC, HTLC, and RTLC. Conditions: 400-atm column inlet pressure; ● and ○: retention factor for 5,6,11,12-tetraphenylnaphthacene at temperatures below and above the boiling point of acetonitrile, respectively; ◆ and ◇: retention factor of 9,10-diphenylanthracene at temperatures below and above the boiling point of acetonitrile, respectively; other conditions are the same as in Fig. 3. (From Ref. 16.)

linear relationship between Ln k and $1/T$ was observed, even when the temperature crossed the boiling point of acetonitrile.

The improved flow and column efficiency at high mobile-phase linear velocity, and reduced solute retention in SGC should speed up separations of samples which have relatively low solubilities in the mobile phase at low temperature. Figure 19 shows the elution of large PAHs under SGC (liquid to vapor) conditions [16]. Using pure acetonitrile as the mobile phase and 400-atm column inlet pressure, the last peak took more than 50 min for elution in RTLC.

The high solvating power of a superheated liquid and its vapor allows the use of selective additives in the mobile phase to achieve special separations. Figure 20 shows a fast SGC separation of enan-

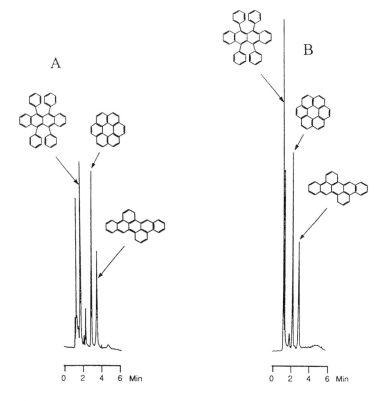

Fig. 19 Fast SGC separations of large PAHs using various mobile phases. Conditions: 55 cm × 250 μm i.d. fused silica capillary column packed with 5-μm polymeric ODS particles (200-Å pores), 300 atm column inlet pressure, temperature program from 65 to 250°C at 25°C min^{-1}. Mobile phases: (A) acetonitrile and (B) hexane. Other conditions are the same as in Fig. 3.

tiomers using permethylated β-cyclodextrin as an additive in hot methanol mobile phase. Highly selective phases, weaker mobile phases, or "diluted" mixed mobile phases such as modified CO_2 can be used to improve the resolution, especially for the separation of complex samples. With the same solute retention (adjusted by changing the composition of the mobile phase), the use of high temperatures provides better resolution [47].

Fig. 20 Fast SGC separation of enantiomers. Conditions: 95 cm × 250 μm i.d. fused silica capillary column packed with 5-μm silica particles (80-Å pores), 0.5% (w/w) permethylated β-cyclodextrin in methanol mobile phase, 380-atm column inlet pressure, 140°C. Other conditions are the same as in Fig. 3.

REFERENCES

1. J. C. Giddings, *Dynamics of Chromatography, Principles and Theory*, Marcel Dekker, New York, 1965.
2. J. C. Giddings, *Unified Separation Science*, Wiley, New York, 1991.
3. M. L. Lee and K. E. Markides, *Science, 235*: 1342 (1987).
4. M. L. Lee, F. J. Yang, and K. D. Bartle, *Open Tubular Column Gas Chromatography, Theory and Practice*, Wiley, New York, 1984.
5. J. C. Giddings, *Anal. Chem., 36*: 741 (1964).
6. D. E. Martire and E. Boehem, *J. Phys. Chem.*, 91: 2433 (1987).

7. M. Roth, *J Phys. Chem.*, *94*: 4309 (1991).
8. M. L. Lee and K. E. Markides (eds.), *Analytical Supercritical Fluid Chromatography and Extraction*, Chromatography Conferences, Provo, Utah, 1990, Chap. 3.
9. L. R. Snyder and J. J. Kirkland, *Introduction to Modern Liquid Chromatography*, Wiley, New York, 1991.
10. J. A. Schmit, R. A. Henry, R. C. Williams, and J. F. Dieckman, *J. Chromatogr. Sci.*, *9*: 645 (1971).
11. J. H. Knox and G. Vasvari, *J. Chromatogr.*, *83*: 181 (1973).
12. F. D. Antia and Cs. Horváth, *J Chromatogr.*, *435*: 1 (1988).
13. S. T. Lee and S. Olesic, *Anal. Chem.*, *66*: 4498 (1994).
14. Y. Shen, Y. J. Yang, and M. L. Lee, *Anal. Chem.*, *69*: 628 (1997).
15. Y. Shen and M. L. Lee, *J Chromatogr A*, *778*: 31 (1997).
16. Y. Shen and M. L. Lee, *Anal. Chem.*, submitted.
17. J. J. van Deemter, F. J. Zuiderwg, and A. Klinkenberg, *Chem. Eng. Sci.*, *5*: 27 (1956).
18. J. H. Knox, *Anal. Chem.*, *38*: 253 (1966).
19. P. J. Schoenmakers, in *Supercritical Fluid Chromatography*, R. M. Smith, ed., RSC Chromatography Monograph Series, The Royal Society of Chemistry, London, 1988, Chap. 4.
20. T. A. Berger and L. M. Blumberg, *Chromatographia*, *38*: 5 (1994).
21. M. N. Myers and J. C. Giddings, *Anal. Chem.*, *37*: 1453 (1965).
22. M. N. Myers and J. C. Giddings, *Anal. Chem.*, *38*: 294 (1966).
23. A. Di. Corcia, A. Liberti, and R. Samperi, *J. Chromatogr.*, *167*: 243 (1978).
24. P. C. Lu, L. M. Zhou, C. H. Wang, G. H. Wang, A. Z. Xia, and F. B. Xu, *J. Chromatogr.*, *186*: 25 (1979).
25. D. R. Gere, R. Board, and D. McManigill, *Anal. Chem.*, *54*: 736 (1982).
26. T. A. Berger and W. H. Wilson, *Anal. Chem.*, *38*: 5 (1994).
27. Y. Shen and M. L. Lee, *Anal. Chem.*, *69*: 2541 (1997).
28. J. H. Knox and J. F. Parcher, *Anal. Chem.*, *41*: 1599 (1969).
29. R. T. Kennedy and J. W. Jorgenson, *Anal. Chem.*, *61*: 1128 (1989).
30. M. J. E. Golay, in *Gas Chromatography*, D. H. Desty, ed., Butterworths, London, 1958.
31. Y. Shen and M. L. Lee, *J Microcol. Sep.*, *9*: 21 (1997).
32. E. Grushka, *Anal. Chem.*, *42*: 1142 (1970).
33. Y. Shen and M. L. Lee, *Chromatographia*, *45*: 67 (1997).
34. A. Kurganov, T. Tssaeva, and K. Unger, *20th International Sympo-*

sium on *High Performance Liquid Phase Separations and Related Techniques*, June 16–21, 1996, San Francisco, p. 184.
35. A. Karmen, I. McCaffrev, and R. L. Bowman, *Nature, 193*: 575 (1962).
36. Y. Shen and M. L. Lee, *Chromatographia, 41*: 665 (1995).
37. Y. Shen, A. Malik, W. Li, and M. L. Lee, *J. Chromatogr. A, 707*: 303 (1995).
38. Y. Shen and M. L. Lee, *J. Microcol. Sep., 8*: 413 (1996).
39. Y. Shen and M. L. Lee, *J. Microcol. Sep., 8*: 519 (1996).
40. Y. Liu, F. Yang, and C. Pohl, *J. Microcol. Sep., 2*: 245 (1990).
41. Y. Shen and M. L. Lee, *Chromatographia*, in press.
42. W. Merk, R. N. Lichtenthaler, and J. M. Prausnitz, *J. Phys. Chem., 84*: 1694 (1980).
43. E. Fernéndez-Sénchez, A. Fernéndez-Torres, J. A. Garcia-Doménguez, and J. M. Santiuste, *J. Chromatogr., 457*:55 (1988).
44. R. A. Keller and L. R. Snyder, *J. Chromatogr. Sci., 9*: 346 (1971).
45. H. Colin, G. Guiochin, and P. Jandera, *Chromatographia, 15*: 133 (1982).
46. A. F. M. Barton, *CRC Handbook of Solubility Parameters and Other Cohesion Parameters*, 2nd ed., CRC Press, Boca Raton, FL, 1991.
47. J. R. Grant, J. W. Dolan, and L. R. Snyder, *J. Chromatogr., 185*: 153 (1979).

4
The Linear-Solvent-Strength Model of Gradient Elution

L. R. Snyder and J. W. Dolan *LC Resources Inc., Walnut Creek, California*

I.	INTRODUCTION	116
II.	LIQUID CHROMATOGRAPHY BASICS	120
	A. Isocratic Elution	120
	B. Gradient Elution	121
	C. Retention Versus Mobile-Phase Composition	122
III.	THE LINEAR-SOLVENT-STRENGTH (LSS) MODEL	123
	A. Quantitative Relationships	123
	B. Comparing Separations by Gradient or Isocratic Elution	131
	C. Large-Molecule Effects	138
	D. Gradient Delay and Elution After the Gradient	144
	E. Mass-Overloaded Separations	145
IV.	OPTIMIZING GRADIENT SEPARATION	150
	A. Gradient Method Development	151
	B. Design and Use of an Initial Gradient Separation	154
	C. Computer Simulation	157

V.	NONIDEAL EFFECTS IN GRADIENT ELUTION	160
	A. Instrumental Effects	161
	B. Non-LSS Conditions; LC Methods Other Than Reversed Phase	163
	C. Nonequilibrium Effects	167
	D. Retention Reproducibility	168
	E. Potential Accuracy of Retention Predictions from the LSS Model	170
	F. Anomalous Band Broadening in Gradient Elution	171
VI.	APPLICATION OF THE LSS MODEL TO GAS CHROMATOGRAPHY	173
VII.	CONCLUSION	175
	REFERENCES	180
	APPENDIX	185

I. INTRODUCTION

The technique of gradient elution, first described in 1952 [1–3], is a solution to certain problems which are often encountered in isocratic liquid chromatography (LC) [4]. Whereas isocratic elution is carried out with a mobile phase of fixed composition, in gradient elution the mobile phase is intentionally varied during a chromatographic run. While this change in mobile-phase composition can take various forms, gradient elution almost always involves an increase in *solvent strength* from the beginning to end of the separation, as a result of mixing a stronger solvent B with a weaker solvent A. Therefore, the retention factor k for each sample component decreases with time after sample injection. Gradients of various kinds are illustrated in Fig. 1; linear gradients as in Fig. 1a are most common.

The effective use of gradient elution requires an understanding of how separation varies with experimental conditions. The gradient itself is defined by the composition of the A- and B-solvents, the steepness of the gradient (rate of change of %B in the mobile phase), and gradient shape (Fig. 1a–d). Column dimensions and mobile-phase flow rate further determine the relationship between the gradient and sample retention. Other factors that affect isocratic separation also play a role in

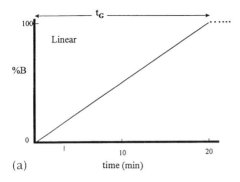

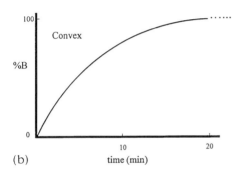

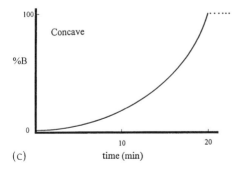

Fig. 1 Examples of different gradient shapes. See text for discussion.

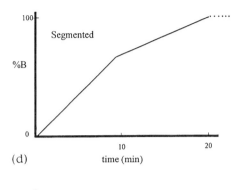

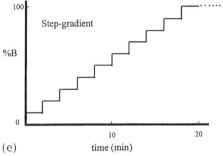

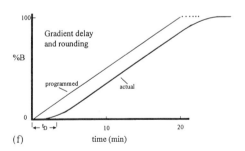

Fig. 1 Continued

gradient elution, especially the dependence of isocratic retention on mobile-phase composition. Gradient elution is more complicated than isocratic elution and more difficult to describe in quantitative terms. Although the essential features of gradient elution were known by the early 1960s (see review of Ref. 5), the widespread application of these principles is only now beginning to take place.

Most workers who use gradient elution are familiar with the principles of isocratic separation. The latter technique is easier to comprehend and has been more widely applied. Since gradient elution can be visualized as a series of a small isocratic steps (cf. Figs. 1a, 1e), our understanding of isocratic elution should be useful as a starting point for a similar description of gradient elution. It is also desirable, so far as possible, to be able to relate gradient separation to "corresponding" isocratic runs. In this way, our knowledge of isocratic method development can be made use of for the selection of optimal gradient conditions. A simple representation of gradient elution is now available, which allows a description and understanding of both isocratic and gradient separations in similar fashion. This *linear-solvent-strength* (LSS) model of gradient elution is based on an approximation of isocratic solute retention as a function of mobile-phase composition %B (Eq. 1), where a stronger solvent B is mixed with a weaker solvent A to form the mobile phase:

$$\log k = a - c(\%B) \tag{1}$$

Here, k refers to an isocratic solute retention factor, and a and c are constants for a particular solute and solvents A and B (column and temperature constant). Equation (1) is usually a good approximation for reversed-phase liquid chromatography (RP-LC) [6], and it can be used for other LC methods (ion-exchange, normal-phase LC) over a limited range in %B [7]. A gradient will be assumed that is linear in the volume fraction ϕ of B (equal to [%B]/100) in the mobile phase entering the column:

$$\phi = a + ct$$
    (a and c constant, t is time after the start of the gradient) \qquad (2)

$$= \phi_0 + \left(\frac{\Delta\phi}{t_G}\right)t \tag{2a}$$

ϕ_0 is the value of ϕ for $t = 0$, $\Delta\phi$ is the change in ϕ during the gradient, and t_G is gradient time (total gradient duration in min).

Equations (1) and (2) give

$$\log k = a' + c't \quad (a' \text{ and } c' \text{ constant}) \tag{3}$$

Gradients which are described (even approximately) by Eq. (3) will be defined as LSS gradients [8]. Compared to non-LSS gradients, LSS gradients are advantageous for the following reasons:

- Overall separation tends to be better for typical samples.
- Relatively simple equations can be derived which describe separation as a function of gradient conditions and facilitate the understanding and optimization of gradient elution.
- These LSS gradient equations are of similar form as corresponding equations for isocratic elution, allowing our knowledge of isocratic elution to be applied to gradient separation.
- A number of questions concerning gradient elution can be more easily described and resolved by means of the LSS model.

This chapter summarizes the development and application of the LSS model for the practical understanding and use of gradient elution.

II. LIQUID CHROMATOGRAPHY BASICS

A. Isocratic Elution

Isocratic retention can be described by

$$t_R = t_0(1+k) \tag{4}$$

or

$$V_R = V_m(1+k) \tag{5}$$

Here, t_R is solute retention time, t_0 is the column dead time, V_R is solute retention volume, and V_m is the column dead volume. For a sufficiently small sample mass and volume, k is usually not a function of sample size (but see Section III.E). The column dead time and volume are related by flow rate F as

$$t_0 = \frac{V_m}{F} \tag{6}$$

Isocratic baseline bandwidth $W = 4\sigma$ (min) is related to column efficiency, expressed as the plate number N as

$$W = 4N^{-1/2} t_R = 4N^{-1/2} t_0 (1+k) \tag{7}$$

N depends on various experimental conditions [9], but it is often assumed to be the same for the different bands in a chromatogram.

Isocratic resolution for two adjacent bands 1 and 2 can be expressed [4] as

$$R_s = \frac{2(t_2 - t_1)}{W_1 + W_2} \tag{8}$$

or

$$R_s = \frac{1}{4}(\alpha - 1)N^{1/2}\left(\frac{k}{1+k}\right) \tag{8a}$$

Here, t_1 and t_2 are the retention times or bands 1 and 2, W_1 and W_2 are their bandwidths W, $\alpha = k_2/k_1$ is the separation factor, and k is the average value of k for the two bands; k_1 and k_2 refer to the values of k for the two bands.

B. Gradient Elution

Figure 1 provides examples of gradients of various types. Our primary interest in this chapter is in *continuous* rather than *stepwise* gradients (e.g., Fig. 1a vs. Fig. 1e). However, in the limit of a large number n of small gradient steps ($n \to \infty$), a stepwise gradient becomes equivalent to a continuous gradient. This is the basis of the fundamental equation of gradient elution retention [10,11]:

$$\int_0^{V_R'} \frac{1}{V_m} \frac{dV}{k_a} = 1 \tag{9}$$

Here, V_R' is the solute retention volume corrected for V_m, V is the cumulative volume of mobile phase that has passed through the column after the start of the gradient, and k_a is the instantaneous value of k for the solute band at any time during the gradient. Equation (9) describes the migration of a solute band through the column. The passage of a differential volume element (dV) of mobile phase through the band center results in a fractional band migration $dx = dV/V_m K_a$. When the total volume of mobile phase passing through the band center equals the corrected retention volume, the sum of these fractional migrations $\Sigma dx = 1$. If k is known as a function of V or time $t = V/F$, then Eq. (9) can be solved (either explicitly or by numerical integration) for the solute retention volume V_R or retention time $t_R = V_R/F$. A major advantage of the LSS model is that it allows an explicit solution of Eq. (9) that is simple and therefore convenient to manipulate.

C. Retention Versus Mobile-Phase Composition
Different LC Methods

Retention for a given solute in RP-LC is usually well approximated [6] by

$$\log k = \log k_w - S\phi \tag{10}$$

where k_w is the value of k for water as mobile phase, and S is a constant when only the volume fraction of organic solvent B (ϕ = %B/100) is varied. Eq. (10) has the same form as Eq. (1); therefore, if a linear gradient is used (Eq. 2), and Eq. (10) is obeyed, an LSS gradient results (Eq. 3). Minor deviations from Eq. (10) are common (e.g., see Figs. 3c, d and discussion of Section V.B), but these deviations seldom lead to significant error in the use of Eq. (10) for gradient elution predictions.

Retention for a given solute in ion-exchange LC (IEC) can be described [12] by

$$\log k = a - z \log c \tag{11}$$

where a and z are constants when only the concentration c of the mobile-phase counterion varies. The constant $z=p/q$, where p is the charge on the solute molecule, and q is the charge on the mobile-phase counterion; in most cases, $q=1$ and therefore $z=p$ will be assumed in the following discussion.

Retention for a given solute in normal-phase LC (NP-LC) can be described [13] by a relationship similar to Eq. (10) for IEC:

$$\log k = c - n \log X_B \tag{12}$$

where c and n are constants, and only the mole fraction X_B of the strong solvent B varies. A more general equation for k as a function of mobile-phase composition is [14]

$$\log k = \log k_p - A_s \varepsilon \tag{12a}$$

where k_p is the value of k for pentane as mobile phase A_s increases with solute molecular size, and ε is a solvent strength parameter that can be calculated from the mobile-phase composition.

Retention for a protein in hydrophobic interaction chromatography (HIC) is given [15,16] by

$$\log k = a + c(S') \tag{13}$$

where a and c are constants, and only the concentration (S') of salt S' in the aqueous mobile phase varies.

III. THE LINEAR-SOLVENT-STRENGTH (LSS) MODEL

Earlier, more detailed accounts of the LSS model can be found in Refs. 17–19. The present review summarizes this treatment in c 'er to highlight important features of the LSS model, as well as incorporates important revisions, additions, and applications since 1986.

A. Quantitative Relationships

An LSS gradient for RP-LC corresponds to a linear gradient (Eq. 2), as may be seen by combining Eqs. (2a) and (10):

$$\log k = \log k_0 - \left(\frac{S \, \Delta\phi}{t_G}\right) t$$

$$= \log k_0 - b\left(\frac{t}{t_0}\right) \tag{14}$$

Here, k_0 is the value of k at the start of the gradient ($t=0$), given by

$$\log k_0 = \log k_w - S\phi_0 \tag{14a}$$

$\Delta\phi = \phi_z - \phi_0$, and t_G is the gradient time (Fig. 1a); ϕ_0 and ϕ_z refer to the values of ϕ at the start and end of the gradient. The quantity b is defined as

$$b = \frac{t_0 \, \Delta\phi \, S}{t_G} = \frac{V_m \, \Delta\phi \, S}{t_G F} \tag{15}$$

b is a fundamental measure of gradient steepness which will facilitate comparisons between isocratic and gradient elution.

In order to create a broadly applicable, yet easily interpreted description of gradient elution, it is useful to begin with the simplest case: samples which are strongly retained at the beginning of the gradient, with no "nonideal" effects of the type discussed in Section V. Because reversed-phase operation represents the most widely used LC procedure, the LSS model will be developed for this case; i.e., Eq. (10) will be assumed to describe isocratic retention. Gradient separation based on other LC methods (Eqs. 11–13) is examined in Section V.B; however, qualitative generalizations for gradient elution which are reported here for RP-LC (based on the LSS model) are similar for each of the different LC methods.

Retention

Equation (9) can be restated as

$$\int_0^{t_R} \frac{1}{t_0} \left(\frac{dt}{k_a} \right) = 1 \tag{9a}$$

$t_R = t_R - t_0$. Equation (9a) with $k_a = k$ from Eq. (14) then yields [17] (see also Refs. 8, 19–21)

$$t_R = \frac{t_0}{b} \log(2.3 k_0 b + 1) + t_0 \tag{16}$$

For solutes that are initially strongly retained (large k_0), Eq. (16) reduces to

$$t_R = \frac{t_0}{b} \log(2.3 k_0 b) + t_0 \tag{16a}$$

Equation (16a) assumes that the sample is injected at the same time that the gradient reaches the column inlet. The assumptions used in the derivation of Eq. (16a) are often imprecise, but deviations from this relationship can be corrected as described below. More important, Eq. (16a) is useful as a simplified model of gradient elution that can provide a number of important generalizations which will be discussed later.

Generally, there will be an equipment hold-up or "dwell" volume V_D that delays the arrival of the gradient at the column inlet by a delay time $t_D = V_D/F$ (see Fig. 1f). Solutes with large values of k_0 will be held at the column inlet until the arrival of the gradient, and for this case t_R is given by

$$t_R = \frac{t_0}{b} \log(2.3 k_0 b) + t_0 + t_D \tag{16b}$$

When the value of k_0 is not large, and t_D/t_0 is not small, "pre-elution" of the solute occurs, and Eq. (16b) becomes less reliable. The more accurate calculation of t_R for this case is described in Section III.D. However, Eq. (16b) will be reliable for most peaks in a gradient chromatogram.

Given values of the experimental conditions (t_G, ϕ_0, ϕ_z, V_m, F, t_D), as well as the isocratic parameters S and k_w, it is possible to predict values of t_R from Eq. (16b) (see Section III.D for the case where k_0 is not large). This in turn is useful for computer simulation and separation optimization (Section IV). Values of S and k_w can be obtained by carrying out two (or more) *isocratic* experiments where only ϕ is varied, and solving the two resulting equations based on Eq. (10) (starting with two values of k and ϕ for each solute). Several studies have verified the accuracy of this approach based on Eq. (16b) for predictions of t_R as a function of experimental conditions [17,21–23].

Values of S and k_w for use with Eq. (16b) are more conveniently obtained from two gradient runs where only b is varied (usually by varying t_G [24], but also by varying F). If the values of b for the two runs are b_1 and b_2, with $b_2 > b_1$ and $\beta = b_1/b_2$, and the retention times for a given solute in the two runs are t_1 and t_2 (each given by Eq. 16b), then solving these two equations for b_1 gives

$$b_1 = \frac{t_0 \log \beta}{t_1 - \frac{t_2}{\beta} - (t_0 + t_D)\frac{(\beta - 1)}{\beta}} \tag{17}$$

and

$$\log k_0 = \frac{b_1(t_1 - t_0 - t_D)}{t_0} - \log(2.3 b_1) \tag{17a}$$

Log k_w can then be obtained from Eqs. (17a) and (14a), and S can be determined from Eqs. (17) and (15). Following the determination of values of S and k_w for each solute from two initial gradient runs, values of t_R can be predicted for other experimental conditions, usually with an accuracy of better than ±1% [7,23] (also see Sect. IV). Eqs. (17) and (17a) assume no pre-elution of the sample (Section III.D) and $2.3 k_0 b >> 1$.

Once values of S and k_w have been derived from two (or more)

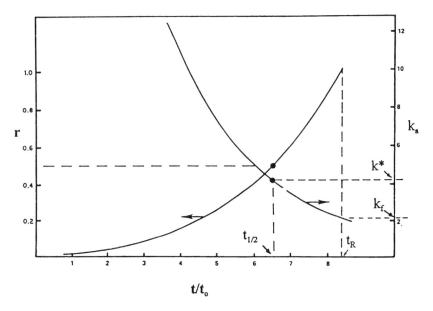

Fig. 2 Illustration of solute migration through the column during gradient elution. See text for details.

initial gradient runs, isocratic retention can also be predicted by means of Eq. (10). It is also possible to compare "average" values of k vs. ϕ for each gradient run with corresponding isocratic k vs. ϕ values, to confirm the same retention relationship (Eq. 10) for *both* isocratic and gradient runs. This also provides a check on the accuracy of Eq. (16b). "Average" or "effective" values of k and ϕ in a gradient run will be defined[‡] (k^* and ϕ^*, respectively) for the solute band when it has migrated halfway through the column; this is illustrated in the hypothetical example of Fig. 2. The instantaneous value of k (k_a) as a function of the fractional distance r migrated by the solute band through the column can be calculated as follows. The integral of Eq. (9a) can be set equal to r (instead of 1 for elution from the column), following which a similar treatment as for Eq. (16) gives

---

[‡]k^* and ϕ^* were previously defined [17,18] as $\bar{k}$ and $\bar{\phi}$.

$$t_r = \frac{t_0}{b} \log(2.3 k_0 br + 1) + rt_0 \tag{18}$$

where t_r is the time at which the band has migrated a fractional distance r along the column. A value of k_a as a function of r can be obtained from Eqs. (14) and (18); $t_r - rt_0$ corresponds to the time t in Eq. (14), so

$$\log k_a = -\log(2.3br + [1/k_0])$$

or

$$k_a = \frac{1}{2.3br + \dfrac{1}{k_0}} \tag{19}$$

For k_0 large and r = 0.5,

$$k^* = \frac{1}{1.15b} \tag{20}$$

$$= \frac{\left(\dfrac{1}{1.5b}\right) t_G F}{V_m \, \Delta\phi \, S} \tag{20a}$$

The quantity k^* can be regarded as the average or effective value of k during gradient elution; k^* is of general importance in understanding gradient elution separation, especially for providing a link between gradient and isocratic separation (Section III.B). With values of S and k_w known for a sample band, insertion of a value of $k^* = k$ from Eq. (20) into Eq. (10) allows the calculation of ϕ^*, equal to the average value of ϕ during gradient elution. Figure 3 illustrates several experimental comparisons of isocratic (k,ϕ) and gradient (k^*,ϕ^*) retention data for the same solute and RP-LC system (only ϕ or b varying). Each data point for (k^*, ϕ^*) in Fig. 3 is the result of two experimental gradient runs. The agreement of these isocratic and gradient data is in each case satisfactory; the average difference in values of ϕ^* vs. ϕ (for the same values of k^* and k) is only ±0.002–0.007 units (1 std. dev.), some of which error appears to be experimental scatter rather than bias (Section V.D). The compounds of Fig. 3 range from small molecules (a, 4,4′-dimethoxybenzophenone; c, diethylphthalate) to an 8-kDa protein (b, insulin) and a 50-kDa molecular weight polystyrene (d). The plots of Figs. 3c, d are seen to be slightly

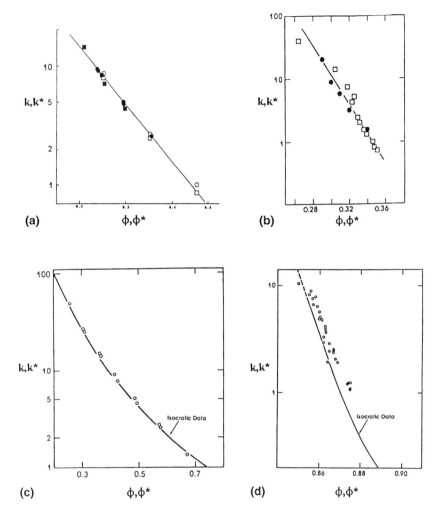

Fig. 3 Comparison of isocratic and gradient retention. Isocratic values of k vs. ϕ measured directly. Gradient values of k^* vs. $ϕ^*$ calculated from gradient runs. (a) 4,4′-dimethoxybenzophenone solute, acetonitrile/water mobile phase; solid data points are derived from gradient data, open data points are isocratic data [18]; (b) insulin solute, acetonitrile/buffer mobile phase; open squares derived from gradient data, closed circles are isocratic data [24]; (c) diethyl phthalate solute, acetonitrile/water mobile phase; circles derived from gradient data, solid line best fit to isocratic data [18]; (d) 50-kDa polystyrene fraction solute, tetrahydrofuran/water mobile phase; circles derived from gradient data, solid line is best fit to isocratic data [24].

curved rather than linear as required by Eq. (10), yet good agreement between values of ϕ^* and ϕ (for $k^* = k$) is still observed. Similar comparisons, with comparable agreement between isocratic and gradient retention, have been reported for several peptides [25]. Correlations as in Fig. 3 serve to confirm the similarity of isocratic and gradient elution, as well as verify the accuracy of the LSS model for the systems studied.

Bandwidth

The final value of k_a (k_f) at the time the band elutes from the column is important in determining band width in gradient elution; k_f can be obtained from Eq. (19) with $r = 1$:

$$\log k_f = -\log\left(2.3b + \frac{1}{k_0}\right)$$

or

$$k_f = \frac{1}{2.3b + \dfrac{1}{k_0}} \tag{21}$$

For bands that are strongly retained at the beginning of the gradient (k_0 large)

$$k_f = \frac{1}{2.3b} \tag{21a}$$

Substitution of k_f into Eq. (7) gives

$$W = 4\, N^{-1/2}\, t_0\, (1+k_f) \quad \text{(no band compression)} \tag{22}$$

However, Eq. (22) is not corrected for band compression during migration through the column as a result of the faster migration of the trailing versus leading edge of the band [20]. Therefore,

$$W = 4\, G N^{-1/2}\, t_0\, (1+k_f) \quad \text{(band compression)} \tag{22a}$$

where the band compression factor G can be derived as a function of b [26]:

$$G^2 = \frac{1 + p + \dfrac{p^2}{3}}{(1+p)^2}$$

with

$$p = \frac{2.3k_0 b}{k_0 + 1} \approx 2.3b \quad \text{(for } k_0 \text{ large)} \tag{23}$$

Values of G (k_0 large) approach 0.6 for $b > 2$ ($k^* < 0.4$), and $G > 0.9$ for $b < 0.1$ ($k^* > 10$). Thus, in most cases band compression results in a 10–40% reduction in W, compared with the value predicted by Eq. (22) (no band compression).

If we know the isocratic value of N, the column dead time t_0, and the gradient steepness b, values of W for gradient elution can be predicted from Eq. (22a). However, this assumes that no other effects contribute to band broadening. In fact, several studies [7,19,23,27–29] have shown that experimental bandwidths are generally larger than those predicted by Eq. (22a), by as much as a factor of 2 for steep gradients ($b > 1$). The origin of this anomalous band broadening is still in question, but see Section V.F. At present it is not possible to accurately predict these deviations from Eq. (22a), which serve to largely cancel the gradient compression effect G. We suggest the use of Eq. (22a) with a fixed value of $G = 1.1$ for estimating bandwidth in gradient elution, because $JG \approx 1.1$ for $0.1 < b < 2.0$, if the solid curve of Fig. 13a is assumed for the bandwidth correction factor J as a function of b. Other workers [22] have suggested Eq. (22a) with $G = 0.85$. Since uncertainty usually exists in the estimation of N as a function of experimental conditions [9,18], the exact choice of an empirical value of G is not critical.

For small molecules, values of S usually fall within narrow limits. For 64 compounds chosen from five different samples that represented widely varying solute structures, one study [30] found $S = 4.2 \pm 0.8$ (1 std. dev.). The variability of S values is typically less for samples composed of compounds of related structure. When values of S are similar for the different compounds in a sample, their values of b (Eq. 15) will also be similar. For a given gradient separation, this means (cf. Eqs. 21 and 22) that the widths of all bands will be about the same, with the exception of bands eluting early in the gradient (k_0 small). Approximately equal bandwidths in gradient elution are commonly observed, as in the example of Fig. 4b. For mixtures of homologous or oligomeric compounds as sample, S commonly increases with compound molecular weight; see the further discussion of Eq. (28).

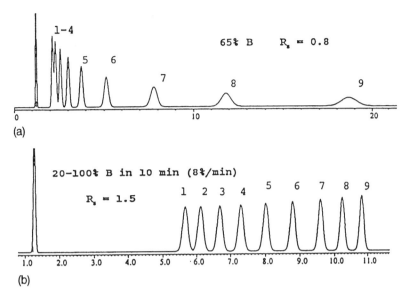

Fig. 4 Comparison of isocratic (a) and gradient elution (b) for a mixture of C_2-C_{10} n-dialkyl phthalates [4]. Computer simulations [32] based on experimental data of Ref. 99.

B. Comparing Separations by Gradient or Isocratic Elution

An often asked question is whether isocratic or gradient elution will provide the "best" separation of a given sample. For samples that contain several components with widely varying values of k (isocratic elution), gradient elution is usually preferred because it avoids bunching of peaks at the beginning of the chromatogram (with loss of resolution) and excessive broadening of peaks (with reduced detectability) at the end of the chromatogram. This is illustrated for a typical sample in Fig. 4, where isocratic and gradient separations for comparable experimental conditions are compared. For the separation and quantitation of a single peak, the relative detection sensitivity for gradient versus isocratic elution is less obvious. In the following section, we will examine this question for separation conditions that give comparable sample resolution.

Bandwidth and Sensitivity

Consider first the question of relative detection sensitivity in gradient versus isocratic separation. Since detector response should be the same for both procedures, bandwidth will in each case determine relative sensititivity. It will be shown in the next subsection that comparable isocratic and gradient resolution results, if the average retention k^* in the gradient run is made equal to k for the isocratic run. If detection sensitivity is defined as S_{iso} and S_{grad} for isocratic and gradient elution, respectively, then

$$\frac{S_{grad}}{S_{iso}} = \frac{W_{iso}}{W_{grad}}$$

where W_{iso} and W_{grad} are corresponding values of W (baseline bandwidths). This relationship, with Equations (7a) and (22) plus $k_f = (1/2)k^*$ (Eqs. 20, 21a), gives

$$\frac{S_{grad}}{S_{iso}} = \frac{2(1+k)}{G(2+k^*)} \qquad (24)$$

which for $k = k^*$ becomes

$$\frac{S_{grad}}{S_{iso}} = \frac{2(1+k)}{G(2+k)} \qquad (24a)$$

where G is given by Eq. (23), and (Eq. 20) $b = 1/1.15k^* = 1/1.15k$. For $1 < k < 10$, $S_{grad}/S_{iso} \approx 2$. Thus, detection sensitivity in gradient elution is expected to be twice as good as for isocratic separation, when sample resolution is comparable and other factors are the same. Because of anomalous band broadening in gradient elution (Section V.F), this sensitivity advantage for gradient elution will generally be somewhat less, especially for steeper gradients (k and k^* small) where detection sensitivity is maximized. Because baseline noise is also generally greater for gradient versus isocratic elution, it is therefore questionable whether gradient elution should be preferred to isocratic elution as a means for maximizing detection sensitivity. However, this argument applies only to the detection and separation of a single peak in the chromatogram and ignores run time (which can be made much shorter in gradient elution when there is a wide range in k-values for the sample components). For the simultaneous detection of several peaks with maximum sensitivity, gradient elution is usually preferred.

Resolution

Concerning resolution for gradient versus isocratic elution, we have noted that values of S for the various components of a sample will tend to be similar; typically $S \approx 4$ for small molecules. However, values of S for adjacent peaks often differ slightly, and this can have a major influence on their resulting isocratic [31] or gradient [32] separation. Therefore, any comparison of isocratic and gradient resolution should distinguish between the case where S is exactly equal for two adjacent bands, versus where the two S-values are significantly different.

Equal Values of S. An equation for resolution R_s in gradient elution can be derived as follows [17]. For gradient conditions, Eq. (16b) for two adjacent solutes 1 and 2 can be written

$$t_1 = \frac{t_0}{b} \log(2.3 k_{01} b) + t_0 + t_D \tag{25}$$

and

$$t_2 = \frac{t_0}{b} \log(2.3 k_{02} b) + t_0 + t_D \tag{25a}$$

Here, subscripts 1 and 2 refer to each solute. Because values of S are assumed equal for each band, the value of b is also the same. The bandwidth W can likewise be assumed the same for each band and is given by Eqs. (21a) plus (22a). Equations (25) and (25a) for t_1, t_2, and W (equal W_1 or W_2) can be substituted into Eq. (8) to give

$$R_s = \frac{2.3}{4} \log\left(\frac{k_{02}}{k_{01}}\right) N^{1/2} [(2.3b + 1)G]^{-1} \tag{26}$$

For small values of x, $2.3 \log x \approx x\text{-}1$, and if S is the same for the two solutes, $\alpha = k_{02}/k_{01}$ for all values of ϕ. Therefore, Eq. (26) can be restated (for closely adjacent bands) as

$$R_s = \left(\frac{1}{4}\right)(\alpha - 1)\ N^{1/2}\ [(2.3b + 1)G]^{-1} \tag{26a}$$

 (i) (ii) (iii)

This has the same form as Eq. (8a) for isocratic elution, with the exception of term (iii), which equals $k/(1 + k)$ for isocratic separation. If k^*

for gradient elution is in fact equivalent to k for isocratic elution so far as resolution, we can write by analogy with Eq. (8)

$$R_s = \frac{1}{4} (\alpha - 1) N^{1/2} \frac{k^*}{(1+k^*)} \quad (26b)$$
$$\quad\quad (i) \quad\quad (ii) \quad\quad (iii)$$

Table 1 compares exact values of R_s (term (iii) only) of Eq. (26a) with "intuitive" values from Eq. (26b) that are based on a mean k-value (k^*) in gradient elution. The accuracy of Eq. (26b) (which is of the same form as Eq. (8a) for isocratic elution) is never worse than ±5% for $k^* >$ 1. As the gradient becomes flatter and more closely resembles isocratic conditions, k^* becomes large. As expected, in the limit of large k^*, Eqs. (26a) and (26b) agree exactly.

Equation (26b) suggests that when isocratic and gradient conditions are normalized (i.e., $k = k^*$), the resolution of two adjacent bands will be similar. This is illustrated in the computer simulations of Fig. 5 (based on Eqs. 16b and 22a) for a hypothetical example, where the same value of $S=4$ was chosen for each solute, $\alpha=1.05$, $N=15000$, and $k_0 = 100$ for the gradient separation. Values of ϕ (isocratic) and b (gradient) have been selected to give values of $k = k^* = 1$, 5 or 20 in these six separations. Resulting values of R_s are essentially the same for each isocratic versus gradient comparison in which $k = k^*$. In the real world, due to the anomalous band broadening often observed in gradient elution (Section V.F), *resolution in gradient elution will usually be somewhat less than in isocratic elution*, when k and k^* are each optimized

Table 1 Comparison of Equations (26a) and (26b) for the Prediction of Resolution in Gradient Elution

b	k^*	$[(2.3b+1)G]^{-1}$ Eq. 26a[a]	$k^*/(1+k^*)$ Eq. (26b)	Error in Eq. (26b)
0.05	17.4	0.945	0.946	0.1%
0.1	8.7	0.895	0.897	0.2%
0.2	4.3	0.808	0.813	0.6%
0.5	1.7	0.620	0.635	2.4%
1.0	0.87	0.444	0.465	4.7%
2.0	0.43	0.281	0.303	7.8%

[a]Values of G calculated from Eq. (23) with $p = 2.3b$.

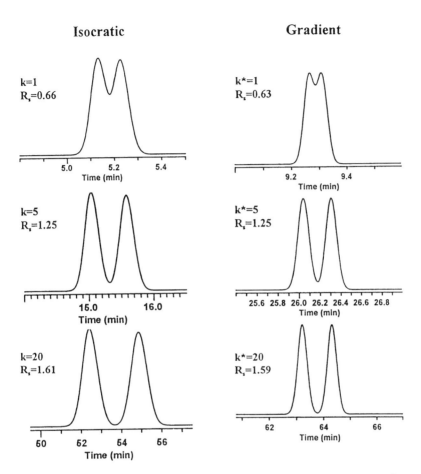

Fig. 5 Comparison of resolution in isocratic and gradient elution for "corresponding" conditions of ϕ and b. Assumes equal S-values for the two solutes. See text for details. Computer simulations based on Eqs. (16b) and (22a). Equation (10) describes isocratic solute retention: for solute 1, $\log k = 2.00-4.00\phi$; for solute 2, $\log k = 2.02-4.00\phi$.

for two adjacent bands in the isocratic and gradient separations. The assumed isocratic plots of k versus ϕ for the two solutes of Fig. 5 are shown in Fig. 6a. The equal values of S for each solute result in parallel plots of log k versus ϕ.

Unequal Values of S. When values of S differ for different solute bands, α will vary with either ϕ (isocratic elution) or b (gradient elution), and *minimum* resolution (other factors equal) may be found for

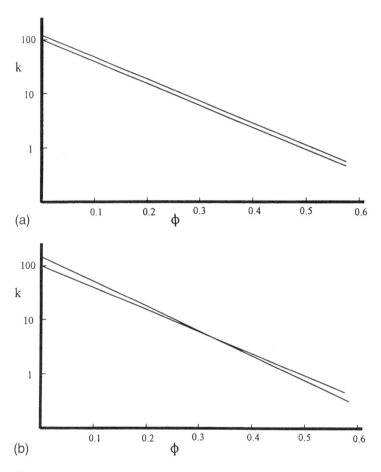

Fig. 6 Assumed retention behavior for two hypothetical solutes used in computer simulations of Figs. 5a and 7b. See text for details.

an intermediate value of ϕ or b [31,32]. This is illustrated for a second set of simulations, based on the retention equations plotted in Fig. 6b and illustrated further in Fig. 7 (cf. Fig. 5 for equal values of S). As in Fig. 5, the comparisons of Fig. 7 show similar resolution in gradient and isocratic elution when $k = k^*$. However, it can be seen that these comparisons of R_s are no longer near-exact, as was the case for Fig. 5 with equal S-values. A rigorous (but complicated) analysis of resolution as a function of b for the case of unequal solute S-values is given in Ref. 22 (Eq. 4.55); see also the discussion in Ref. 17. Computer simulations (unreported data) suggest a simpler treatment: the value of k^* for mini-

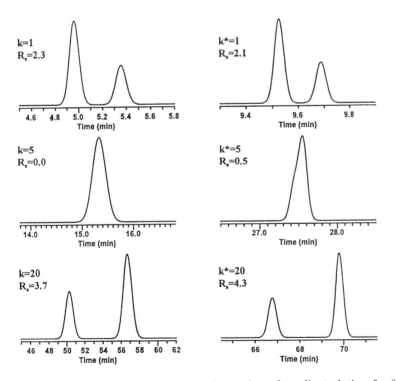

Fig. 7 Comparison of resolution in isocratic and gradient elution for "corresponding" conditions of ϕ and b. Assumes unequal S-values for the two solutes. See text for details. Computer simulations as in Fig. 5; for solute 1, log k = 2.15–4.40ϕ; for solute 2, log k = 2.02–4.00ϕ.

mum resolution ($R_s = 0$) will be generally smaller than k for $R_s = 0$ by an empirical factor γ, which is in turn given approximately by

$$\gamma = [8.7 - 1.3 \, (\log k)] \frac{\Delta S}{S*} \tag{27}$$

Here, ΔS refers to the difference in S values for the two solutes, and $S*$ is the average value of S. Thus, if the two S-values differ by 10% (as in Fig. 7; a large, but not uncommon difference), the value of $k*$ for $R_s = 0$ will equal about 0.8 times the value of k for $R_s = 0$. This is further illustrated by comparing isocratic versus gradient values of R_s from the examples of Fig. 7 with gradient predictions for $k* = 0.8k$:

$k, k*$	Isocratic	Resolution R_s Gradient ($k*=k$)	Gradient ($k* = 0.8k$)
1	2.3	2.1	2.1
5	0.0	0.5	0.0
20	3.7	4.3	3.7

Values of R_s for these isocratic separations agree better with the gradient separations where $k* = 0.8k$, rather than with those for $k = k*$. However, in most cases the discrepancies in resolution observed for $k = k*$ will have little practical significance. A rigorous derivation of γ (Eq. 27) has been suggested by a reviewer and is included in the Appendix.

C. Large-Molecule Effects

Samples composed of synthetic or natural polymers often require the use of gradient elution, especially for RP-LC separation. As will be seen, the LSS model has proved useful in understanding and optimizing these separations. In some ways, the gradient separation of macromolecules seems to follow different rules than are observed for the separation of compounds with molecular weights <10 kDa (Table 2). For very large molecules (>100 kDa), isocratic elution with $k \gg 0$ may be difficult to achieve; i.e., for small changes in %B, retention seems to change rapidly from $k \sim 0$ to no elution at all. These observations suggest that gradient elution differs fundamentally for large versus small molecules,

Table 2 Characteristics of Large-Molecule versus Small-Molecule RP-LC[a]

Gradient Elution
1. Change in flow rate or column length has little effect on sample resolution
2. Change in flow rate or column length has little effect on ϕ at elution (ϕ_e)
3. Changes in column type or surface area have little effect on ϕ at elution (ϕ_e)

Isocratic Elution
1. For very large molecules, isocratic elution with $k > 0$ is not possible

[a]See the second, third, and fourth subsections of Section III.C.

and various alternative descriptions of large-molecule gradient elution have been suggested:

1. The "on-off" model [33–36] assumes that each solute band is retained strongly at the column inlet until ϕ reaches a certain "critical" value ϕ_c; when $\phi = \phi_c$, the value of k for the band decreases from a very large value to 0, and the solute band then moves through the column with $k_a = 0$ and is eluted in a mobile-phase composition $\phi_e = \phi_c$.
2. The "precipitation" model, applicable mainly to synthetic polymers, assumes that the solute is initially insoluble, but for some mobile-phase composition ϕ_c the solute dissolves and is then carried through the column in solvent of composition $\phi_c = \phi_e$, i.e., without retention [37,38].
3. The "critical solution" model [39,40] is based on a detailed statistical-thermodynamic theory of chromatographic retention and arrives at similar predictions as for the "on-off" model.

Large molecules have some special characteristics which occasionally affect both their isocratic and gradient separation in such a way as to lead to possible discrepancies in comparisons of isocratic and gradient elution:

1. Experimental values of isocratic retention k sometimes vary with flow rate [24].
2. Experimental values of S may differ for isocratic versus gradient separation [41].
3. Solute conformation may vary for small changes in ϕ, and these changes in conformation may be slow [42–45].

4. Solute retention may deviate markedly from Eq. (10) for small values of k [46] (in some cases such deviations may be due to inaccurate values of t_o).

We will not pursue these and other large-molecule retention anomalies here (see Section V.C), but simply note that it is rare that they have a significant effect on gradient separation, Instead, we will next examine the predictions of conventional chromatographic theory for the separation of large versus small molecules in the absence of any special effects as above. The LSS model is well suited for such an examination.

Several early studies noted that values of the parameter S increase with molecular weight M, for the separation of synthetic polymers [47] and proteins [48]. As a rough approximation for RP-LC [49].

$$S \approx 0.5 M^{1/2} \tag{28}$$

Thus, values of S greater than 100 are possible for solute molecular weights >100 kDa. In such cases, an increase in ϕ by 0.01 unit (1% B) can result in a 10-fold reduction in k. Large values of S for big molecules can be rationalized in terms of different models of the RP-LC retention process [50,51]. "Critical solution" behavior as defined by Boehm and co-workers implies values of S that are effectively infinite, so that isocratic elution becomes impossible; k for the solute is then either 0 or ∞. Likewise, critical solution behavior leads directly to the on-off model of large-molecule gradient elution. A number of reports from different groups have presented arguments in favor of either the critical solution or LSS model for the gradient separation of large molecules. In fact, the two models address different aspects of large-molecule gradient elution and are actually complementary rather than oppositional (one possible exception is whether critical solution behavior represents a continuum or discontinuity in the retention process). We will first examine the consequences of large S-values in terms of the LSS model. Then we will attempt to reconcile various differences in interpretation of experimental results. The next three sections address the issues of Table 2. For a fuller discussion of "critical solution" vs conventional chromatographic behavior, the reader is referred to several summary accounts that present differing points of view [40,45,49,52].

Gradient Retention of Large-Solute Molecules: On-Off Versus Conventional Elution

The combination of Eqs. (15) and (20) leads to

$$k^* = \frac{1}{1.15}\left(\frac{t_G}{t_0 \Delta\phi}\right)\frac{1}{S} \tag{29}$$

For typical "small" molecules, $S \approx 4$, and for typical values of t_o (determined by column size and flow rate) a gradient steepness $\Delta\phi/t_G$ will be selected to give a reasonable value of k^*: $1<k^*<10$ (as for isocratic elution, where the usual goal is $1<k<10$). If similar gradient conditions (values of t_G/t_o, $\Delta\phi$) are used for a large-molecule sample, where S can be greater than 100, resulting values of k^* will be very small ($k^*<0.1$). This means that during most of the time the sample moves through the column, there is very little retention. Under these conditions, the on-off model represents a good description of the actual situation. However, the use of a gradient steepness $\Delta\phi/t_G$ that is similar for both large and small molecules is a poor use of gradient elution. In order to maintain desirable values of k^* ($1<k^*<10$) for the separation of large molecules, it is necessary to reduce gradient steepness (Eq. 20a; Ref. 53). When this is done, large molecules will move through the column in similar fashion as small molecules; i.e., with $k_a >> 0$. This also means that solute migration rate will increase from beginning to end of elution (as in Fig. 2). This sample migration pattern within the column has been observed experimentally for the gradient separation of high-molecular-weight protein samples [54], in confirmation of theoretical predictions based on the LSS model [53] for samples that are retained during their passage through the column.

Gradient Retention of Large-Solute Molecules: Resolution as a Function of Column Length and Flow Rate

A number of workers [55–58] have observed that gradient separations of protein samples (large molecules) give similar resolution, regardless of changes in flow rate or column length. This is not the case for isocratic elution, where sample resolution usually increases if flow rate is decreased or column length is increased (because of the dependence of N on column length and flow rate). This apparently anomalous behavior for protein samples in gradient elution has been argued as additional evidence for the on-off model. This comparison is faulty, however, because when flow rate or column length are changed for gradient elution, there is a corresponding change in k^* (Eqs. 15,20), as well as in N:

$$k^* = \frac{1}{1.15}\left(\frac{\Delta\phi S}{t_G}\right)\frac{V_m}{F} \tag{29a}$$

Because V_m is proportional to column length, an increase in column length or a decrease in flow rate will increase N but decrease k^*. R_s increases with increase in either N and k^* (Eq. 26b), so changes in column length or flow rate will tend to cancel so far as their net effect on resolution. This analysis has been carried further [53,59], with the quantitative conclusion that resolution will not change much when column length or flow rate is varied (holding t_G constant) for the gradient separation of large molecules. Resolution can usually be improved by increasing column length or reducing flow rate, *if k^* is held constant by a corresponding increase in gradient time t_G*. For a further discussion of the maximization of resolution for the gradient separation of large molecules in terms of the LSS model, see Ref. 18.

Gradient Retention of Large-Solute Molecules: Mobile-Phase Composition ɸ at Elution

For large-molecule solutes, the mobile-phase composition at the time the solute elutes from the column ($\phi = \phi_e$) changes very little for changes in experimental conditions. If values of k change from essentially infinite to zero for a very small change in ϕ (critical solution model), the on-off model would apply and this behavior would be expected. However, the same behavior is predicted by the LSS model for large values of S, even when $k_a >> 0$ during migration of the sample through the column. This can be seen as follows. From Eq. (2a).

$$\phi_e = \phi_0 + \frac{\Delta\phi}{t_G}(t_R - t_0 - t_D) \tag{30}$$

Note that the value of ϕ at the end of the column is delayed relative to Eq. (2a) by the column dead time t_0 and the equipment delay time t_D. Eqs. (15), (16b), and (30) then yield

$$\phi_e = \phi_0 + \frac{1}{S}\log\ 2.3k_0 b \tag{31}$$

$$= \phi_0 + \frac{1}{S}\log 2.3k_0 - \frac{1}{S}\log\left(\frac{F}{V_m}\right) - \frac{1}{S}\log\left(\frac{t_G}{\Delta\phi S}\right) \quad (31a)$$

(i)   (ii)   (iii)   (iv)

The effect of a change in column type or surface area is described by term (ii) of Eq. (31). If a large change in k_o (e.g., by a factor of 10) results from this change in the column, the relative change in ϕ_e is predicted to be $\log(10)/S$, which for $S>100$ will be less than 0.01 units in ϕ (1% B). Likewise, a change in ϕ_e as a result of change in t_G, flow rate or column length is described by terms (iii, iv) of Eq. (31). Again, when $S>100$, the change in ϕ_e for a 10-fold change in t_G, flow rate or column length will be less than 0.01 unit.

It can also be seen that ϕ_e does not depend on ϕ_o, as long as k_o is large (which will be the case, whenever $\phi_o < \phi_e$). Thus, Eqs. (14a) and (31) give

$$\phi_e = \frac{1}{S}\log k_w - \frac{1}{S}\log 2.3b \quad (31b)$$

For a given solute, values of k_w and S are constant. Similarly, the above discussion of Eq. (31a) is equivalent to arguing that ϕ_e does not depend on b. Therefore, as long as S is large and the solute is retained initially with k_o large, ϕ_e will change very little when ϕ_o or b is changed.

Isocratic Retention of Large-Solute Molecules

It was originally observed [60] that synthetic polymers with molecular weights >50 kDa cannot be eluted under isocratic conditions with $k>0$. This was argued as evidence for critical solution behavior and an effectively infinite value of S, so that k equals 0 or ∞. Such a discontinuous change in S with increasing molecular weight does not seem reasonable, based on the known applicability of Eq. (28) for $M<10^5$ kDa. More recently, it has been shown [44,45] that isocratic elution of very large polymer molecules with $k>0$ can in fact be achieved, by means of sample conditioning procedures that allow full equilibration of the sample with the mobile phase prior to injection.

We can summarize the above discussion as follows. The critical solution model of Boehm et al. [40,50] agrees with other models [51] and numerous experimental studies in predicting that S increases for the RP-LC separation of large solute molecules (approximately in accordance with Eq. 28). For some (undesirable steep) gradient conditions, large values of S can result in very small values of k_a during migration of the sample through the column, and the on-off model then applies. For preferred gradient conditions, this will not be the case, and the separation of large molecules does not differ in principle from the separation of small molecules. The LSS model is applicable for both cases ($k^* \sim 0$, or $k^* > 1$). The sample-precipitation model is a separate case, discussed more fully in Refs. 38 and 52. Sample precipitation will hardly ever apply to the analytical separation of biomolecules, but can be important for the separation of synthetic polymers.

D. Gradient Delay and Elution After the Gradient

If the solute is not strongly retained at the beginning of gradient elution (k_0 not large), and if the equipment hold-up volume V_D is significant, then the solute can migrate part way through the column before it is overtaken by the gradient. This effect, referred to as *solute pre-elution*, will introduce an error into the value of t_R calculated from Eq. (16b). This error can be determined and used to correct Eq. (16b) as follows [61]. The solute fractional migration x (cf. r in Fig. 2) during pre-elution is given as

$$x = \frac{t_D}{t_0 k_0} \tag{32}$$

The corrected solute retention time is then

$$t_R = t_D + t_{1-x} \tag{33}$$

where t_{1-x} is the time required to migrate through the remaining distance ($[1-x]L$) within the column, and L is column length. For a column of fractional length $1-x$, the correct retention time is given by Eq. (18) with $1-x = r$, or

$$t_{1-x} = \frac{t_0}{b} \log[2.3 k_0 b (1-x) + 1] + (1-x) t_0 \tag{34}$$

with x given by Eq. (32). Substitution of Eqs. (32) and (34) into (33) then gives the correct value of t_R for the case where solute pre-elution occurs. This treatment is equivalent to the model of Jandera and Kucerova based on the division of the column into two connected segments [62].

If x is assumed $\ll 1$ (a common case), Eq. (34) can be simplified [61] and related to a change in ϕ_e ($\delta\phi_{pe}$):

$$\delta\phi_{pe} = \frac{\dfrac{V_D}{V_m}}{2.3Sk_0} \tag{35}$$

For example, assume typical values of $V_D = 4$ mL, $V_m = 1.5$ mL (15 × 0.46-vm column), $S = 4$ (solute molecular weight = 100–400), and $k_0 = 10$. The resulting error in Eq. (16) expressed as a change in ϕ_e, is then $\delta\phi_{pc} = 0.03$, which can be related to a positive error in t_R from Eq. (16) of $-0.03(t_G/\delta\phi)$ min.

If the solute is not eluted prior to the end of the gradient ($t_R < t_G + t_0 + t_D$), then Eq. (16) will again be in error. In this case, the fractional migration r through the column is given by Eq. (18) with $t_r = t_G + rt_0 + t_D$. For k_0 large, corresponding to late elution from the column, the fractional migration r can be determined as

$$r = \frac{10^g}{2.3k_0 b} \tag{36}$$

where $g = bt_G/t_0$. The total retention time is then

$$t_R = t_G + t_0 + t_D + (1-r)k_z t_0 \tag{37}$$

where k_z is the value of k_a when $\phi = \phi_z$ (at the end of the gradient).

E. Mass-Overloaded Separations

Isocratic Elution

The theory of preparative HPLC separation is now well developed. For the case of lightly overloaded separation, simple, reliable equations can be derived which describe bandwidth and retention time as a function of sample weight and experimental conditions. To a reasonable approximation, band broadening as a function of sample weight can be represented by a family of overlapping bands of right-triangle shape. This is illustrated in Fig. 8a for a typical example. Retention time and bandwidth for a single solute can be defined as in Fig. 8b. Note that the retention time t_R for a small sample defines the end of the band in Fig. 8b, so that only the bandwidth W is required to define the band and relate this to resolution of one band from another. Bandwidth as a function of sample weight w is given as [64]

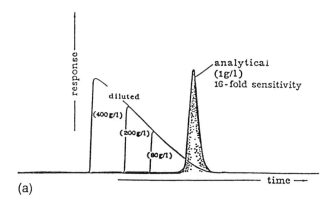

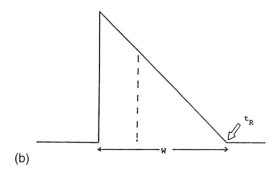

Fig. 8 Bandwidth as a function of sample weight in isocratic elution. (a) Overlapped elution bands for different sample sizes [63]; (b) definition of retention time and bandwidth for separation with a large sample weight [64].

$$W = \left\{ \left(\frac{16}{N}\right) t_0^2 (1+k)^2 + \frac{6 t_0^2 k^2 w}{w_s} \right\}^{1/2} \tag{38}$$

(i)   (ii)

Here, N and k refer to values for a small sample, and w_s is the saturation capacity of the column (maximum possible uptake of solute by the stationary phase). Note that term (i) of Eq. (38) is the band broadening that results for a small sample, while term (ii) represents the additional band broadening as a result of mass overload. Equation (38) has been shown to be reliable for several different RP-LC systems [64,65]. The column capacity w_s can be determined from measurements of W for a small sample weight and a large sample weight. From the chromatogram of a small sample, values of N, t_0, and k can be determined. Values of w and W for the large-sample-weight experiment plus small-sample values of k, t_0, and N can then be substituted into Eq. (38) to solve for w_s (see also the discussion of Ref. 64).

Gradient Elution

For a small sample, equivalent resolution is expected for isocratic versus gradient elution when experimental conditions are selected so that $k = k^*$ (Section III.B, "Resolution" subsection). It can be argued [66] that comparable resolution should also be found for overloaded separation, when w/w_s is the same in isocratic and gradient elution, and $k = k^*$. This hypothesis is tested in Fig. 9 for the separation of two compounds as a function of sample size and corresponding conditions for isocratic versus gradient separation ($k = k^*$). Three elution curves are overlapped for each of the examples in Fig. 9, for the separate injection of each compound ("1-solute") and the mixture ("2-solute"). The agreement between isocratic and gradient separation in each comparison (for the same sample weight) is almost exact, when it is recognized that band shape in mass-overloaded gradient elution is distorted by the gradient to give "shark-fins" instead of right triangles as in Fig. 9.

A further analysis of gradient elution under overloaded conditions [66,67] suggests that W in gradient elution will be given by Eq. (38), provided k is replaced by k_f. Since k_f can be calculated from experimental conditions (Eqs. 15,20), it is then possible to calculate W and sample resolution in mass-overloaded gradient elution:

$$W = \left\{ \left(\frac{16}{N}\right) t_0^2 G^2 (1+k_f)^2 + 6 t_0^2 k_f^2 \left(\frac{w}{w_s}\right) \right\}^{1/2} \tag{38a}$$

$$\qquad\qquad\quad \text{(i)} \qquad\qquad\quad \text{(ii)}$$

Term (i) of Eq. (38a) can be determined from small-sample experiments, and this allows the evaluation of term (ii) (referred to as W_{th}^2) when the sample size is large enough to cause additional band broadening. Term (ii) can be restated (Eqs. 21, 38a) as $W_{th}^2 = 1.1(t_0^2 G^2/b^2)(w/w_s)$; therefore, a log-log plot of $W_{th}(b/t)_0$ versus w/w_s should give a straight line of slope 0.5.

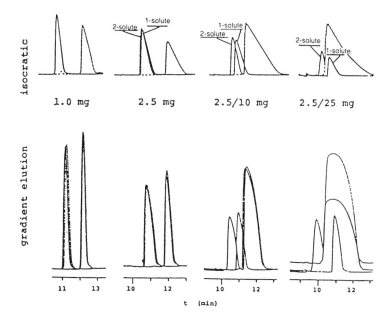

Fig. 9 Comparison of isocratic and gradient separation as a function of sample weight, when corresponding conditions are used ($k = k^*$). Sample is a mixture of two xanthines; sample weights for each compound indicated in fig. (1.0 mg each, 2.5 mg each, 2.5 mg of first and 10 mg of second compound; 2.5 mg of first and 25 mg of second compound [66]). Elution curves for the individual solutes ("1 solute") and mixture are overlapped in each case.

Term (ii) of Eq. (38a) is tested in Fig. 10 for the separation of two different solutes as a function of sample weight w. The predicted linear relationship is observed in Fig. 10 for both solutes, and the data fall close to the solid curve from Eq. (38a) with a value of $G = 1.0$ (see discussion of Section III. A, "Bandwidth" subsection). This represents reasonable agreement of experiment with LSS theory, especially when the possible importance of anomalous band broadening (Section V.F) is considered.

The equivalence of k and k^* for mass-overloaded separation can be used in similar fashion as for small-sample separations (Section III.B). Thus, as in the latter case, the separation of two adjacent bands is usually best carried out isocratically, since gradient resolution is not expected to be better than for isocratic separation. Likewise, the same

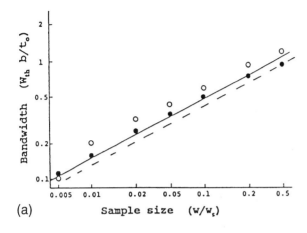

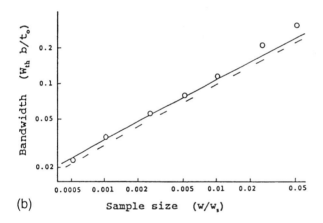

Fig. 10 Comparison of experimental and predicted (Eq. 38a, solid curve) bandwidths in gradient elution as a function of sample weight [67]. Solid curve is for $G = 1$. (a) Caffeine solute, 0–50% methanol/buffer gradient; o, t_G = 30 min; •, t_G = 60 min; (b) benzyl alcohol solute, 0–50% methanol/water gradient in 30 min.

considerations that are used to optimize mass-overloaded isocratic separation will be applicable for corresponding gradient separations, when gradient elution is preferred because of other factors (more than two sample components having widely different values of k, the separation of large-molecule samples, etc.).

IV. OPTIMIZING GRADIENT SEPARATION

The LSS model has proved especially helpful in facilitating the development of gradient elution procedures. HPLC method development can have several goals [4], but the achievement of adequate sample resolution is usually paramount. Effective gradient method development recognizes that equal resolution is found for "corresponding" isocratic and gradient separations, i.e., where $k = k^*$. This means that method development strategies which are suitable for isocratic separation can be applied in similar fashion to the development of gradient methods. Since most chromatographers have more experience with *isocratic* method development, this offers a means for using this experience in the development of *gradient* methods. The related nature of isocratic and gradient method development is discussed further in Section IV.A.

Method development is best begun by carrying out an initial separation of the sample with a full-range gradient (0–100% B; or better, 5–100% B; see Section V.C). Application of the LSS model allows appropriate conditions for this initial experiment to be predicted in advance (e.g., for $1<k^*<10$), as opposed to the trial-and-error selection of an acceptable mobile phase for isocratic elution. From this initial gradient run, it is possible to select conditions for the next method development experiment, which can be either an isocratic or gradient run. The design and interpretation of this initial gradient separation are discussed in Section IV.B.

The equations that were derived from the LSS model in Section III allow the quantitative prediction of either isocratic or gradient separation as a function of certain experimental conditions. Two initial gradient runs are required for the determination of values of k_w and S for each sample component. It is then possible to predict isocratic separation as a function of %B (equal to 100ϕ), and gradient separation as a function of gradient conditions (t_G, ϕ_0, ϕ_z). Although this has not yet been discussed, it is also possible to extend these predictions to changes in gradient shape (e.g., segmented gradients) and changes in "column conditions": column length and diameter, particle size, and flow rate. These predictive capabilities are best realized by *computer*

Table 3 Experimental Conditions Which Can Be Changed to Improve Resolution in Reversed-Phase Gradient Elution (Eq. 8a)[a]

Retention (k^*)	*Column plate number (N)*
Gradient steepness b	Column length
	Particle size
	Flow rate
Selectivity (α)	
Neutral and ionic solutes	*Ionic solutes*
Gradient steepness (k^*)	pH
Solvent type	Ion-pair-reagent concentration
Column type	Concentration of additives or buffers
Temperature	Column source (manufacturer)

[a]See discussion of Refs. 4, 68, and 69.

simulation—the use of a computer to carry out the necessary calculations and present them to the user in the form of chromatograms, tables, and graphical plots that summarize one or more simulated separations. Computer simulation is discussed briefly in Section IV.C.

A. Gradient Method Development

A detailed discussion of isocratic and gradient method development, as well as their interrelationship, is given in Ref. 4. To summarize that treatment, sample resolution can be systematically improved by changing one or more of the separation conditions listed in Table 3. Since the resolution of a "critical" band pair[†] (that band pair in the chromatogram having the smallest value of R_s) depends on any one variable in the same way for isocratic or gradient elution, it is convenient to express this dependence in terms of the simpler isocratic system (for values of $k = k^*$). The following discussion is organized according to the three subdivisions of Table 3 (k^*, N, α).

Retention (k^*)

Usually the first step in method development is to optimize sample retention k^*, as represented by term (iii) of Eq. (26b): $k^*/(1+k^*)$. Condi-

---

[†]There is no connection between "critical" resolution and "critical solution" theory.

tions should be adjusted initially to provide a value of k^* that provides a good compromise among resolution, run time, and detection sensitivity (peak height). As in isocratic elution, where the usual aim is $1<k<10$ for all sample bands, it is initially desirable to have $1<k^*<10$. Selection of conditions for $1<k^*<10$ is easier in gradient elution, since for a linear gradient all bands will have similar values of k^*. For a small-molecule sample, S can be assumed equal to about 4, and Eq. (20a) then becomes

$$k^* \approx \frac{0.2 t_G F}{V_m \Delta\phi} \quad (S = 4) \tag{39}$$

For the usual 4.6-mm ID column, V_m (mL) can be approximated by $L/10$ (cm), which gives

$$k^* \approx \frac{2 t_G F}{L \Delta\phi} \quad (S = 4, \text{ column ID} = 0.46 \text{ cm}) \tag{39a}$$

Usually the flow rate and column length will be selected for other reasons (plate number, pressure drop, run time [68]), and initially it is best to select an intermediate value of k^*; e.g., $k^* = 5$. A maximum gradient range ($\Delta\phi \approx 1.0$) is also desirable for an initial gradient separation of a sample whose elution characteristics are not known. This leaves gradient time as the only quantity in Eq. (39a) that has not been specified. Solving for t_G gives

$$t_G \approx \frac{5}{2}\left(\frac{L}{F}\right) \quad (S = 4, \text{ column ID} = 0.46 \text{ cm}, \Delta\phi = 1) \tag{40}$$

For typical conditions [68], e.g., $L = 15$ cm and $F = 2$ mL/min, a recommended gradient time of about 20 min can be obtained from Eq. (40). The value of k^* chosen for the initial separation can vary by as much as a factor of 2, without a large effect on average sample resolution. Smaller values of k^* and t_G mean shorter run times and sharper peaks (better detection sensitivity), while larger values of t_G result in longer run times and broader peaks.

Selectivity (α)

Once a suitable value of k^* has been determined (in terms of average resolution, run time and peak width), it is recommended [4] to next optimize selectivity or band spacing. Generally it is desired to separate all adjacent bands with some minimum resolution; e.g., $R_s = 1.5$ for baseline separation. Typically, one or more ("critical") band pairs will overlap and require an increase in R_s. This can be achieved by varying one or more of the conditions of Table 3 ("selectivity [α]"). The first variable which should be investigated is gradient steepness or the value of k^*. Thus, the initial selection of k^* (see preceding subsection) aims at achieving an average resolution for the entire sample which is generally adequate. However, an individual band pair may be unresolved for this *average* optimum value of k^*. This is illustrated in the example of Fig. 7 for $k^* = 5$. In such cases, gradient steepness and k^* can be varied with the aim of improving the resolution of various critical band pairs. At the same time, it is necessary that adequate resolution ($R_s > 1.5$) be maintained for the other band pairs in the chromatogram, which is most conveniently effected by using computer simulation (Section IV.C).

With the exception of gradient steepness, the other selectivity variables of Table 3 can be varied in the same way as for isocratic elution. The effect of such changes in conditions for a gradient separation will be exactly the same as for isocratic elution, as long as corresponding conditions ($k^* = k$) are compared. For a further discussion of selectivity effects in either isocratic or gradient elution, see Refs. 4, 69, and 70.

Plate Number (N)

A plate number cannot be calculated from a gradient chromatogram by means of Eq. (7) for isocratic elution. If a value of N is calculated from an isocratic separation, however, the same value of N can be assumed for gradient elution. Since N can vary significantly as a function of k [9], this means that a comparison of N for isocratic versus gradient elution should be made under conditions where $k = k^*$. A value of N calculated from a gradient run can be obtained from Eqs. (21a) and (22a):

$$N = 16\left\{\frac{(2.3b+1)Gt_0}{2.3bW}\right\}^2 \tag{41}$$

A major difference between isocratic and gradient separation concerns the effect of changes in flow rate or column length on retention and band spacing. In isocratic elution, a change in F or L does not affect values of k or α. This is not the case for gradient elution, as can be seen from Eq. (39a). If it is desired to change N in gradient elution, in order to either improve resolution or shorten run time, it is usually necessary to maintain k^* approximately constant so as to retain the optimum band spacing or selectivity achieved in the preceding subsection. While run time and detection sensitivity will usually not change much when L or F are varied (holding other conditions including t_G fixed) [53], selectivity can change dramatically. To maintain constant selectivity when making changes in L or F, t_G should be changed at the same time in such a way as to hold k^* constant (see Eq. 39). Thus, if column length or flow rate is increased by some factor, then gradient time must be increased (L) or decreased (F) by this same factor. For a column of fixed size, this requirement for constant selectivity can also be met by maintaining gradient volume = $t_G F$ constant [71, 22, Section 4.5].

There is one exception to the rule that selectivity will not change if b (Eq. 15) is held constant when varying column length or flow rate in gradient elution. If there is a significant gradient delay t_D, the average value of k_0 for two adjacent solutes is not large, and values of S for the two solutes are significantly different, then holding b fixed while varying F or L may not ensure that selectivity remains the same [71a]. If a critical band pair elutes early in the chromatogram, a slight adjustment in b may be necessary to maintain a previously optimized selectivity after changing F or L while holding b constant.

B. Design and Use of an Initial Gradient Separation

Each experiment during method development can be used to design the next experiment. This is especially true of an initial gradient separation, which can provide the following information:

- Is isocratic or gradient elution best for subsequent experiments and the final method?
- If isocratic elution is recommended, what value of ϕ should be used for the second (isocratic) experiment?
- If gradient elution is recommended, what values of ϕ_0 and ϕ_z should be used for the second experiment?

Isocratic or Gradient Elution?

Consider first whether isocratic or gradient elution will be best for the final method. This is determined by the allowable range in k for isocratic separation. Let values of k for the first (a) and last (z) eluted peaks be k_a and k_z. It is then required that the ratio k_z/k_a be less than some maximum value; e.g., $k_z/k_a < 10$, if $1 < k < 10$. Retention times t_{Ra} and t_{Rz} for peaks a and z in a gradient separation can be approximated by Eq. (16); assume that S for each band is approximately the same, then

$$t_{Rz} - t_{Ra} = \frac{t_G}{\Delta\phi S} \log\left(\frac{k_z}{k_a}\right) \qquad (42)$$

or

$$\left[\log\left(\frac{k_z}{k_a}\right)\right]_{max} = \frac{\Delta\phi S}{t_G}(t_{Rz} - t_{Ra}) \qquad (42a)$$

and for a full-range gradient ($\Delta\phi=1.0$)

$$\left[\log\left(\frac{k_z}{k_a}\right)\right]_{max} = S\frac{t_{Rz} - t_{Ra}}{t_G} \qquad (42b)$$

Here, $[\log(k_z/k_a)]_{max}$ corresponds to the maximum allowable value of k_z/k_a, which in turn is determined by the maximum allowable range in isocratic k-values for all the sample peaks. Equation (42b) can be rearranged to

$$f = \frac{t_{Rz} - t_{Ra}}{t_G} = \frac{1}{S}\left[\log\left(\frac{k_z}{k_a}\right)\right]_{max} \qquad (42c)$$

which relates the fraction f of the gradient chromatogram that is filled with sample bands to the allowable range in isocratic k-values and to S.

Table 4 summarizes the application of Eq. (42c) for different values of k_z/k_a and samples of differing molecular weight (and S). Depending on the range in isocratic k-values that is acceptable, different maximum values of $f = (t_{Rz} - t_{Ra})/t_G$ are allowable in the initial gradient run (for which $\Delta\phi = 1.0$). Note that higher molecular weight samples have smaller allowable values of f, which is another reason why gradient elution is usually required for such samples.

What Value of ϕ for Isocratic Separation?

Since large-molecule samples will usually require gradient elution, we will restrict our analysis to small-molecule samples, for which $S \approx 4$. This case has been analyzed in detail [30]. If the value of S can be assumed, then b can be calculated from the experimental conditions (Eq. 15), and a value of k_0 can be obtained from (17a). This then allows the

Table 4 Isocratic or Gradient elution? Interpretation of an Initial Gradient Separation (see text for details a value of $\Delta\phi = 1.0$ is assumed)

Allowable range in isocratic retention	Maximum value of $f = (t_{Rz} - t_{Ra})/t_G$			
	$M<500$[a] ($S = 4$)	$M = 500–2000$ ($S = 10$)[b]	$M = 2000–10{,}000$ ($S = 19$)[b]	$M = 10{,}000–100{,}000$ ($S = 47$)[b]
$1<k<10$	0.25	0.10	0.05	0.02
$1<k<20$	0.33	0.13	0.06	0.03
$0.5<k<20$	0.40	0.16	0.08	0.03

[a]Average value from Ref. 29.
[b]Calculated from a more precise form of Eq. (28) for peptides and proteins [23]: $S = 0.48 M^{0.44}$.

calculation of values of k as a function of ϕ (Eqs. 10, 14a). Depending on the allowable value of k for the first or last band, or the desired average value of k (e.g., $k = 5$), the required value of ϕ in isocratic elution can be calculated from Eq. (10). It has been shown [30] that the predicted value of ϕ for isocratic separation will be most accurate (considering error in the assumed value of $S = 4$), when the initial gradient run is carried out under conditions of $k^* = k$ for the peak whose desired k-value in isocratic elution has been specified. See (Refs. 4 and 29) for tables which facilitate the prediction of ϕ for the second (isocratic) experiment in method development. The above approach based on the LSS model can be compared with other procedures for estimating isocratic retention from a single gradient run [72].

What Values of ϕ_0 and ϕ_z for Gradient Elution?

This question has been addressed in Ref. 30. The value of ϕ_e in gradient elution is given by

$$\phi_e = \frac{(t_R - t_0 - t_D)\Delta\phi}{t_G} + \phi_0 \qquad (43)$$

If gradient steepness (b) will not be changed, it is usually safe to take $\phi_0 = \phi_e - 0.05$ for the first band in the initial gradient, and $\phi_z = \phi_e + 0.05$ for the last band. In this way, a "safety margin" of 0.05 unit in ϕ is allowed at the beginning and end of the separation. However, if b is to be varied (usually desirable in order to optimize selectivity), the effect of a change in b on the value of ϕ_e needs to be considered. From Eq. (16b), the change in ϕ_e (δ_e) as a result of a change in b from a value b_1 to b_2 can be shown to be

$$\delta\phi_e = \frac{1}{S} \log\left(\frac{b_2}{b_1}\right) \tag{44}$$

Thus, for sharper gradients (larger b), ϕ_e will increase; i.e., all peaks will be shifted toward the end of the chromatogram. In method development for gradient elution, it is desirable to vary b in order to optimize selectivity (Section IV.A). If we allow k^* to vary between 0.5 and 20 (probably a maximum practical variation), this means a maximum variation from some central value of b by a factor of about 6. For small-molecule samples with $S \approx 4$, Eq. (44) predicts a maximum change in $\delta\phi_e$ by about 0.2 unit. To be safe, therefore, a value of ϕ_0 for the second and subsequent gradient experiments should be set equal to 0.2 unit less than the value of ϕ_e for the first band in the first gradient experiment. Likewise, a value of ϕ_z for later experiments should be set equal to 0.2 unit more than the value of ϕ_e for the last band in the first experiment. For large-molecule samples, this safety factor can be reduced in proportion to the value of S. For example (see Table 4), the adjustment in ϕ for the first and last peaks need be only ±0.02 unit for a 10–100 kDa sample ($S \approx 47$) (e.g., a typical protein).

C. Computer Simulation

The preceding relationships derived from the LSS model allow the prediction of gradient retention as a function of linear-gradient conditions (t_G, $\Delta\phi$, etc.,) which combine to determine a value of b (Eq. 15). This treatment can be extended to calculations of retention for nonlinear gradients, by dividing solute migration through the column into individual steps for each gradient segment. It is also possible to estimate values of N as a function of experimental conditions, including column length, particle size, and flow rate [9,18]. These calculations can be combined within a computer program to predict (and optimize) gradient separation as a function of several variables:

- Initial and final values of ϕ in the gradient (ϕ_0 and ϕ_z)
- Gradient time
- Gradient shape (segmented gradients or curved gradients approximated by segments)
- Column conditions: column length, particle size, flow rate

It is only required to begin with values of k_w and S for each solute in the sample. These values can in turn be obtained from two initial gradient runs where only b is varied (Eqs. 14,17,17a). This approach to the computer simulation of gradient elution was first reported in 1987 [32,73] and later described in further detail [74–76].

Computer simulation can be used to simultaneously optimize gradient conditions such as t_G, ϕ_0, and gradient shape (e.g., see also the discussion of Refs. 77 and 78). More recently, computer simulation has been extended to the prediction of separation as a function of the simultaneous variation of gradient conditions and an additional variable; e.g., gradient steepness and temperature [70,79]. This approach is based on the modeling of values of t_R in gradient elution as a function of some second variable (e.g., temperature), with b held fixed. For the case of temperature as the second variable, a total of four initial experiments are required before computer simulation can begin: two experiments at one temperature T_1 with b varied, and two experiments at a second temperature T_2, again with b varied. This approach to gradient elution optimization is illustrated in Fig. 11 for the separation of a 20-component sample (tryptic digest of recombinant human growth hormone). Figure 11a describes the four experiments required for computer simulation, Fig. 11b shows a map of resolution versus gradient time and temperature for this separation, and Fig. 11c shows a predicted separation (from Fig. 11b) that provides baseline resolution (R_s = 1.5) and minimum run time for a linear gradient. Run time is further reduced in Fig. 11d, by the use of a segmented gradient plus a smaller-particle column. The results of Figs. 11b–d are provided by computer simulation using commercial software (DryLab, LC Resources), based on the four experiments of Fig. 11a.

The application of the LSS model for a change in separation temperature proceeds as follows. It is assumed that (a) isocratic retention for each solute can be represented by

$$\log k = a + \frac{c}{T} \tag{45}$$

(b) values of S do not change with temperature T, and (c) the coefficients a and c in Eq. (45) do not vary appreciably with small changes in ϕ. This then gives [79]

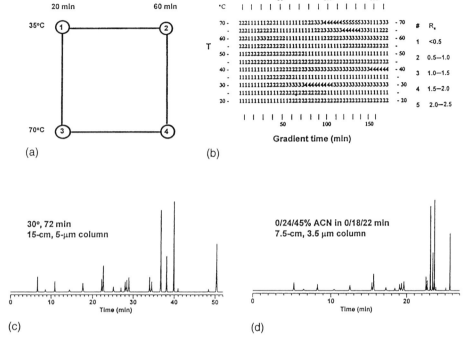

Fig. 11 Computer simulation for the simultaneous optimization of gradient steepness and temperature [70]. (a) initial four experiments required; other values of T and t_G can be substituted; (b) resolution map for separation of tryptic digest of recombinant human growth hormone (initial four experiments for 20 and 60°C, 30- and 120-min gradients); (c) preferred linear-gradient separation of sample predicted by (b); (d) predicted segmented-gradient separation for a decrease in run time.

$$t_R = a' + \frac{c'}{T} \qquad (46)$$

where t_R refers to retention time in gradient elution with b held fixed. Experimental plots of t_R versus $1/T$ exhibit slight curvature [79], presumably because of the failure of some of the assumptions involved in the derivation of Eq. (46). On the other hand, plots of t_R versus T are more nearly linear. Therefore, the empirical relationship

$$t_R = a' - c'T \qquad (46a)$$

appears more suitable for computer simulation. Using Eq. (46a), two gradient runs with only temperature varying allow values of t_R to be predicted as a function of T (b fixed). Two additional runs at a different gradient time with only temperature varying allow values of t_R to be predicted as a function of temperature for this second gradient time. Values of S and k_w can then be determined as a function of temperature, from predicted retention times at each temperature for the two runs varying in gradient time (Eqs. 17, 17a). Both Eqs. (46) and (46a) can be unreliable for temperatures <20°C, precluding their use for computer simulation at lower temperatures.

The application of computer simulation for developing gradient methods has been described in detail in previous publications, and no attempt will be made here to repeat this discussion. However, it should be stressed that multicomponent samples often benefit from just the right gradient steepness (value of b), and in some cases segmented gradients can be decidedly advantageous. Finding an optimum gradient (either linear or segmented) is in many cases only practical with the aid of computer software. For a recent summary of computer simulation for gradient method development, see Chapter 10 of Ref. 4. Several studies have been reported of the application of computer simulation using DryLab to "real" samples composed of small molecules [27,32,70,80–83], peptides [74,84–88], or proteins [75,87,90–93]. The accuracy of computer simulation has been corroborated by numerous comparisons of predicted versus actual separations for a wide variety of samples [27,32,73–75,79,80,82–85,88,89,91,93–96a]. Potential errors in computer simulation are discussed further in Section V, but errors in simulated retention times are usually insignificant (i.e., errors in ϕ_e <0.01).

V. NONIDEAL EFFECTS IN GRADIENT ELUTION

The phenomena discussed in this section represent exceptions to the "ideal" gradient elution behavior assumed in Sections III and IV. In most cases, these "nonideal" effects apply to any theory of gradient elution and can therefore lead to error in the predictions of the LSS model. In some cases, the further application of the LSS model has enabled the identification and characterization of these nonideal effects. This in turn has allowed estimates of the importance of these effects (so far as predictions based on the LSS model), and has pointed the way to their recognition and correction in specific cases, such as improving the predictions of computer simulation (although in many cases these errors cancel in computer simulation).

A. Instrumental Effects

Nonideal instrumental effects can be divided into the following categories:

- Error in the formation of the gradient
- Error in flow rates
- Rounding of the gradient by dispersion within the HPLC system
- Delay of the gradient by the hold-up volume from the mixer to the column inlet

Equipment Errors

Malfunctioning of a gradient system can lead to error in the gradient supplied to the column inlet. The diagnosis and cure of such problems has been discussed [97]. The first step in checking for equipment error is to measure the gradient delivered to the column. This is easily done [97] by removing the column from the system and carrying out a blank gradient in which a UV-absorber has been added to solvent B. The recorder display will then resemble that of Fig. 1f. If the system is performing properly, the gradient will be linear over the range 5–95% B within ±1% or better. Gradient accuracy can be further verified by running a series of step gradients as in Fig. 1e. with steps at 10,20,30%, . . .

Errors in flow rate can occur for both isocratic and gradient systems [98]. For a low-pressure mixing system, the error in predicted values of t_R can be estimated as follows. A change in F (δF) results in a proportional change in t_0, equal to $(\delta t_0)_F$ (Eq. 6), and b (Eqs. 6,15). The resulting change in ϕ_e due to a change in F can then be calculated (Eqs. 15,16a)

$$(\delta\phi_e)_F = \frac{\left(\dfrac{1}{2.3S} + \dfrac{\Delta\phi t_0}{t_G}\right)}{\dfrac{\delta F}{F}} \tag{47}$$

Here, $\delta F/F$ represents the fractional error in flow rate, and $(\delta\phi_e)_F$ is the resulting error in ϕ_e. As an example, assume an error in F of 1% and values of $S = 4$, $\Delta\phi = 1$, a 25×0.46 cm column, a flow rate of 1 mL/min, and a gradient time of 20 min. For these conditions, $t_0 \approx 2.5$ min, and the error in ϕ_e (due to error in F) is then 0.002 (Eq. 47). Typically, significant errors in predicted values of ϕ_e require major errors in F. In the case of computer simulation, when all experiments are carried out on the same equipment and any errors in flowrate are the same (bias, not random error), these errors will largely cancel and their effect on predicted values of t_R will be negligible. Likewise, because values of S for the various components of a sample tend to be similar, errors predicted by Eq. (47) will tend to be the same for all bands. The net result of a small change or error in flow rate is therefore to shift the entire chromatogram slightly toward lower or higher retention times, with little change in band spacing or resolution.

Gradient Rounding

Gradient rounding at the beginning and end of the gradient (illustrated in Fig. 1f), will reduce the retention times of bands eluting near the start of the gradient, and increase the retention times of bands eluting near the end of the gradient [61]. Because gradient rounding will vary from one HPLC gradient system to another, a gradient method that is affected by gradient rounding will be less rugged and therefore less desirable. For linear gradients, gradient rounding will normally have little effect on sample resolution, if values of ϕ_0 and ϕ_z are selected which result in elution of all bands well after ϕ_0 and before ϕ_z. A similar rounding will also occur for segmented gradients at the point where two segments join [96]. Gradient rounding for segmented-gradient separations can have a more serious effect on separation and lead to error in computer simulation, because the segment juncture will often occur in the middle of the chromatogram and in close proximity to bands within the chromatogram. One study has shown [96] that the effect of gradient rounding on t_R is most pronounced for bands that elute just after the juncture of the two segments. For a more detailed discussion of gradient rounding, see Refs. 61 and 96.

Gradient Delay

The effect of gradient delay (Fig. 1f) on sample retention is described by Eqs. (16b)(k_0 large) and 34 (k_0 small). Two problems arise as a result of gradient delay. First, the accurate application of Eq. (16b) or (34) for the prediction of t_R requires that the gradient hold-up volume V_D be known. Second, when a gradient method is transferred from one LC system to another, the hold-up volumes for the two systems are likely to be different, which can cause a change in separation. These two problems do not have much to do with LSS theory and have been discussed previously [4,100].

B. Non-LSS Conditions; LC Methods Other Than Reversed Phase

Solute retention for ion-exchange and normal-phase LC (Section I.C,) does not obey Eq. (1), so the use of linear gradients with these LC methods does not result in LSS conditions. Plots of log k versus ϕ for RP-LC also often exhibit some curvature, as seen in Figs. 3c and d. Modest failure of Eq. (1) when linear gradients are used does not preclude the application of the LSS model to these situations, especially for the qualitative application of generalizations obtained in earlier sections for reversed-phase separation. Some loss in accuracy for predictions of retention based on the LSS model can result, but most of the generalizations arrived at in this chapter are still valid for these non-LSS separations. Likewise, while an LSS separation generally provides the best overall resolution of a sample, linear gradients with ion-exchange or normal-phase LC do not result in noticably poorer separations for most samples. It is also important to recognize that applicability of Eq (1) is not required over a wide range in values of k or %B (e.g., $0.1 < k < 100$) in order to use the LSS model. For most gradient separations, significant migration through the column only occurs for a narrow range of k_a-values (e.g., $1 < k_a < 10$) that depends on the value of b for the separation.

Ion-Exchange LC (IEC)

For IEC and $1 < k < 10$, plots of log k versus salt concentration (%B or c) become increasingly linear as the charge z on the solute molecule increases [7]. For values of $z > 2$, the failure of Eq. (1) for IEC is probably no greater than for typical RP-LC separations, and predictions for linear-gradient separations based on the LSS model should be acceptable. Separations of protein samples are often carried out using IEC, and z is often >2 for these multiply charged molecules. For solutes with $z > 2$, the LSS relationships used for RP-LC can be used in similar fashion for predicting IEC retention as a function of gradient conditions. As in RP-LC, we will define ϕ equal to the volume fraction of the B-solvent; e.g., containing 0.5 M salt. The quantity $\Delta\phi$ is then the value of ϕ for the B-solvent at the end of the gradient ($\phi = \phi_z$) minus the value of ϕ at the start of the gradient ($\phi = \phi_0$). Other parameters (t_G, F, V_m, S, b) are defined in the same way as for RP-LC.

If two linear-gradient IEC separations are carried out with only b varying (e.g., with different values of t_G), values of k_0 and b for each solute can be obtained from Eqs. (17) and (17a). Retention time t_R can then be predicted for other values of b, ϕ_0, ϕ_z, etc., from Eq. (16b). Values of a and z in Eq. (11) can be calculated by measuring values of k^* and ϕ^* for each of the two gradient experiments as follows. The value of k^* is determined exactly as for RP-LC (Eq. 20). The value of ϕ^* can be related to the time $t_{1/2}$ at which the band has migrated halfway through the column, which is given by Eq. (18) ($t_{1/2} = t_r$) with $r = 0.5$. The value of ϕ^* is then (see Eq. 2a)

$$\phi^* = \phi_0 + (t_{1/2} - 0.5 t_0 - t_D) \frac{\Delta \phi}{t_G} \qquad (48)$$

Equation (48) recognizes that the value of ϕ from Eq. (2a) is measured at the column inlet (not the midpoint) and allows for a gradient delay time t_D which is not considered by Eq. (2a). If k_0 is assumed large, Eqs. (16b) and (18) then allow $t_{1/2}$ to be expressed as

$$t_{1/2} = t_R + \frac{t_0}{b} \log 0.5 - 0.5 t_0$$

which with Eq. (48) gives

$$\phi^* = \phi_0 + \frac{\Delta \phi}{t_G} \left[t_R - t_0 - t_D - 0.3 \left(\frac{t_0}{b} \right) \right] \qquad (49)$$

This value of ϕ^* can then be converted to a value c^*, which represents the mobile-phase counterion concentration at the column midpoint (corresponding to k^*).

Values of k^* and c^* can be obtained from each of the two gradient experiments used initially to measure k_0 and b, following which Eq. (11) can be solved for values of a and z. The accuracy of this LSS approach to modeling IEC retention has been verified by several studies involving the IEC separation of protein samples [41,101]. Some examples are shown in the log-log plots of Fig. 12. The data points calculated from gradient experiments are seen to agree well with experimental isocratic

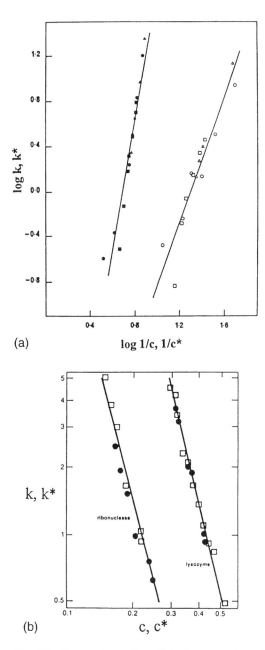

Fig. 12 Comparison of predicted vs. experimental isocratic ion-exchange retention for four proteins. (a); ovalbumin, • experimental isocratic data; other points, calculated from gradient data; (b) carbonic anydrase, ○, experimental isocratic data; other points, calculated from gradient data [41]; (b) ribonuclease and lysozyme, □, experimental isocratic data; •, calculated from gradient data [101].

data, confirming the reliability of the above approach. Straight-line plots as predicted by Eq. (11) are also observed. For $z < 3$, application of the LSS model as above can be made more accurate by first determining z and then empirically correcting this value by subtracting 0.3 unit [7,18]. A more accurate value of a can then be calculated from Eq. (11), using derived values of k^* and ϕ^* for an experimental run used initially to calculate k_0 and b.

It is possible to predict gradient IEC retention in similar fashion as above, but without relying on the LSS approximation [102,103]. It has been claimed that this more rigorous (and complicated) approach results in more accurate predictions of solute retention. However, a comparison of retention time predictions from the rigorous and LSS models gave the following results for 32 oligonucleotides and 34 polysaccharides [102]:

Solute type	Mean error in predicted retention times	
	Rigorous model	LSS approximation
Oligonucleotides	2.77%	2.83%
Polysaccharides	0.63%	0.66%

Clearly, either the rigorous or LSS-approximation approach can give accurate predictions of retention, and the difference in accuracy may not be significant. However, with the exception of one of the above compounds (out of a total of 66), $z > 2$ for all of the solutes studied, so accurate predictions from the LSS model are not unexpected.

For a further discussion of the LSS model applied to IEC retention, see Refs. 7 and 18.

Normal-Phase LC (NP-LC)

Equations (11) (IEC) and (12) (NP-LC) are similar form, especially if it is noted that volume fraction ϕ can be substituted for mole fraction X in Eq. (12) with little effect on the linearity of plots of log k versus log ϕ. Therefore, application of the LSS model to NP-LC can proceed in much the same way as for IEC (see preceding subsection). The LSS model was in fact first proposed [8] and experimentally verified [104] for the case of NP-LC separation (using Eq. 12a to describe isocratic retention). However, little application of this model to NP-LC has since been reported. Possible reasons for this lack of interest are (a) values of the coefficient n in Eq. (12) are typically small ($n < 3$), so that the accuracy of the LSS model is thereby reduced, and (b) the popularity of normal-phase HPLC is currently limited. See also the important discussion of Jandera et al. for the more precise prediction of gradient retention for NP-LC and IEC [62,105,106], as well as experimental verification of this theory by Baba et al. [107].

Hydrophobic Interaction Chromatography (HIC)

Because Eq. (13) for HIC is similar in form to Eq. (10) for RP-LC, the application of the LSS model for quantitative predictions of HIC retention will be the same as for RP-LC, and most of the above relationships for RP-LC can be used directly. Several studies [15,16,108] have confirmed some of these predictions for the HIC separation of proteins.

C. Nonequilibrium Effects

Nonequilibrium effects in gradient elution arise as a result of the change in solvent composition with time. Several different phenomena of this kind have been observed:

- Incomplete equilibration of the column with the starting mobile phase (ϕ_0) prior to beginning the gradient
- Solvent "demixing" during the gradient
- Slow change in stationary-phase conformation as a result of change in mobile-phase composition
- Slow change in the conformation of solute molecules (e.g., proteins) as a result of change in mobile-phase composition

Column equilibration prior to the start of gradient elution is ensured by washing the column with a sufficient volume of the initial mobile phase prior to starting gradient elution. When the column is fully equilibrated, retention times for early bands will not change when the volume of the initial column wash is increased further. Column equilibration can take longer for gradients that begin with 0%B (water), and this problem can be minimized by instead starting at 2–5%B.

Solvent "demixing" refers to the selective uptake of the strong solvent from the mobile phase, e.g., sorption of the organic solvent during RP-LC separation. As a result, the actual gradient is depleted of the strong solvent during the early part of gradient elution and solute retention is increased. The effect of solvent demixing on RP-LC gradient retention has been analyzed [99,109]. Changes in retention as a result of solvent demixing are best measured in terms of change in ϕ* or ϕ_e ($\delta\phi_e$), and these changes are greater for smaller values of the gradient volume ($t_G F$) and larger values of column surface area (or the phase ratio). Solvent demixing in NP-LC [62] and IEC [110] has also been discussed.

The conformation of RP-LC stationary phases (e.g., C_{18}) is known to change with mobile-phase composition, and these changes can be slow [111]. For low values of ϕ, the alkyl chains are in a "collapsed" confirmation; at high values of ϕ, the alkyl chains are more extended. The effect of slow changes in stationary-phase conformation on RP-LC gradient retention has not been investigated.

Slow change in confirmation of peptide or protein solutes as a function of mobile-phase composition has been reported for both RP-LC and IEC separation, with resulting changes in solute retention (Section III.C). Such effects usually have a more serious effect on band shape and width [112]. See also the discussion of (113,114).

D. Retention Reproducibility

Even in the absence of nonideal effects as described in the previous three sections), every gradient separation is subject to small errors that result in some variability in measured t_R values for identical experimental conditions. The gradient equipment may be free of bias (Section V.A), but some variability in flow rate or gradient delivery over time cannot be avoided. Also, the column retention characteristics can change with time as determined by its history (column manufacture and prior usage). If random changes in conditions over time result in a change in isocratic sample retention, the resulting effect on variability of gradient retention can be calculated as follows. From Eq. (16a), a change in k_0 by an amount δk_0 results in a change in t_R by

$$\delta t_R = \frac{t_0}{2.3b}\left(\frac{\delta k_0}{k_0}\right) \tag{50}$$

The corresponding change in ϕ_e (assume S does not change) is then (Eqs. 2,15)

$$\delta\phi_e = \left(\frac{\Delta\phi}{t_G}\right)\delta t_R = \frac{1}{2.3S}\left(\frac{\delta k_0}{k_0}\right) \tag{51}$$

If a change in k_0 results in a corresponding change in k (Eq. 10, no change in S), the change in isocratic retention ($\delta k/k$) can replace $\delta k_0/k_0$ in Eq. (51). Note that some fractional change in k_0 results in a defined change in ϕ_e and relative retention time (as long as $\Delta\phi$ is held constant). Expressing the reproducibility of gradient retention in terms of values of $\delta\phi_e$ (rather than changes in t_R) has the added advantage that errors in isocratic and gradient retention can be compared directly (see following discussion).

In one study [99], five neutral molecules (dialkyl phthalates) were used as sample for the study of retention variability in gradient elution. Over a period of two months, 30 isocratic and 30 gradient runs were carried out, with the results of Table 5. The variability of isocratic retention (CV = $\delta k/k$) was 1.2%, and the average value of S for the solutes of Table 5 is $S = 2.9$. Equation (51) then predicts an average variability

Table 5 Reproducibility of Isocratic and Gradient Elution [99]. Retention Times for Five n-Dialkyl Phthalates; 75% Acetonitrile/water (isocratic) or 10–100% Acetonitrile/water; 25 × 0.46-cm Column; 2 mL/min; 35°C; 30 Measurements Each Separation over a Two-Month Measurement Period

	(a) Data for individual solutes					
Solute	Isocratic retention data			Gradient retention data		
	t_R (min)	k	CV[a]	t_R (min)	k*	Std. dev. ϕ_c
di-C$_1$	1.74 ± 0.01	0.6	1.8%	8.84 ± 0.03	2.0	0.003
di-C$_2$	2.26 ± 0.01	1.0	1.0	10.32 ± 0.02	1.8	0.002
di-C$_3$	2.53 ± 0.015	1.3	1.1	10.92 ± 0.02	1.6	0.002
di-C$_4$	5.56 ± 0.045	4.0	1.0	12.84 ± 0.02	1.5	0.002
di-C$_5$	9.90 ± 0.11	8.0	1.2	13.75 ± 0.02	1.4	0.002
avg.			± 1.2%			± 0.002
(b) Average data for other gradient runs as a function of b						
t_G (min)		k*			Std. dev. ϕ_c	
5		0.8			0.002	
10		1.6			0.002[b]	
20		3.2			0.002	
40		6.4			0.004	

[a]Coefficient of variation.
[b]Data of Table 5a.

of gradient retention ($\delta\phi_e$) of $\pm(1/2.3 \times 2.9)(0.012) = \pm 0.0018$. This is very close to the observed variability of gradient retention (avg. std. dev. of $\phi_e = 0.002$ in Table 5a). This comparison suggests that variability in the column, resulting in variation in isocratic retention, accounted for most of the observed variation in gradient retention. It should be noted that the HPLC system used for the study of Table 5 was designed for an accuracy in gradient delivery (%B) of $\pm 0.1\%$ absolute. Other gradient systems may be less reliable in this respect, leading to greater variation in gradient retention.

Equation (51) is not a function of gradient steepness b, therefore the variability of ϕ_e over time should be the same when gradient steepness is varied. This was observed to be the case for the example of Table 5b. Retention variability from one run to the next might be expected to cancel to some extent, when using computer simulation. However, this will be less true when the time between various experimental runs is increased, as was observed in one study [89] of the gradient separation of a mixture of 30S ribosomal proteins. When all experiments for computer simulation and verification of predictive accuracy were carried out within two days with a column that had been "conditioned" prior to use, the average error in predicted values of t_R was $\pm 0.4\%$, corresponding to an error of ± 0.0004 unit in ϕ_e. "Conditioned" as opposed to "new" columns were also found to provide more reproducible retention times. When the experimental runs were carried out over a longer period and/or with "new" columns, the average error in predicted values of t_R (and ϕ_e) was as much as 10-fold greater. However, it should be noted that the conditions for these experiments (low-pH mobile phase) were likely to lead to some column degradation and change in sample retention.

E. Potential Accuracy of Retention Predictions from the LSS Model

One study [99] has attempted to correct for some of the nonideal effects summarized in this section and to compare values of gradient retention that are predicted from the application of Eq. (16b) with use of independent isocratic data to define values of k_w and S. Twenty-six different run conditions, comprising differences in gradient conditions (t_G, F, L) or the C_{18} packings were investigated, using the five-component sample of Table 5. The average error in predicted gradient retention was less than 0.01 unit in ϕ^*. A systematic trend in these errors was observed, with negative errors of as much as 0.01 unit in ϕ^* for low and high values of ϕ^*, and positive errors of as much as 0.01 unit for $\phi^* = 0.5$–0.7. An explanation for these residual errors has so far not been suggested.

F. Anomalous Band Broadening in Gradient Elution

Section III.A called attention to the fact that bandwidths in gradient elution are usually greater than predicted by Eqs. (21a)–(23), if isocratic values of N are used (suggesting smaller values of N for gradient elution). If W_{exp} refers to the measured value, and W_{calc} is the predicted or isocratic value, the excess band broadening can be defined by $J = W_{exp}/W_{cal}$. Figure 13a summarizes values of J as a function of b for a number of solutes from several different studies. While there is considerable scatter in these data, it is clear that anomalous band broadening as measured by J increases with b. Several possible effects might explain $J > 1$ for larger b:

- Uncorrected extra-column effects
- Distortion of the gradient by the equipment
- Variation of N with k

Extra-column band broadening always contributes to the final width of an elution band. In the case of gradient elution, extra-column band-broadening caused by the sample valve and lines between the sample loop and the column is usually negligible, because the sample is strongly retained during injection. However, any additional band broadening that occurs past the column outlet will add to the value of W predicted by Eq. (22a). As b increases, W from Eq. (22a) decreases, and the effect of extra-column band broadening will become progressively larger. Figure 13b illustrates the calculated dependence of J versus b for different relative contributions of extra-column band broadening (cf. experimental behavior of Fig. 13a).

Distortion of the gradient by the equipment (Section V.A) is another possible explanation for the J effect, in that such distortion can lead to excess band broadening, and gradient distortion increases for steeper gradients (larger values of b). However, we consider this effect less likely. For linear gradients, distortion will be most serious at the beginning and end of the gradient. Initial distortion of the gradient should have little effect on bandwidth, because the solute will be retained strongly at the column inlet (as long as k_0

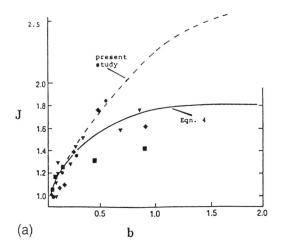

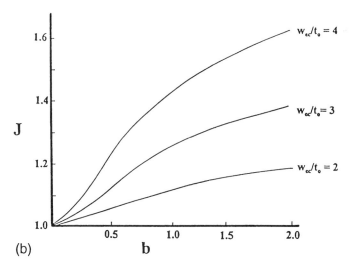

Fig. 13 Anomalous band broadening in gradient elution as measured by the correction factor J. (a) summary of data for J vs. b from several studies as summarized in Ref. 27; two curves (- - - and ———) are from different studies; (b) calculated values of J vs. b for different contributions from extra-column band broadening.

is large), regardless of such distortions. In the case of the insulin solute of Fig. 3b, the initial value of k_0 is estimated to be very large ($k_0 \sim 10^6$), yet the value of J for insulin is similar to values for other (smaller) solute molecules. Distortion at the end of the gradient will not affect the elution of typical solutes (which leave the column well before the gradient ends).

The final possibility seems the most likely, namely the general tendency of N to increase as k increases [115,116]. Thus the value of N assumed in Eq. (22a) corresponds to the isocratic value for $k = k^*$. This value of N for a given solute is the same throughout an isocratic separation, because k does not change during isocratic elution. However, in gradient elution the value of k decreases during migration of the band through the column (Fig. 2). This means that N likewise decreases during elution. This variation of N along the column during gradient elution is equivalent in isocratic elution to combining columns in series that have different values of N. For that case, it has been shown [117] that the value of N for the combined columns is always less than the average N (for $k = k^*$) for the column set. The result of this dependence of N on k^* in gradient elution will be an underestimation of W by Eq. (22a). Because the change of N with k^* is most pronounced for small values of k [115], this in turn means that J should increase with b as observed.

It is in principle possible to calculate the effective value of N in gradient elution from a knowledge of N versus k, but the latter relationship will vary with the column, the solute and other conditions. This is further shown in the scatter of the data of Fig. 13a. Consequently, a precise correction for the J effect (as a result of variation of N during separation) seems impractical. The data of Fig. 13a suggest that $JG \approx 1.0–1.2$, which is equivalent to replacing G in Eq. (22a) by the constant factor 1.1. In principle, it is possible to eliminate excess band broadening in gradient elution that arises from extra-column or gradient distortion effects. However, this is not possible for band broadening due to the variation of N with k.

VI. APPLICATION OF THE LSS MODEL TO GAS CHROMATOGRAPHY

The present chapter deals with liquid chromatography, but it is worth noting that the LSS concept has been extended to gas chromatography (GC). Whereas in LC changes in ϕ are most commonly used to control retention, changes in temperature are most convenient in GC. Temperature-programmed GC [118] has many obvious similarities to gradient elution for LC, as is evident from a superficial comparison of these two techniques.

Isothermal retention in GC as a function of temperature is given by Eq. (45). If this relationship is approximated by

$$\log k = a - cT \tag{52}$$

and a linear temperature ramp is used (common practice)

$$T = a' + c't \tag{53}$$

then

$$\log k = a'' - c''t \tag{54}$$

which can be regarded as of the same form as an LSS gradient.

The fundamental equation for GC retention (cf. Eq. 9a) can be written as [118]

$$\int_0^{t_R} \frac{1}{t_0} \frac{dt}{1+k_a} = 1 \tag{55}$$

and insertion of Eq. (54) into Eq. (55) gives an expression for t_R in temperature-programmed GC with a linear temperature ramp [119]

$$t_R = \frac{t_0}{2.3b'} \ln[e^{2.3b'}(k_0+1) - k_0] \tag{56}$$

where

$$b' = \frac{t_0 \Delta T c'}{t_P} \tag{57}$$

ΔT refers to the change in temperature during the temperature program, and t_p is the time of the program. These relationships, and corresponding expressions for bandwidth, allow the use of computer simulation for the prediction and optimization of GC separations as a function of the temperature program [119–122]. The logic of optimizing a GC temperature program has been shown to be quite similar to the optimization of an LC gradient.

VII. CONCLUSION

Sections I–V illustrate the ability of the LSS model to simplify our understanding and use of gradient elution. As a result, many previously unappreciated aspects of this technique are today more easily grasped and taught to a new generation of chromatographers. Until the implications of the LSS model were fully appreciated, the practical improvement of gradient separations was necessarily a trial-and-error procedure for all but the most experienced chromatographers. Today, any gradient separation can be readily optimized by application of the same principles that apply for isocratic elution. The use of computer simulation (based on the LSS model) has made gradient method development even more efficient.

The initially surprising behavior of large–sample molecules in gradient elution has created further confusion, which in turn led to incorrect conclusions concerning the role of the column in these separations. Application of the LSS model to the latter question has proved valuable in terms of both theory and practice. The LSS model has also enabled us to use gradient elution more effectively as a probe of various other fundamental features of chromatographic retention. Finally, the LSS model has helped untangle some of the principles of mass-overloaded (i.e., preparative) separations using gradient elution. This has made it possible for practical chromatographers to apply a rational approach to method development for these previously hard-to-understand separations (Chapter 13 of Ref. 4).

In retrospect, it appears remarkable that this simple concept (the LSS model) has proved to have such widespread and significant application to our understanding and use of gradient elution. Having said this, we would like to emphasize that the LSS model is only an approximation to the more complex general case, which is described rigorously by Eqs. (9) or (9a). Past discussions of gradient elution theory have sometimes overlooked this approximate nature of the LSS model, and we hope to have corrected this misimpression here. Finally, many of the conclusions which can be obtained from the application of the LSS model are also derivable (if less obviously) from classical gradient elution theory based on Eqs. (9) and (9a). Pavel Jandera in his excellent book [21] and numerous publications over the past 25 years was in fact the first to recognize some of these generalizations, as noted a few times in this chapter.

ACKNOWLEDGMENTS

We are very much indebted to the following people for their painstaking review of the present chapter and for many corrections and improvements: Dr. Charles Lochmuller, Dr. Pavel Jandera, Dr. Peter Schoenmakers, and Dr. Tim Schunk.

SYMBOLS

Units are given for nondimensionless terms; defining equation given in parentheses

$a, c; a', c'; a'', c''$	generally refer to constants in various equations (1–3,12,13,45–46a,52–54)
A, B	refers to weaker and stronger mobile-phase solvents, respectively; in gradient elution, A- and B-solvents refer to the two solutions (with compositions ϕ_0 and ϕ_z) that are mixed to form the gradient
b	gradient steepness parameter (Eq. 15)
b'	temperature steepness parameter in GC (Eq. 57)
b_1, b_2	values of b for two runs that will be used for computer simulation and/or to determine values of k_w and S for each solute in the sample
c	concentration of counterion (salt) in the mobile phase used for IEC (M)
c^*	average or effective value of c in IEC gradient elution (similar to ϕ^*)
f	$(t_{Rz} - t_{Ra})/t_G$; see Table 4 and discussion of Eq. (42c)
F	flowrate (mL/min)
G	gradient compression factor (Eqs. 22a, 23)
GC	gas chromatography
HIC	hydrophobic interaction chromatography
HPLC	high-performance liquid chromatography
IEC	ion-exchange chromatography
k	retention factor
kDa	kiloDalton
k_a	"actual" or instantaneous value of k in gradient elution (see Fig. 2);
k_f	final value of k (or k_a) when a solute band reaches

	the end of the column; i.e., at the time of elution (Eqs. 21, 21a)
k_0	value of k (or k_a) at start of gradient elution (Eq. 14a)
k_{01}, k_{02}	values of k_0 for two adjacent solutes (Eq. 26)
k_w	extrapolated value of k for water as mobile phase (RP-LC) (Eq. 10)
k_z/k_a	ratio of isocratic k-values for first (a) and last (z) peaks
$(k_z/k_a)_{max}$	maximum value of k_z/k_a for which isocratic elution is practical (Eqs. 42a–c)
k^*	value of k_a when a solute band has migrated halfway through the column (Fig. 4, Eq. 29a); k^* was formerly called $\bar{k}$ [17,18]
L	column length (cm)
LC	liquid chromatography
LSS	linear solvent strength
M	solute molecular weight (Daltons)
n	constant in Eq. (12); roughly equal to ratio of molecular sizes of solute and B-solvent molecules in NP-LC
N	column plate number (Eq. 7)
NP-LC	normal-phase chromatography
p	contribution to gradient compression factor (Eq. 23); also, charge on the solute molecule
q	charge on mobile-phase counterion IEC
r	fractional distance migrated by a solute through the column (Fig. 2)
RP-LC	reversed-phase liquid chromatography
R_s	resolution (Eqs. 8, 8a)
S	solute parameter equal to $d(\log k)/d\phi$ (Eq. 10)
S'	salt concentration in Eq. (13)
S^*	average value of S for two adjacent bands (Eq. 27)
S_{grad}, S_{iso}	relative band sensitivity (peak height) for gradient and isocratic elution (Eqs. 24, 24a)
t	time (min)
T	temperature (K)
t_D	gradient delay time, equal to V_D/F (see Fig., 1f)
t_G	gradient time (min); time from beginning to end of gradient

t_0	column dead time (min)(Eq. 6)
t_r	time at which band has migrated a fractional distance r through the column
t_R	solute retention time (min)
t_{Ra}, t_{Rz}	retention times for first (a) and last (z) bands in a gradient chromatogram (Eq. 42a)
t_1, t_2	retention times (min) of adjacent solutes 1 and 2 (Eqs. 8,25,25a); also, values of t_R for a given solute in two runs where b is varied (Eqs. 17,17a)
$t_{1/2}$	time t at which a solute has migrated halfway through the column ($r = 0.5$)
t_{1-x}	the time required to migrate a fractional distance $1-x$ within the column (Eq. 34)
V	volume of mobile phase eluted from column in time t
V_D	equipment hold-up or "dwell" volume (mL)
V_m	column dead volume (mL), equal to $t_0 F$
V_R	solute retention volume (mL)
w	weight of a solute that is injected onto a column (mg)
w_s	saturation capacity of a column (mg); maximum weight of a solute that can be retained by the stationary phase of a column
W	solute baseline bandwidth (min) (Eqs. 7, 22a)
W_{calc}, W_{exp}	calculated and experimental values of W in gradient elution
W_{grad}, W_{iso}	bandwidths for gradient and isocratic elution
W_1, W_2	value of W for solutes 1 and 2
x	fractional solute migration through the column (equal to r) during pre-elution as a result of the equipment hold-up volume V_D
X_B	mole fraction of B in the mobile phase for NP-LC (Eq. 12)
z	ratio of charge for solute (p) and counterion (q) molecules (Eq. 11); discussion here assumes that $q = 1$ so that $z = p$
α	isocratic separation factor for two adjacent solute bands, equal to ratio of their k-values
β	b_2/b_1 (Eq. 17)

γ	correction factor for case of unequal S-values for two adjacent bands (Eq. 27)
δF	an error in flow rate F (Eq. 47)
δk_0	a change in k_0 due to some change in the column (Eqs. 50,51)
$\delta \phi$	a change in ϕ_e due to some "nonideal" effect
$\delta \phi_e$	a change in ϕ_e (Eqs. 44, 51)
$(\delta \phi_e)_F$	an error in δ_e due to an error in F (Eq. 47)
$\delta \phi_{pe}$	error in Eq. (16b) (expressed in terms of $\delta\phi$) due to solute-pre-elution (Eq. 35); also, a change in δ_e as a result of a change in b (Eq. 44)
$\Delta \phi$	change in ϕ during the gradient, equal to $\phi_z - \phi_0$
ϕ	volume fraction of strong solvent B in the mobile phase, equal to %B/100 (Eqs. 2,2a for gradient elution)
ϕ_c	critical value of ϕ such that $k = 0$, where a small decrease in ϕ leads to significant solute retention
ϕ_e	value of ϕ at the column exit when the solute elutes from the column
ϕ_0, ϕ_z	value of ϕ at the start (0) and end (z) of the gradient
ϕ^*	value of ϕ at the column midpoint, when a solute band has reached the midpoint (Eq. 49); ϕ^* is calculable from Eq. (10) for $k = k^*$; ϕ^* was formerly called $\bar\phi$

IN REMEMBRANCE

One of the authors (LRS) would like to acknowledge Cal Giddings as a friend and inspiration for the past 30 years. Cal was unique among modern chromatographers in his theoretical insight and ability to reduce complicated problems to manageable calculations. His influence on the research of other chromatographers has been enormous. He was also highly innovative, as witnessed by his invention and theoretical description of the various FFF methodologies.

A few personal examples where Cal's prior work was of critical importance to the author are worth noting (full references follow): the efficiency of columns in series [1], the development of HPLC [2], the importance of stationary phase diffusion in HPLC [3], and an understanding of dispersion in segmented flow [4]. In each of these published studies, an initially puzzling observation or technical challenge was reduced to a simple follow-up of a clear analysis in his book, *Dynamics of Chromatography*.

We will miss him.

READINGS

1. J. Kwok, L. R. Synder, and J. C. Sternberg, *Anal. Chem.*, *40* (1968) 118.
2. L. R. Snyder, *Anal. Chem.*, 39 (1967) 698.
3. R. W. Stout, J. J. DeStefano, and L. R. Snyder, *J. Chromatogr.*, *282* (1983) 263.
4. L. R. Snyder and H. J. Adler, *Anal. Chem.*, *48* (1976) 1022.

REFERENCES

1. L. Hagdahl, R. J. P. Williams, and A. Tiselius, *Ark. Kemi*, *4* (1952) 193.
2. K. O. Donaldson, V. J. Tulane, and L. M. Marshall, *Anal. Chem.*, *24* (1952) 185.
3. H. Busch, R. B. Hurlbert, and V. R. Potter, *J. Biol. Chem.*, *196* (1952) 717.
4. L. R. Snyder, J. L. Glajch, and J. J. Kirkland, *Practical HPLC Method Development*, 2nd ed., Wiley-Interscience, New York, 1997, Chap. 8.
5. L. R. Snyder, *Chromatogr. Rev.*, *7* (1965) 1.
6. K. Valko, L. R. Snyder, and J. L. Glajch, *J. Chromatogr.*, *656* (1993) 501.
7. M. A. Quarry, R. L. Grob, and L. R. Snyder, *Anal. Chem.*, *58* (1986) 907.
8. L. R. Snyder, *J. Chromatogr.*, *13* (1964) 415.
9. S. G. Weber and P. W. Carr, in *High Performance Liquid Chromatography*, P. R. Brown and R. A. Hartwick, eds., Wiley-Interscience, New York, 1989, Chap. 1.
10. B. Drake, *Ark. Kemi*, *8* (1955) 1.
11. E. C. Freiling, *J. Am. Chem. Soc.*, *77* (1955) 2067.

12. F. C. Nachod, ed., *Ion Exchange. Theory and Application*, Academic Press, New York, 1949.
13. E. Soczewinski, *Anal. Chem.*, *42* (1969) 179.
14. L. Snyder, *Principles of Adsorption Chromatography*, Marcel Dekker, New York, 1968, Chap. 8.
15. N. T. Miller and B. L. Karger, *J. Chromatogr.*, *326* (1985) 45.
16. G. Rippel and L. Szepesy, *J. Chromatogr. A*, *664* (1994) 27.
17. L. R. Snyder, in *High-performance Liquid Chromatography, Advances and Perspectives*, Vol. 1, Cs. Horvath, ed., Academic Press, Orlando, FL, 1980, p. 208.
18. L. R. Snyder and M. A. Stadalius, in *High-performance Liquid Chromatography. Advances and Perspectives*, Vol. 4, Cs. Horvath, ed., Academic Press, Orlando, FL, 1986, p. 195.
19. L. R. Snyder, J. W. Dolan, and J. R. Gant, *J. Chromatogr.*, *165* (1979) 3.
20. L. R. Snyder and D. L. Saunders, *J. Chromatogr. Sci.*, *7* (1969) 195.
21. R. A. Hartwick, C. M. Grill, and P. R. Brown, *Anal. Chem.*, *51* (1979) 34.
22. P. Jandera and J. Churacek, *Gradient Elution in Column Liquid Chromatography*, Elsevier, Amsterdam, 1985.
23. J. W. Dolan, L. R. Snyder, and J. R. Gant, *J. Chromatogr.*, *165* (1979) 31.
24. M. A. Stadalius, H. S. Gold, and L. R. Snyder, *J. Chromatogr.*, *296* (1984) 31.
25. M. T. W. Hearn and M. I. Aguilar, *J. Chromatogr.*, *359* (1986) 31.
26. H. Poppe, J. Paanakker, and M. Bronckhorst, *J. Chromatogr.*, *204* (1981) 77.
27. J. D. Stuart, D. D. Lisi, and L. R. Snyder, *J. Chromatogr.*, *485* (1989) 657.
28. M. T. W. Hearn and M. I. Aguilar, *J. Chromatogr.*, *397* (1987) 47.
29. M. T. W. Hearn and M. I. Aguilar, *J. Chromatogr.*, *352* (1986) 35.
30. L. R. Snyder and J. W. Dolan, *J. Chromatogr.*, *721* (1996) 3.
31. L. R. Snyder, M. A. Quarry, and J. L. Glajch, *Chromatographia*, *24* (1987) 33.
32. J. W. Dolan, L. R. Snyder, and M. A. Quarry, *Chromatographia*, *24* (1987) 261.
33. F. E. Regnier, *Science*, *222* (1983) 245.
34. G. Lindgren, B. Lundstrom, I. Kallman, and K.-A. Hansson, *J. Chromatogr.*, *296* (1984) 83.

35. J. Koyama, J. Nomura, Y. Shiojima, Y. Ohtsu, and I. Horii, *J. Chromatogr.*, *625* (1992) 217.
36. N. I. Dubinina, O. I. Kurenbin, and T. B. Tennikova, *J. Chromatogr. A*, *753* (1996) 217.
37. G. Glockner and J. H. M. van den Berg, *J. Chromatogr.*, *352* (1986) 511.
38. M. A. Quarry, M. A. Stadalius, T. H. Mourey, and L. R. Snyder, *J. Chromatogr.*, *358* (1986) 1.
39. R. E. Boehm and D. E. Martire, *Anal. Chem.*, *61* (1989) 471.
40. A. Alhedai, R. E. Boehm, and D. E. Martire, *Chromatographia*, *29* (1990) 313.
41. A. N. Hodder, M. I. Aguilar, and M. T. W. Hearn, *J. Chromatogr.*, *476* (1989) 391.
42. M. T. W. Hearn, A. N. Hodder, and M. I. Aguilar, *J. Chromatogr.*, *327* (19850 47.
43. A. N. Hodder, K. J. Machin, M. I. Aguilar, and M. T. W. Hearn, *J. Chromatogr.*, *517* (1990) 31.
44. C. H. Lochmuller and M. B. McGranaghan, *Anal. Chem.*, *61* (1989) 2449.
45. C. H. Lochmuller, C. Jiang, Q. Liu, and V. Antonucci, *Crit. Rev. Anal. Chem.*, *26* (1996) 29.
46. B. G. Belenkii, A. M. Podkladenko, O. I. Kurenbin, V. G. Mal'tsev, D. G. Nasledov, and S. A. Trushin, *J. Chromatogr.*, *645* (1993) 1.
47. J. P. Larmann, J. J. DeStefano, A. P. Goldberg, R. W. Stout, L. R. Snyder, and M. A. Stadalius, *J. Chromatogr.*, *255* (1983) 163.
48. S. Terabe, H. Nishi, and T. Ando, *J. Chromatogr.*, *212* (1981) 295.
49. M. A. Stadalius, M. A. Quarry, T. H. Mourey, and L. R. Snyder, *J. Chromatogr.*, *358* (1986) 17.
50. R. E. Boehm, D. E. Martire, D. W. Armstrong, and K. H. Bui, *Macromolecules*, *16* (1983) 466.
51. X. Geng and F. E. Regnier, *J. Chromatogr.*, *296* (1984) 15.
52. T. C. Schunk, *J. Chromatogr. A*, *656* (1993) 591.
53. L. R. Snyder, M. A. Stadalius, and M. A. Quarry, *Anal. Chem.*, *55* (1983) 1412A.
54. J. M. DiBussolo and J. R. Gant, *J. Chromatogr.*, *327* (1985) 67.
55. J. L. Meek and Z. L. Rosetti, *J. Chromatogr.*, *211* (1981) 15.
56. R. Eksteen, Supelco Inc., Bellefonte, PA, Bulletin 795, 1983.
57. N. H. C. Cooke, B. G. Archer, M. J. O'Hare, E. C. Nice, and M. Capp, *J. Chromatogr.*, *255* (1983) 115.
58. R. M. Moore and R. R. Walters, *J. Chromatogr.*, *317* (1984) 119.

59. S. Yamamoto, M. Nomura, and Y. Sano, *J. Chromatogr.*, *409* (1987) 101.
60. D. W. Armstrong and R. E. Boehm, *J. Chromatogr. Sci.*, *22* (1984) 378.
61. M. A. Quarry, R. L. Grob, and L. R. Snyder, *J. Chromatogr.*, *285* (1984) 1.
62. P. Jandera and M. Kucerova, *J. Chromatogr. A*, *759* (1997) 13.
63. F. Eisenbeiss, S. Ehlevding, A. Wehrli, and J. F. K. Huber, *Chromatographia*, *20* (1985) 657.
64. L. R. Snyder, G. B. Cox, and P. E. Antle, *Chromatographia*, *24* (1987) 82.
65. J. E. Eble, R. L. Grob, P. E. Antle, and L. R. Snyder, *J. Chromatogr.*, *405* (1987) 45.
66. J. E. Eble, R. L. Grob, P. E. Antle, and L. R. Snyder, *J. Chromatogr.*, *405* (1987) 51.
67. G. B. Cox, L. R. Snyder, and J. W. Dolan, *J. Chromatogr.*, *484* (1989) 409.
68. J. A. Lewis, L. R. Snyder, and J. W. Dolan, *J. Chromatogr. A*, *721* (1996) 15.
69. L. R. Snyder, *J. Chromatogr. B*, 688 (1997).
70. L. R. Snyder, J. W. Dolan, I. Molnar, and N. M. Djordjevic, *LC.GC*, *15* (1997).
71a. J. W. Dolan and L. R. Snyder, *J. Chromatogr. A*, in press.
71. H. Engelhardt and H. Elgass, *Chromatographia*, *22* (1986) 31.
72. P. J. Schoenmakers, A. Bartha, and H. A. H. Billiet, *J. Chromatogr.*, *550* (1991) 425.
73. J. W. Dolan and L. R. Snyder, *LC.GC*, *5* (1987) 970.
74. J. W. Dolan, D. C. Lommen, and L. R. Snyder, *J. Chromatogr.*, *485* (1989) 91.
75. B. F. D. Ghrist, L. R. Snyder, and B. S. Cooperman, in *HPLC of Biological Macromolecules*, K. M. Gooding and F. E. Regnier, eds., Marcel Dekker, New York, 1990, p. 403.
76. N. Lundell, *J. Chromatogr.*, *639* (1993) 97.
77. P. Jandera and J. Churacek, *J. Chromatogr.*, *192* (1980) 19.
78. P. Jandera and M. Spacek, *J. Chromatogr.*, *366* (1986) 107.
79. P. L. Zhu, L. R. Snyder, J. W. Dolan, N. M. Djordjevic, D. W. Hill, L. C. Sander, and T. J. Waeghe, *J. Chromatogr. A*, *756* (1996) 21.
80. N. G. Mellisch, *LC.GC*, *9* (1991) 845.
81. L. Wrisley, *J. Chromatogr.*, *628* (1993) 191.
82. T. Bonfichi, *J. Chromatogr. A*, *678* (1994) 213.

83. H. Fritsch, I. Molnar, and M. Wurl, *J. Chromatogr.*, *684* (1994) 65.
84. L. R. Snyder, J. W. Dolan and D. C. Lommen, in *HPLC of Peptides and Proteins: Separation, Analysis and Conformation*, R. S. Hodges, ed., CRC Press, Boston, 1991, p. 725.
85. R. Chloupek, W. S. Hancock, and L. R. Snyder, *J. Chromatogr.*, *594* (1992) 65.
86. W. Hancock, R. C. Chloupek, J. J. Kirkland, and L. R. Snyder, *J. Chromatogr. A*, *686* (1994) 31.
87. R. C. Chloupek, W. S. Hancock, B. A. Marchylo, J. J. Kirkland, B. Boyes, and L. R. Snyder, *J. Chromatogr. A*, *686* (1994) 45.
88. L. R. Snyder, in *New Methods in Peptide Mapping for the Characterization of Proteins*, W. S. Hancock, ed., CRC Press, Boca Raton, FL, 1996, p. 31.
89. B. F. D. Ghrist, B. S. Cooperman, and L. R. Snyder, *J. Chromatogr.*, *459* (1989) 1.
90. B. F. D. Ghrist and L. R. Snyder, *J. Chromatogr.*, *459* (1989) 25.
91. B. F. D. Ghrist and L. R. Snyder, *J. Chromatogr.*, *459* (1989) 63.
92. L. R. Snyder, in *Methods in Enzymology*. Vol. 271. *High Resolution Separation and Analysis of Biological Macromolecules*, W. S. Hancock and B. L. Karger, eds., Academic Press, Orlando, FL, 1996, p. 151.
93. R. Dappen and I. Molnar, *J. Chromatogr.*, *592* (1992) 133.
94. T. H. Dzido and H. D. Smolarz, *J. Chromatogr.*, *679* (1994) 59.
95. J. D. Stuart and D. D. Lisi, *J. Chromatogr.*, *550* (1991) 77.
96. D. D. Lisi, J. D. Stuart, and L. R. Snyder, *J. Chromatogr.*, *555* (1991) 1.
96a. T. H. Dzido and A. Sory, *Chem. Anal. (Warsaw)*, *41* (1996) 113.
97. T. Culley and J. W. Dolan, *LC,GC*, *13* (1995) 456.
98. J. P. Foley, J. A. Crow, B. A. Thomas, and M. Zamora, *J. Chromatogr.*, *478* (1989) 287.
99. M. A. Quarry, R. L. Grob, and L. R. Snyder, *J. Chromatogr.*, *285* (1984) 19.
100. L. R. Snyder and J. W. Dolan, *LC,GC*, *8* (1990) 524.
101. R. W. Stout, S. I. Sivakoff, R. D. Ricker, and L. R. Snyder, *J. Chromatogr.*, *353* (1986) 439.
102. T. Sasagawa, Y. Sakamoto, T. Hirose, T. Yoshida, Y. Kobayashi, and Y. Sato, *J. Chromatogr.*, *485* (1989) 533.
103. E. S. Parente and D. B. Wetlaufer, *J. Chromatogr.*, *355* (1986) 29.
104. L. R. Snyder and H. D. Warren, *J. Chromatogr.*, *15* (1964) 344.

105. P. Jandera and J. Churacek, *J. Chromatogr.*, *91* (1974) 223.
106. P. Jandera and J. Churacek, *J. Chromatogr.*, *104* (1975) 9.
107. Y. Baba, N. Yoza, and S. Ohashi, *J. Chromatogr.*, *350* (1985) 119.
108. M. De Frutos, A. Cifuentes, and J. C, Diez-Masa, *J. High Resolut. Chromatogr.*, *19* (1996) 521.
109. A. Velayudhan and M. R. Ladisch, *Anal. Chem.*, *63* (1991) 2028.
110. A. Velayudhan and M. R. Ladisch, *Ind. Eng. Chem. Res.*, *34* (1995) 2805.
111. R. K. Gilpin, *J. Chromatogr. A*, *656* (1993) 217.
112. K. Nugent, W. Burton, and L. R. Snyder, *J. Chromatogr.*, *443* (1988) 381.
113. M. Hanson, K. K. Unger, J. Schmid, K. Albert, and E. Bayeer, *Anal. Chem.*, *65* (1993) 2249.
114. Y. Liu, D. J. Anderson, and J. R. Shainoff, *J. Chromatogr. A*, *753* (1996) 63.
115. R. W. Stout, J. J. DeStefano, and L. R. Snyder, *J. Chromatogr.*, *282* (1983) 263.
116. J. H. Knox and H. P. Scott, *J. Chromatogr.*, *282* (1983) 297.
117. J. Kwok, L. R. Snyder, and J. C. Sternberg, *Anal. Chem.*, *40* (1968) 118.
118. W. E. Harris and H. W. Habgood, *Programmed Temperature Gas Chromatography*, Wiley, New York, 1967.
119. D. E. Bautz, J. W. Dolan, W. D. Raddatz, and L. R. Snyder, *Anal. Chem.*, *62* (1990) 1561.
120. D. E. Bautz, J. W. Dolan, and L. R. Snyder, *J. Chromatogr.*, *541* (1991) 1.
121. J. W. Dolan, L. R. Snyder, and D. E. Bautz, *J. Chromatogr.*, *541* (1991) 21.
122. L. R. Snyder, D. E. Bautz, and J. W. Dolan, *J. Chromatogr.*, *541* (1991) 35.

APPENDIX: ISOCRATIC VERSUS GRADIENT RESOLUTION FOR CASE OF UNEQUAL S-VALUES

For the case where the S-values for two adjacent bands are equal and experimental conditions are selected so that $k = k^*$, Section III.B, has shown that resolution R_s is the same in isocratic and gradient elution. When the S-values are unequal, it was shown in Section III.B, that plots of R_s versus either k or k^* are no longer exactly equal. In the latter case, similar values of R_s are obtained if values of k are compared with values of γk^*, rather than with k^*. The correction factor γ, equal to k^*/k for $R_s = 0$, was determined empirically as summarized by the approximate relationship Eq. (27).

One of the reviewers of this chapter (Peter Schoenmakers) has now derived a more exact relationship for γ as a function of values of S for the two adjacent solutes 1 and 2. With his permission, this derivation is given below. There will be a value of ϕ or b for which the two bands overlap completely and $R_s = 0$; similarly, there will be values of k and k^* for $R_s = 0$. In the following derivation, these values (i.e., for $R_s = 0$) of ϕ, b, k and k^* will be understood. Likewise, values of k_0 and b for solutes 1 and 2 will be designated as k_{01}, k_{02}, b_1, and b_2, respectively.

For the case of $R_s = 0$, the retention times for the two bands will be the same. Therefore, from Eq. (16b) we can write

$$\frac{1}{b_1}\log(2.3 k_{01} b_1) = \frac{1}{b_2}\log(2.3 k_{02} b_2) \tag{58}$$

Also, from Eq. (15)

$$b_2 = \left(\frac{S_2}{S_1}\right) b_1 \tag{59}$$

Combination of Eqs. (58) and (59) gives

$$b_1 = \frac{1}{2.3}\left(\frac{k_{02} S_2}{S_1}\right)^{S_1/(S_2-S_1)} k_{01}^{S_2/(S-S_2)} \tag{60}$$

Now, $\gamma = k^*/k$; which with Eq. (15) provides

$$\gamma = \frac{1}{1.15 b_1 k} \tag{61}$$

Eq. (14a) (for $R_s = 0$) gives

$$\log k = \log k_{o1} - S_1 \phi = \log k_{o2} - S_2 \phi$$

or

$$\phi = \frac{\log\left(\frac{k_{02}}{k_{01}}\right)}{S_1 - S_2} \tag{62}$$

Substitution of Eq. (62) into Eq. (14a) (for either solute) yields

$$k = k_{01}^{-S_2/(S_1-S_2)} k_{02}^{S_1/(S_1-S_2)} \tag{63}$$

Substitution of Eqs. (60) and (63) into (61) then gives

$$\gamma = \left(\frac{2.3}{1.15}\right) k_2^{S_1/(S_1-S_2)} k_1^{-S_2/(S_1-S_2)} k_1^{S_2/(S_1-S_2)} k_2^{-S_1/(S_1-S_2)}$$
$$= 2(S_2/S_1)^{S_1/(S_1-S_2)} \tag{64}$$

For the conditions of Fig. 7 ($S_1 = 4.4$, $S_2 = 4$), Eq. (64) predicts a value of $\gamma = 0.7$, which is close to the empirical value of 0.8.

5
High-Performance Liquid Chromatography-Pulsed Electrochemical Detection for the Analysis of Antibiotics

William R. LaCourse and Catherine O. Dasenbrock   *University of Maryland Baltimore County, Baltimore, Maryland*

I.	INTRODUCTION	190
	A. Background	190
	B. Historical Perspective of Pulsed Electrochemical Detection	191
	C. Electrocatalysis at Noble Metal Electrodes	193
II.	AMINOGLYCOSIDES	196
	A. Voltammetry	196
	B. The PAD Waveform	198
	C. Waveform Optimization	199
	D. Review of Applications	205
III.	PENICILLINS AND CEPHALOSPORINS	209
	A. Voltammetry	210
	B. The IPAD Waveform	212
	C. IPAD Waveform Optimization	214
	D. Review of Applications	217
IV.	FUTURE RESEARCH	223

189

A.	Capillary Electrophoresis—PED	224
B.	Future Applications	225
V.	CONCLUSIONS	225
	REFERENCES	228

I. INTRODUCTION

The use of antibiotics to combat bacterial infections is one of the most significant medical advances in history. Countless lives are saved or improved everyday with the almost routine prescription of antibiotics. The analysis of antibiotics is crucial to the proper administration of these drugs for human and animal consumption. Methods to speciate and quantitate antibiotics are invaluable to the discovery, development, and formulation of antibiotic-based drugs and products. Antibiotics can be found in numerous delivery forms, including tablets, suspensions, creams, ointments, feeds, etc. Even more demanding analytically is the determination of antibiotics within biological matrices to monitor the activity of the pharmaceuticals, their metabolism, fate, and of significant importance in veterinary pharmaceuticals, the presence and quantity of drug residues in animal tissues and fluids. As an example, antibiotic residues in milk resulting from treatment for mastitis in dairy cows may induce allergic reactions in certain individuals.

A. Background

Traditionally, antibiotics have been analyzed by some form of nonspecific bioassay, without specificity, which allows interferences from active metabolites, degradants, and other antibiotics. Methods employing high-performance liquid chromatography (HPLC) with ultraviolet (UV) detection suffer from poor detection limits and extensive sample preparation [1]. Typical detection limits for penicillins reported are in the range of 0.5 to 20 ppm [2]. Substantially lower detection limits have been reported [3], but the lower detection limits are due to lengthy pre-analysis concentration steps rather than an advancement in detection. Aminoglycosidic antibiotic detection limits by short-wavelength UV detection are worse, and, as a result, derivatization is often used to improve the detection properties of aminoglycosides [4,5]. Derivatization agents exploit free primary amine functionalities for O-phthalaldehyde [6,7] and 2,4-dinitrophenylhydrazine [8] chemistries. Derivatization procedures increase sample preparation, limit analyte applicability to reactive compounds, and may be difficult to apply in complex biological matrices without extensive sample preparation.

Since many antibiotics are not electroactive at constant (dc) applied potentials [1], only a limited number of compounds have been studied by HPLC-electrochemical detection (EC). Chloramphenicol is an antibiotic which has been readily determined by HPLC-EC [9,10]. Other antibiotics determined by HPLC-EC have included azithromycin [11,12], erythromycin and related macrolides [13–15], clarithromycin [16,17], amoxicillin [18], gentamicin [19], cephalosporins [20], and streptomycin [21]. Amperometric detection at a fixed potential is limited generally to compounds containing an aromatic moiety [22] or specific/uniquely electroactive substances [23]. HPLC-hv-EC [24,25], which photolyzes suitable penicillins/cephalosporins to electroactive products, and indirect detection using the loss of electrogenerated bromine [26], have been pursued in an effort to take advantage of the benefits of electrochemical detections.

The high sensitivity and selectivity of electrochemical detection are desired for antibiotic determinations in complex matrices. Unfortunately, as stated earlier, it is generally accepted by many that the majority of antibiotics are electroinactive. In reality, the loss of activity at solid anodes (e.g., Au, Pt, and C) under fixed-potential conditions is attributable to fouling of the electrode surface by adsorption of reaction products and intermediates and not electroinactivity. Even for reversible redox couples which are considered to be well-behaved, dc amperometry is often accompanied by a daily ritual of disassembling the electrochemical cell and mechanically polishing the working electrode. In this manner, fouling from nonspecific adsorption processes and/or mechanistic consequences is physically removed from the electrode surface. An alternative approach is to combine electrochemical detection with "on-line" cleaning. Hence, in order to maintain uniform and reproducible electrode activity at noble metal electrodes for polar aliphatic compounds, *pulsed electrochemical detection* (PED) was developed [27].

B. Historical Perspective of Pulsed Electrochemical Detection

The use of alternating positive and negative potential pulses to clean and reactivate noble metal electrodes that have become fouled by adsorption of solution impurities has been employed since early in this century. These early pulsed waveforms were applied to Pt electrodes by Hammett in 1924 [28] and by Armstrong in 1934 [29]. In the 1950s and 1960s, the development of hydrocarbon fuel cells inspired research on pulsed waveforms to maintain the activity of their noble metal electrodes [30–33]. As a consequence, virtually every publication concerning voltammetric data obtained on noble metal electrodes describes a procedure for electrode pretreatment to maintain a high state of electrode activity and to give reproducible results. Preparation and reactivation of noble metal electrodes is commonly achieved by the repetitive application of cyclic potential scans or alternated positive and negative potential pulses.

In the 1970s, interest in the application of pulsed waveforms to facilitate electroanalytical detection at noble metal electrodes intensified with the advent of electrochemical detection in flow-injection and liquid chromatographic systems. Clark et al. [34], MacDonald and Duke [35], and Stulik and Hora [36] all showed greater reproducibility for electrochemical detections of either organic or inorganic compounds at Pt electrodes by using pulsed waveforms.

Pulsed amperometric detection (PAD) was first introduced in 1981 for the detection of aliphatic alcohols at Pt electrodes [27,37]. The Johnson group at Iowa State University pioneered research on pulsed waveforms at noble metal electrodes for the detection of polar aliphatic compounds. In conjunction with Dionex Corporation (Sunnyvale, CA), the first commercial detector dedicated to PAD was joined with high performance anion-exchange chromatography (HPAEC) for the direct determination of carbohydrates [38,39]. The application of triple-step potential-time waveforms in electrochemical detection at noble metal electrodes was known as *pulsed amperometric detection* [22,38–41]. PAD was soon followed by *pulsed coulometric detection* (PCD) [42,43], which integrated the amperometric signal. Regardless of the specific form of signal measurement, all of these techniques are now grouped under the title of PAD. *Potential sweep-pulsed coulometric detection* (PS-PCD) [43,44], which incorporated a triangular potential scan to the detection step of PCD has become known as *integrated pulsed amperometric detection* (IPAD) [45]. The "integrated" description refers to the coulometric rejection of the oxide background. All detection strategies based on the application of multistep potential-time waveforms at noble metal electrodes for electrochemical detection in HPLC fall under the term *pulsed electrochemical detection* [22,46,47]. The analytical significance of PED is best reflected in the fact that PED instrumentation has become available from at least eight commercial manufacturers.

Although increased sensitivity and reproducibility has also been reported for pulsed potential cleaning at carbon electrodes by several researchers [48–52], those electrodes have not generally been successful for the detection of polar aliphatic compounds. This effect is attributable to the absence of appropriate electrocatalytic properties of carbon surfaces to support the anodic oxygen-transfer reaction mechanisms of polar aliphatic compounds.

C. Electrocatalysis at Noble Metal Electrodes

The majority of easily detected compounds at solid anodes under constant applied potentials are aromatic compounds (e.g., phenols, aminophenols, catecholamines, etc.). These compounds have greater electroactivity than polar aliphatic compounds due to the beneficial decrease in the activation barrier for the reaction mechanism as a consequence of stabilization via π-resonance of the free-radical products of one-electron oxidations. Therefore, a desirable characteristic of electrodes in dc amperometry is inertness. The electrode serves as a sink to provide and remove electrons with no involvement in the reaction mechanism.

Since π-resonance does not exist in polar aliphatic compounds (e.g., aminoglycosides) for stabilization of reaction intermediates, the heterogeneous rate constants for the anodic mechanisms for these compounds at inert electrodes are typically much smaller than for aromatic compounds. Hence, an alternative mode of stabilization is needed to lower the activation barriers of polar aliphatic compounds to enhance reaction rates. Such stabilization is possible at electrodes that have unsaturated d-orbitals such as clean noble metal electrodes. Faradaic processes that benefit from electrode surface interactions are described as "electrocatalytic." Unfortunately, an undesirable consequence of this approach is the accumulation of adsorbed carbonaceous materials, which eventually foul the electrode surface. Historical conclusions of electroinactivity for many compounds, including many antibiotics, at noble metal electrodes using dc amperometry is attributable to fouling of the electrode surface as a result of *high* electrocatalytic activity.

Noble metals (i.e., Pt and Au) are commonly considered to be inert, but under electrochemical conditions these electrodes are quite active. Figure 1A shows the current-potential (i-E) plot for a Au rotating disk

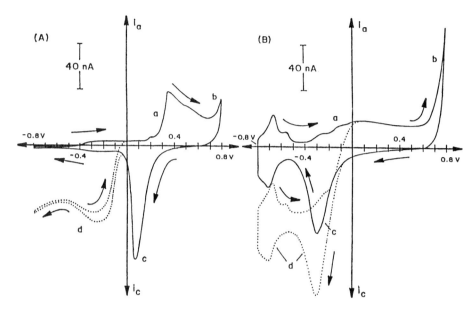

Fig. 1 Cyclic voltammetric response (i-E) for (A) Au and (B) Pt RDE. Conditions: rotation speed, 900 rpm; scan rate, 200 mV s^{-1}; Ag/AgCl reference electrode. Solutions: (———) 0.1 M NaOH, de-aerated; (·····) 0.1 M NaOH. (Reprinted from Ref. 47 with kind permission of Elsevier Science.)

electrode (RDE) in 0.1 M NaOH with (·····) and without (———) dissolved O_2. The anodic signals for the positive scan correspond to the formation of surface oxide (wave a) and the breakdown of water to O_2 (wave b). The cathodic peak on the following reverse scan corresponds to the cathodic dissolution of the surface oxide (wave c). If dissolved O_2 is present, a cathodic wave (wave d) is observed during the positive and negative scans. Except for hydrogen adsorption, reduction, and oxidation at potentials less than –500 mV, all other features of Pt electrodes are similar to that of Au electrodes in 0.1 M NaOH (Fig. 1B). Note that wave d for O_2 reduction is well resolved from wave a for a Au electrode; whereas, there is significant overlap of O_2 reduction and oxide formation for a Pt electrode. This difference accounts for the general preference of Au over Pt electrodes for the majority of PED applications.

From the voltammetric data, it is apparent that surface oxide is reversibly formed and dissolved by the application of alternating positive and negative potentials, respectively. Oxide-free, or "clean," noble metal surfaces have an affinity to adsorb organic compounds. Upon changing to a more positive potential, electrocatalytic oxidation of adsorbed compounds is promoted via anodic oxygen transfer from H_2O by transient, intermediate products in the surface-oxide formation mechanism (i.e., AuOH and PtOH). Any fouling, which results as a consequence of the catalytic detection process or adsorbed compounds is oxidatively desorbed from the electrode surface by application of a large positive potential causing the formation of a stable surface oxide (i.e., AuO and PtO) [30–33, 53]. These surface oxides are quite inert and must be cathodically dissolved by a negative potential excursion to restore the native reactivity of the oxide-free metal surface.

There are three steps: (1) electrocatalytic oxidation for detection, (2) oxide formation to "clean" the electrode surface, and (3) oxide removal to reactivate the electrode and preadsorb analyte for the next cycle which form the basis of all pulsed electrochemical detection techniques. Three modes of anodic electrocatalytic detection can occur at noble metal electrodes.

Mode I: Direct Detection at Oxide-Free Surfaces. At potentials less than ca. +200 mV (Fig. 1A) oxidation of the compound can occur with little or no concurrent formation of surface oxide. The surface-stabilized oxidation results in a product which may leave the diffusion layer, readsorb for further oxidation, or foul the electrode surface. The baseline signal originates primarily from double-layer charging, which decays quickly to a virtual zero value. All alcohol-based compounds are detected by Mode I at Au electrodes in alkaline solutions and Pt electrodes in acidic solutions [27,37–41,54–58]. The carbohydrate moieties of aminoglycosides will undergo Mode I detections.

Mode II: Direct Oxide-Catalyzed Detection. In contrast to Mode I detections, Mode II detections require the concurrent formation of surface oxide. Hence, Mode II detections occur at potentials greater than about +150 mV (Fig. 1A). Oxidation of preadsorbed analyte is the primary contributor to the analytical signal; however, simultaneous catalytic oxidation of analyte in the diffusion layer is not excluded. The oxidation products may leave the diffusion layer or foul the electrode surface. Readsorption of analyte and detection products is attenuated by the surface oxide. The background signal, resulting from anodic formation of surface oxide, is large and has a deleterious effect on detectability in LC-PAD. Aliphatic amines and amino acids [44,59,60] are detected by Mode II at Au and Pt electrodes in alkaline solutions. Numerous sulfur compounds [61–65] are also detected by Mode II at Au and Pt electrodes in both alkaline and acidic solutions.

Mode III: Indirect Detections at Oxide Surfaces. Essential to Mode I and Mode II detections is the preadsorption of the analyte at oxide-free surfaces at negative potentials prior to electrocatalytic oxidation of the analyte itself. Species which adsorb strongly to the electrode surface and are electroinactive interfere with the oxide formation process. Preadsorbed species reduce the effective surface area of the electrode surface, and the analyte signal originates from a suppression of oxide formation. Since the baseline signal results from anodic currents from surface oxide formation at a "clean electrode" surface, a negative peak results. Analyte detection as a result of the suppression of surface oxide formation is known as a Mode III detection. Sulfur-containing and inorganic compounds have been detected by Mode III [46,47]. Although indirect detection can be achieved via suppression of the response of weakly adsorbing PAD-active compounds or dissolved O_2 reduction, neither of these indirect detection processes is considered to be a Mode III detection, which must originate from the nature of the noble electrode itself.

Since electrocatalytic-based detection of various members within a class of compounds is controlled primarily by the dependence of the catalytic surface state on the electrode potential rather than by the redox potentials ($E°$) of the reactants, selectivity is achieved via chromatographic separation. Although voltammetric resolution is restricted, a limited degree of selectivity can be obtained from control of detection parameters.

II. AMINOGLYCOSIDES

The detection of aminoglycoside antibiotics are best realized at electrode surfaces that are virtually free of oxides (Mode I), and these detections are best implemented using PAD waveforms. In order to understand the electrochemical nature of aminoglycosides, the voltammetric response of glucosamine will be studied in detail.

A. Voltammetry

The current-potential (i-E) response is shown in Fig. 2 for a Au RDE in 0.1 M NaOH with (——) and without (----) glucosamine in the absence

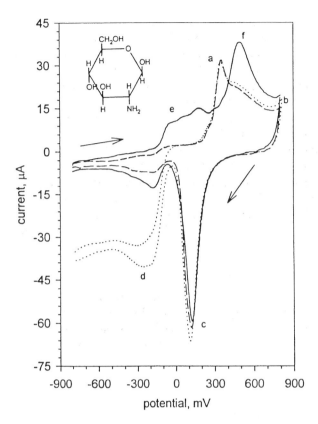

Fig. 2 Cyclic voltammetric response (i-E) for glucosamine at a Au RDE. Conditions: rotation speed, 900 rpm; scan rate, 250 mV s^{-1}; Ag/AgCl reference electrode. Solutions: (——) 100 mM glucosamine, de-aerated; (- - - -) 0.1 M NaOH de-aerated; (·····) 0.1 M NaOH.

of dissolved O_2. With the presence of glucosamine, an anodic wave is observed for the oxidations of the alcohol groups in the region of ca. –200 to +200 mV (wave e). As expected for a Mode I detection, the anodic signal is attenuated abruptly during the positive scan with onset of oxide formation (wave a). Concomitantly with the leading edge of oxide formation, the anodic detection of the amine group (Mode II) commences during the positive scan at ca. +200 mV (wave f). Upon reversal of the potential scan, monolayer coverage of the stable oxide is complete and oxide formation ceases. When oxide formation ceases, so does the amine oxidation, which supports the conclusion that it is a oxide-catalyzed detection. In addition, the absence of signal in the region of ca. +800 to ca. +100 mV indicates the absence of activity for the oxide-covered electrode surface for the alcohol group as well as the amine. Anodic waves e and f are both observed to increase in signal intensity with increases in glucosamine concentration. Although PAD for aminoglycosides can be performed at either Au or Pt electrodes in alkaline media, the use of Au electrodes has the distinct advantage that detection can be achieved without simultaneous reduction of dissolved O_2 (wave d). Dependent upon the individual structure of each aminoglycoside, the extent/magnitude of alcohol and amine oxidation will vary correspondingly to the number and nature of groups present in the molecule. The residual response characteristics correspond to those discussed in Fig. 1A.

B. The PAD Waveform

The PAD waveform is shown in Fig. 3. The detection potential (E_{det}) in the waveform is chosen to be appropriate for the desired surface-catalyzed reaction, which is usually in the oxide free region (ca. −50 to 200 mV; Fig. 2) for aminoglycosides and the current is sampled during a

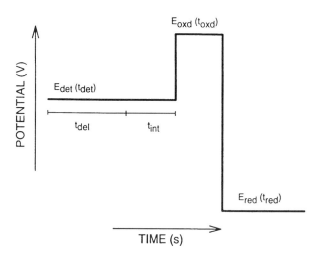

Fig. 3 Generic pulsed amperometric detection waveform.

short time period (t_{int}) after a delay of t_{del}, near the end of the detection period (t_{det}). Current sampling in PAD involves some form of signal averaging over the period of one 60-Hz oscillation. Since there is no contribution to signal strength from the 60-Hz noise, the time integral of the 60-Hz sinusoidal noise signal (16.7 ms) is zero. Extension of this strategy to the integration of an integral number (m) of 16.7-ms periods results in a significant increase in analytical signal strength while maintaining a 60-Hz noise signal of zero. Typically, the integration period (t_{int}) is 200 ms, which is $m = 12$. In addition, since 200 ms is an integral multiple of one period of 50-Hz oscillation (i.e., 20 ms), 50-Hz line noise is also minimized. Following the detection process, the electrode is "cleaned" by oxidative desorption, which is concurrent with formation of surface oxide during a positive step to the value E_{oxd} for a time period of t_{oxd}. The "clean" electrode surface is then reactivated by a subsequent negative potential step to E_{red} (t_{red}) to achieve cathodic dissolution of the surface oxide prior to the next cycle of the waveform.

Anodic detection of the alcohol functionalities in carbohydrate-based compounds occurs in a potential region where there is only a very small background signal for the concurrent formation of surface oxide (Mode I). The experimental basis of chromatographic peaks for carbohydrates by Mode I in LC-PAD is illustrated in Fig. 4 by the chronoamperometric (i-t) response curves generated following the potential step E_{red} to E_{det} in the PAD waveform. The residual current (curve a) decays quickly, and the baseline signal in LC-PAD is minimal for $t_{del} >$ ca. 100 ms. Curve b represents the transient i-t response for the presence of the reactant and the peak shown is representative of the corresponding anodic signal expected in HPLC-PAD for the value of t_{del} indicated. For a specified time period (t_{int}), the output from sampling the signal is often either the average of the current in amperes or its integrated equivalent in coulombs. Since the potential is always changing in PAD, the detection process takes into effect capacitive discharge currents, and as a consequence, the baseline for PAD is stable almost immediately (i.e., usually in less than 5 min). In contrast, stabilization of capacitive currents in dc amperometry often requires a significant amount of time to settle down (i.e., minutes to hours).

C. Waveform Optimization

Selection of potential values for the PAD waveform has been discerned traditionally from the (i-E) response obtained for the analyte(s) of interest using cyclic voltammetry (CV) generated using a triangular po-

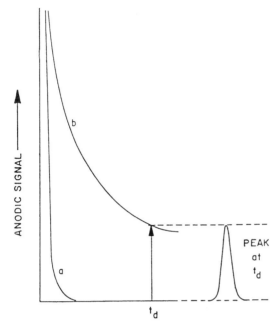

Fig. 4 Chronoamperometric response (i-t) following a potential step E_{red} to E_{det} in the PAD waveform without (a) and with (b) analyte to illustrate the origins of chromatographic peaks of Mode I detections. t_d is the same as t_{del}.

tential-time (E-t) waveform. Recently, LaCourse and Johnson have shown that the analyte response using CV, in which potential and time are coupled, is biased by the electrode surface at previous potential values in the scan [66]. This effect is particularly noticeable for Mode I detections of strongly fouling compounds. Aside from inappropriate potential selection of PAD waveforms from CV data, CV is virtually useless for determining optimum time parameters. PAD waveform optimization is best performed using a voltammetric response obtained when a PAD waveform is applied at a hydrodynamically controlled electrode (i.e., RDE) with small incremental changes in one of the parameters of the waveform (e.g., E_{det}) for each cycle of a multistep waveform. The experiment is performed in the presence and absence of the analyte of interest, and the data files are subtracted to produce a "background-corrected" response. Such an approach is known as *pulsed voltammetry* (PV). A detailed description of PV has been published [66]. PV has proven to be the definitive method for the optimization of PAD waveforms.

Figure 5 shows the "background-corrected" i-E_{det} response (positive scan direction) at a Au RDE for (a) 50, (b) 20, and (c) 10 µM glucosamine in 0.1 M NaOH. As expected, the anodic wave that begins at ca. −700 mV and peaks at ca. +200 mV corresponds to the oxidation of the alcohol groups in glucosamine. The anodic wave for oxidation of the amine group, which is concomitant with the onset of oxide formation, is the anodic wave from ca. +300 to +700 mV. The residual response (·····) is also shown. A maximum response is obtained at E_{det} = +200 mV for glucosamine. The advantage of PV is clearly evident when

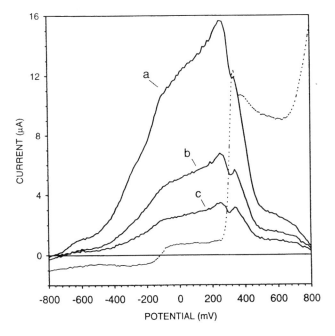

Fig. 5 Pulsed voltammetric response of glucosamine at a Au RDE in 0.1 M NaOH. This plot is "background-corrected". ¹Conditions: rotation speed, 900 rpm. Waveform: E_{det}, variable; t_{det}, 450 ms; t_{del}, 240 ms; t_{int}, 200 ms; E_{oxd}, +800 mV; t_{oxd}, 180 ms; E_{red}, −800 mV, t_{red}, 360 ms. Solutions: (——) glucosamine, (a) 50 µM, (b) 20 µM, and (c) 10 µM; (·····) 0.1 M NaOH, de-aerated.

one compares the PV in Fig. 5 with the cyclic voltammogram in Fig. 2 of glucosamine. In addition to the optimum value of E_{det}, all other potential and time parameters can be easily and accurately chosen on the basis of PV data at a RDE. The pulsed voltammograms in Fig. 6 for streptomycin (----) and spectinomycin (——) are very similar to the PV of glucosamine, and, as a rule, optimized PAD waveform parameters are applicable to the universal and highly sensitive detection of virtually all aminoglycosides.

In PAD, the mechanism of detection requires preadsorption of the analyte to the electrode surface prior to detection at E_{det}. Hence, detection of weakly adsorbing functional groups (e.g., alcohol moieties) of aminoglycosides can be enhanced by the presence of a more strongly adsorbing functional groups (e.g., amine- or sulfur-based moieties) on the same molecule. Figure 7 shows the pulsed voltammogram of lincomycin. Note the enhanced signal for alcohol oxidation due to the

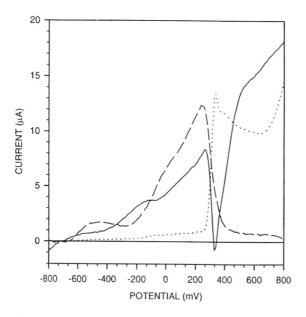

Fig. 6 Pulsed voltammetric responses of streptomycin and spectinomycin at a Au RDE in 0.1 M NaOH. These plots are "background-corrected." Conditions: same as Fig. 5. Solutions: (----) 100 μM streptomycin; (——) 100 μM spectinomycin; (·····) 0.1 M NaOH, de-aerated.

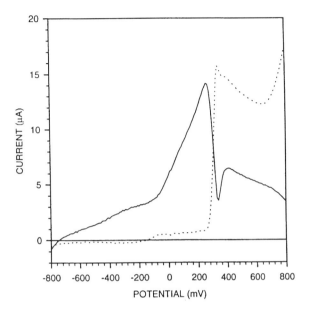

Fig. 7 Pulsed voltammetric response of lincomycin at a Au RDE in 0.1 M NaOH. This plot is "background-corrected." Conditions: same as Fig. 5. Solutions: (———) 200 µM lincomycin; (·····) 0.1 M NaOH, de-aerated.

presence of the sulfur-containing group, which has a very strong affinity for the electrode surface. This fortuitous combination of functional groups also facilitates the detection of aminoglycosides in the presence of organic modifiers, which may have an affinity for the electrode surface. Figure 8 shows the separation of a carbohydrate mixture containing sorbitol (a), fucose (b), galactosamine (c), glucose (d), *N*-acetyl-galactosamine (e), fructose (f), and sucrose (g). Chromatogram A represents the post-column addition of pure 0.2 M NaOH, whereas chromatogram B is for post-column addition of 0.2 M NaOH/20% acetonitrile (ACN). The response for the majority of the sugars is severely attenuated by the presence of the ACN, with a decrease as large as 97% for sucrose. The single exception for this mixture is galactosamine whose signal was not attenuated by the ACN. The persistence of the signal for galactosamine is the beneficial result of the ability of the amine group to adsorb in spite of the presence of ACN on the oxide-free surface. In the case of *N*-acetyl-galactosamine, the loss of signal in

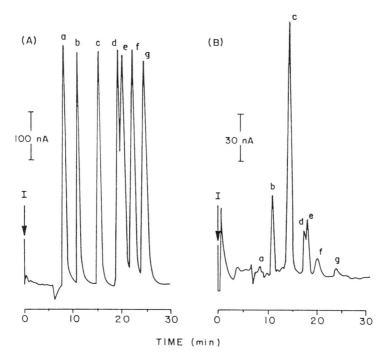

Fig. 8 Chromatograms for carbohydrates with (A) and without (B) the post-column addition of acetonitrile. Conditions: see Ref. 91. Peaks: (a) D-glucitol, 1.6 nmole; (b) fucose, 1.9 nmole; (c) galactosamine, 1.5 nmole; (d) D-glucose, 1.9 nmole; (e) N-acetyl-galactosamine, 2.5 nmole; (f) D-fructose, 3.1 nmole; (g) sucrose, 2.3 nmole. (Reprinted from Ref. 91 with kind permission of Elsevier Science, Amsterdam, The Netherlands.)

the presence of ACN is indicative of the weakened amine adsorption as a result of steric hindrance from the acetyl group.

Effect of pH and Organic Modifiers. It is observed that the rates of anodic mechanisms at Au electrodes decline with increasing solution acidity and virtually no response for alcohol groups is obtained for pH < 12. The enhancement of oxidation rates by high alkalinity is concluded to result because H^+ is produced in reaction steps leading to the rate determining step [67,68]. In contrast to the case of Au, useful analytical signals for alcohols and carbohydrates can be obtained at Pt electrodes even in concentrated acidic media [69]. This is explained by the stronger adsorption of the intermediate reaction products at Pt which has a lower electronic occupancy in surface d-orbitals than Au.

Chromatographic separations are controllable via changes in pH, ionic strength, and organic modifiers. The consideration of pH-gradient elution must recognize the effect of pH change on the background signal as well as the choice of E_{det} for maximum sensitivity in HPLC-PAD. The potential for onset of oxide formation at Au electrodes shifts to more negative values with increases in pH at a rate of ca. –60 mV pH^{-1}. The effect of pH on the oxide formation process is attributable to the pH-dependent nature of Au oxide formation, i.e., $Au(H_2O)_{ads} \rightarrow Au\text{-}OH + H^+ + e^-$. Because of optimal choice of E_{det} corresponds approximately to the value for onset of oxide formation, values of E_{det} should be adjusted by the amount –60 mV pH^{-1} from the value of +200 mV recommended for 0.1 M NaOH. The negative shift in oxide formation with increasing pH can be reflected by a large baseline change in LC-PAD under pH-gradient elution when E_{det} remains constant throughout the gradient [70]. This effect can be alleviated to a great extent by substitution of a pH-sensitive glass-membrane electrode for the Ag/AgCl reference electrode in the PAD cell. Because the response of the glass-membrane electrode is ca. –60 mV pH^{-1}, the value of E_{det} is automatically adjusted during execution of pH gradients [70]. Under ionic strength conditions suitable for electrochemical detection (i.e., $\mu > 50$ mM), the effect of changing ionic strength is reflected as minor perturbations in the background signal from variations in the rate of oxide formation. This effect is not noticed under isocratic LC conditions. Under gradient conditions (e.g., increasing acetate concentration), both positive and negative baseline drifts have been observed. In comparison to ionic strength effects, changes in the concentration of organic modifiers can have a much greater effec on the baseline signal in LC-PAD [45]. This can occur, even for electroinactive organic additives, because the modifiers are frequently adsorbed at the electrode surface with a resulting suppression of the oxide formation process.

D. Review of Applications

Aminoglycosides

The interest in the analysis of the components of the fermentation broth and the determination of aminoglycoside antibiotics in biological fluids has created a need for a sensitive method of detection for these compounds. The aminoglycosides lack a UV chromophore, which eliminates the ability of direct detection by ultraviolet spectrometry. They do, however, contain several available hydroxyl groups which make them detectable by PAD.

Aminoglycoside antibiotics were first determined by pulsed amperometric detection at a platinum electrode in 1985 by Polta and Johnson [71]. Aminoglycoside antibiotics include a group of compounds called the nebramycin factors, which are produced by fermentation of *Streptomyces tenebrarius*. Polta and Johnson focused on the nebramycin factors apromycin and tobramycin. For the analysis, they used a "dual-column technique" in which they used a cation-exchange guard column as a preconcentrator column and a neutral polystyrene analytical column. The sample was injected onto the preconcentrator column with a 10 mM phosphate buffer pH 5.2 and then the preconcentrator column was backflushed with NaOH onto the analytical column. The dual-column technique was developed to improve the detection limits by increasing the amount of material injected onto the column This technique was compared with a single column technique in which the sample was simply injected onto the analytical column. The detection limit achieved for tobramycin was 0.8 ppm in a clean standard using the single-column technique, whereas the dual-column technique offered lower detection limits by ca. one order of magnitude. Tobramycin and apromycin were detectable in fermentation broth and spiked blood serum by LC-PAD using this dual column technique (Fig. 9).

This initial work was extended by Statler in 1990 [72]. Tobramycin was determined by anion exchange chromatography followed by PAD at a Au electrode under alkaline conditions. The detection limit for tobramycin was improved to 0.2 ppb. The chromatographic resolution is improved in this report, largely because of the improvement in the HPLC column technology by this time. The detection sensitivity was improved by using a Au electrode and also by detection under alkaline conditions. Post-column base was added to increase the ionic strength and raise the pH to improve the sensitivity of the analysis.

McLaughlin and Henion [73] compared PAD with ion-spray mass spectrometry detection for spectinomycin, hygromycin B, streptomycin, and dihydrostreptomycin. They used a C-2 reversed phase col-

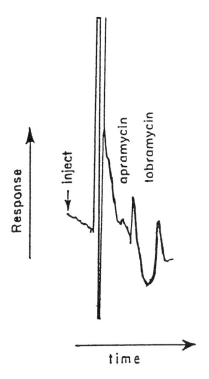

Fig. 9 Chromatogram of spiked blood serum using the dual-column technique. Conditions: 6-min wash, flow rate 0.6 mL/min. Analyte concentrations are 0.8 ppm. (Reprinted from Ref. 71 with permission from Elsevier Science, Amsterdam, The Netherlands.)

umn with pentafluoropropionic acid (PFPA) and heptafluorbutyric acid (HFBA) as ion pairing agents. The PAD analysis was accomplished at a Au electrode with the addition of post-column NaOH. In Fig. 10, the analysis of the target compounds in extract of control bovine kidney fortified at the 20 ppm level is shown. McLaughlin and Henion found that PAD detection was compatible with the ion pairing agent and more versatile than MS detection. The MS detection was more sensitive than PAD, but was incompatible with the ion-pairing agent.

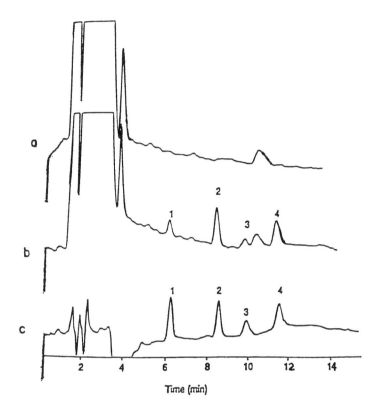

Fig. 10 HPLC-PAD of (a) matrix solid phase dispersion (MSPD) extract of control bovine kidney, (b) MSPD extract of bovine kidney fortified at the 20 ppm level and (c) synthetic mixture of standards at levels of 15 ng per component injected, representative of 100% recovery. Peaks: (1) spectinomycin; (2) hygromycin; (3) streptomycin; (4) dihydrostreptomycin. (Reprinted from Ref. 73 with permission from Elsevier Science, Amsterdam, the Netherlands.)

Phillips and Simmonds [74] determined spectinomycin and its degradation products using cation-exchange chromatography with PAD. Spectinomycin will degrade to actinamine under acidic conditions and to actinospectinoic acid under basic conditions. A 150 mM sodium acetate buffer pH 5–6 was optimal for the separation, and the addition of post column base was necessary for the detection. The addition of base post-column did not cause degradation of the spectinomycin to antinospectinoic acid before the PAD detection. Spectinomycin and its degradation products were easily detectable under the conditions of the assay. The limit of detection for spectinomycin was 0.4 ppm. The limit of quantitiation for spectinomycin was reported as 1.2 ppm with a between-day RSD of 0.3–0.9%.

Kaine and Wolnik [75] developed an assay for the determination of the four major components of gentamicin to aid in the identification of the source of pharmaceutical preparations of gentamicin. These four components are known as gentamicin C_1, C_{1a}, C_2, and C_{2a} and are used to construct fingerprints of the preparations from different manufacturers. This assay was accomplished using anion-exchange chromatography with PAD detection at a Au electrode. For the separation, gradient chromatography was employed with 3 mM NaOH initially, and then ramping to 5 mM NaOH over 14.9 min with the addition of post-column 0.5 M NaOH to raise the pH and ionic strength of the eluant and to maintain baseline stability. For this work, both bulk product and injectable preparations were analyzed. This analysis was sufficient to determine the four major components of gentamycin as well as some minor unidentified components. These minor components proved most useful for the identification of the source of the preparation for both the bulk product and the injectable preparations. The detection limit for gentamicin was 1 ppm and the peak area relative standard deviation (RSD) was 1–4%.

Analysis of neomycin A, B, and C in topical lotions was accomplished using HPAEC with PAD at a Au electrode [76]. The separation was accomplished with a NaOH gradient that started with 100% water and was ramped linearly to 6 mM NaOH over 10 min and the addition of 0.5 M NaOH post-column. Figure 11 shows a chromatogram of a topical neomycin preparation in which neomycin B and C are clearly identifiable. The detection limit for this assay was 0.2 ppm for each compound.

III. PENICILLINS AND CEPHALOSPORINS

Aminoglycoside detection relies upon the detection of the alcohol groups to take advantage of the simplicities of PAD waveforms for Mode I detections, and the amine or sulfur (e.g., lincomycin) moieties are exploited to increase the adsorption of the analyte to the electrode

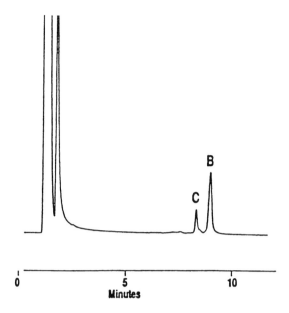

Fig. 11 Neomycin B and C in topical lotion. (Reprinted from Ref. 76 with permission from Dionex Corp.)

surface. In the case of compounds without alcohol groups (e.g., penicillins, cephalosporins, and peptide-based antibiotics), only Mode II detections are often available. The majority of work in this area has been performed on penicillins and cephalosporins of which the sulfur moieties dominate the electrochemical response. Therefore, penicillamine, which contains both an amine group and a thio group, will be examined in detail.

A. Voltammetry

Although amine compounds can be detected directly at Pt and Au electrodes only under alkaline conditions, numerous organic and inorganic sulfur-containing compounds can be detected across the entire pH range at both electrodes. Both functional groups are detected by the oxide-catalyzed mechanism of Mode II. The Au electrode is preferred over Pt to minimize interference from dissolved O_2.

Adsorption is a prerequisite to detection and therefore at least one nonbonded electron pair must reside on the S-atom. The kinetics for detection of adsorbed S-compounds are quite favorable at pH's from 2 to 7. Since alcohol and amine groups are detected only under highly acidic and/or alkaline conditions, the detection of sulfur compounds under mildly acidic conditions is highly selective. In addition, these pH conditions are compatible with reversed-phase separations.

The current-potential (i-E) response is shown in Fig. 12 for a Au RDE in 250 mM acetate buffer (pH = 3.75) with (——) and without (----) penicillamine. With the presence of 10 mM penicillamine (——), an an-

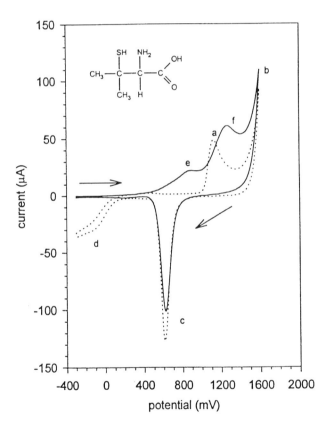

Fig. 12 Cyclic voltammetric response (i-E) for penicillamine at a Au RDE. Conditions: rotation speed, 900 rpm; scan rate, 250 mV s^{-1}; Ag/AgCl reference electrode. Solutions: (——) 75 mM glucosamine, de-aerated; and (·····) 250 mM NaOAc buffer, pH 3.75.

odic wave is observed on the positive scan at ca. +400 mV to ca. +1000 mV (wave e). This wave corresponds to oxidation of the amine moiety and, to a limited degree, the thiol. The bulk of thiol detection is the peak at ca. ca +1000 mV to +1600 mV (wave f). Anodic wave e and f are observed to increase in signal intensity with increases in penicillamine concentration. The adsorption of pencillamine is so strong that the onset of oxide formation is pushed to higher potentials. This perturbation of the oxide in the presence of analyte leads to deleterious consequences (e.g., negative peaks) when using PAD waveforms. The majority of penicillins and cephalosporins are dominated by the voltammetric behavior of the sulfur group and, otherwise, behave similarly to penicillamine. Reduction of dissolved O_2 occurs for E < ca. +100 mV (wave d) during both the positive and negative scans. The residual response characteristics correspond to those discussed in Fig. 1A except that the overall response is shifted to more positive potentials as a consequence of the pH dependence of oxide formation.

B. The IPAD Waveform

Anodic detection of amine- and sulfur-containing compounds occurs in potential regions where there is a significant signal for the concurrent formation of surface oxide (i.e., oxide-catalyzed or Mode II detections). The origins of detection peaks using PAD waveforms are depicted in Fig. 13. The residual i-t response (curve a) corresponds to double-layer charging and to the formation of surface oxide. The current from double-layer charging decays quickly, but the surface-oxide formation current decays much more slowly. Baseline signals for Mode II typically have nonzero values over a large range of t_{del} values. The i-t response corresponding to the presence of adsorbed analyte is represented as curve b. For small $t_{del,1}$, "negative" peaks can be obtained in LC-PAD because of initial inhibition of the oxide formation process by the adsorbed analyte. For larger values of $t_{del,2}$, "positive" peaks are obtained when sufficient oxide has been produced to catalyze the anodic reaction of the adsorbate. In the case of an unfortunate choice of an intermediate value of t_{del} ($t_{del,c}$), a detection peak might not be obtained.

The baseline signals for Mode II detections are frequently observed to drift to large anodic values, especially for new or freshly polished electrodes. This drift is the consequence of a slow growth in the true electrode surface area as a result of surface reconstruction caused by

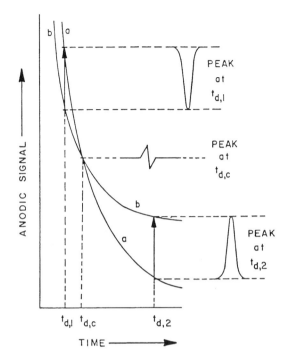

Fig. 13 Chronoamperometric response (i-t) following a potential step E_{red} to E_{det} in the PAD waveform without (a) and with (b) analyte to illustrate the origins of chromatographic peaks of Mode II detections.

the oxide on-off cycles in the applied multistep waveforms. As listed in Table 1, Mode II detections performed with PAD waveforms are subject to a number of disadvantages due to the presence of the forming surface oxide.

Although the concomitant formation of surface oxide is required for the detection of penicillins and cephalosporins, the problem of baseline anomalies and drift can be significantly diminished by the use of the waveform in Fig. 14. Here, the electrode current is integrated electronically throughout a rapid cyclic scan of the detection potential (E_{det}) within the pulsed waveform. The potential scan proceeds into (positive scan) and back out of (negative scan) the region of the oxide-catalyzed reaction for detection by Mode II. The anodic charge for ox-

Table 1 Disadvantages of PAD for Oxide-Catalyzed Detections

Baseline/background sensitivities	
pH organic modifiers ionic strength temperature	Any changes or gradients in any of these variables will lead to baseline drifting. The baseline drift is attributable to variations in the extent of surface oxide formation
Sample related effects	
poor signal-to-noise ratio	The sample current is only a fraction of the total signal. The majority of signal for Mode II detections is derived from surface oxide formation. Oxide-formation signal tends to be noisy.
postpeak 'dips'	The presence of the analyte at the electrode surface interfers with surface oxide formation. Hence, the background is different in the presence and absence of the analyte, which often results in a dip after the chromatographic peak.

Source: From Ref. 47 with permission.

ide formed on the positive sweep tends to be compensated by the corresponding cathodic charge (opposite polarity) for dissolution of the oxide on the negative sweep. As a consequence, reactions accompanied by the formation of noble metal oxide can be coulometrically monitored with automatic rejection of the surface oxide background signal.

C. IPAD Waveform Optimization

There are several requirements pertaining to the cyclic sweep of E_{det} in IPAD which must be satisfied to achieve maximum success for applications to LC. These criteria are as follows:

1. The cyclic scan of E_{det} must begin (E_{dst}) and end (E_{dnd}) at a value for which the electrode is free of surface oxide.
2. The value of E_{dnd} should not extend into the region for cathodic detection of dissolved O_2.
3. The positive scan maximum (E_{dmx}) must not extend beyond the value for anodic solvent breakdown.

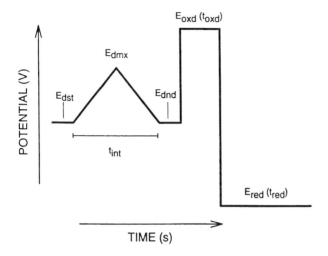

Fig. 14 Generic integrated pulsed amperometric detection waveform.

At present, the IPAD waveform should be optimized by utilizing the criteria presented above in relation to CV data and using PV to optimize the pulsed potential steps as described for optimizing PAD waveforms. Figure 15 shows the PV of penicillamine under the same conditions of Fig. 12. Note that the analytical signal can be found at the onset of oxide formation where analyte coverage versus rate of oxide formation is most optimal. As evidenced by the response concomitantly with oxide formation at high positive potentials, the anodic wave corresponds mostly to the oxidation of the thiol. The negative dip is an artifact of background correction, which results from the background signal in the presence of the analyte (i.e., suppression of the onset of oxide formation) being different from the background signal in the absence of the analyte. The pulsed voltammograms in Fig. 16 for amoxicillin (———) and cephapirin (----) are representative of virtually all penicillins and cephalosporins, respectively. Note how strongly these compounds are adsorbed to the electrode surface as reflected in the magnitude of the negative-dipping phenomenon. Although the third criterion specifies that one should not scan beyond the potential at which O_2 evolution occurs, a greater S/N is often achieved by scanning only 100 to 200 mV beyond the onset of oxide formation. Presently, work is underway to develop an IPAD optimization program similar to that of PV.

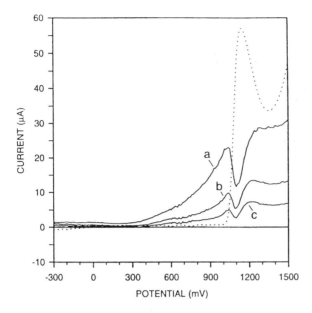

Fig. 15 Pulsed voltammetric response of penicillamine at a Au RDE in 250 mM acetate buffer, pH 3.75. This plot is "background-corrected." Conditions: rotation speed, 900 rpm. Waveform: E_{det}, variable; t_{det}, 450 ms; t_{del}, 240 ms; t_{int}, 200 ms; E_{oxd}, +1500 mV; t_{oxd}, 180 ms; E_{red}, –300 mV, t_{red}, 360 ms. Solutions: (—) penicillamine, (a) 50 μM, (b) 20 μM, and (c) 10 μM; (·····) residual, de-aerated.

Since IPAD coulometrically rejects the oxide background, the use of an IPAD waveform results in a significant decrease in baseline shifts for changes/gradients in pH, organic modifier concentration, ionic strength, and temperature [45]. In other words, IPAD results in greater compatibility with gradient chromatography. Figures 17A and 17B show a comparison of PAD and IPAD, respectively, at a Au electrode for the isocratic separation of three amino acids on an anion-exchange column [47]. The baseline drift is attributable to room temperature fluctuations to which the oxide is sensitive. The three compounds in Fig. 17B have detection limits of 1–10 pmol. Clearly, IPAD is preferred over PAD to minimize baseline offset and drift. Similar results are observed for thiocompounds. For Mode I detections, IPAD offers no advantages and the use of PAD waveforms for aminoglycosides is highly recommended.

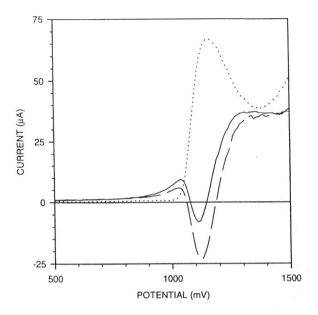

Fig. 16 Pulsed voltammetric response of of amoxicillin (——) and cephapirin (- - -) at a Au RDE in 250 mM acetate buffer, pH 3.75. These plots are "background-corrected." Conditions/Waveform: see Fig. 15. Solutions: 100 μM each compound; (·····) residual, de-aerated.

D. Review of Applications
Penicillins

Because of their ubiquitous use in modern medical practices, analysis of penicillin antibiotics in biological matrices and preparations is important. Many attempts have been made to formulate a sensitive assay for penicillins within complex matrices. PAD/IPAD is a good choice for the analysis of penicillins because this mode of detection is very sensitive for the sulfur moiety contained on the β-lactam ring.

Koprowski, Kirchmann, and Welch [77] showed that penicillin antibiotics can be detected by PAD following a flow injection experiment in either a direct detection mode or an indirect detection mode. In Fig. 18, the output from a direct PAD waveform and an indirect PAD waveform are demonstrated. For this work, eight penicillin antibiotics were tested and all showed similar electrochemical behavior at a Au elec-

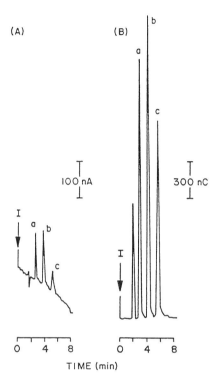

Fig. 17 Comparison of (A) PAD and (B) IPAD for the detection of amino acids. Conditions: column, CarboPAC-PA1; mobile phase, 0.1 M NaOH; flow rate, 0.9 mL min^{-1}, injection volume, 50 µL; (I) injection time. Peaks: (a) lysine, 280 pmol; (b) asparagine, 320 pmole; and (c) 4-hydroxyproline, 351 pmol. (Reprinted from Ref. 47 with kind permission from Elsevier Science.)

trode, implying that each undergoes a similar reaction at the electrode surface. This conclusion supports the idea that the sulfur moiety is responsible for the electrode activity. Koprowski et al. also investigated the use of a two-step waveform. Penicillin G was used as a model compound and a detection limit of 0.2 ppm was achieved for the two step PAD waveform, and for three-step PAD, detection limits of 0.7 ppm and 0.1 ppm for direct and indirect detection, respectively.

This work was continued by Kirchmann and Welch [78] when they reported on an HPLC-PAD method for the separation and quantitation

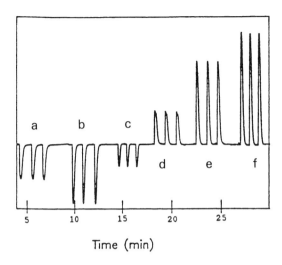

Fig. 18 Crossover from indirect to direct PAD by variation of detection potential. Waveform: $t_{det} = 0.333$s, $E_{oxd} = 1700$ mV, $t_{oxd} = 0.167$s $E_{red} = -200$ mV, $t_{red} = 0.167$s. Solvent: 0.2 M acetate buffer (pH 4.7), flow rate = 0.8 mL/min. Sample: 380 penicillin G. (a) $E_{det} = 1000$ mV, (b) $E_{det} = 1300$ mV, (c) $E_{det} = 1200$ mV, (d) $E_{det} = 1300$ mV, (e) $E_{det} = 1400$ mV, and (f) $E_{det} = 1500$ mV. (Reprinted from Ref. 77 with permission of VCH publishers.)

of penicillin antibiotics. The separation of the antibiotics was accomplished using a C-18 reversed phase column with a 0.2 M acetate buffer pH 4.75 and an acetonitrile/methanol organic modifier mixture. Again, they investigated the use of both direct and indirect detection of the penicillins of interest. With an isocratic separation, there was poor resolution of amoxicillin and ampicillin, the two earliest eluting compounds. Since PAD detection has limited compatability with organic modifier gradients, a traditional gradient could not be used. Instead, they kept the total amount of organic constant, but varied the composition of the organic portion of the mobile phase. This served to preserve the baseline stability while also allowing for better separation of the early eluting compounds. The best gradient for the separation of nine penicillins was initially 15:10:75 acetonitrile:methanol:acetate buffer with a linear ramp to 30:0:70 acetonitrile:methanol:acetate buffer over 15 min with a flow rate of 2 mL/min. The detection limit for ampicillin under the gradient conditions was reported to be 0.2 ppm. All other compounds had detection limits within a factor of 2 of the ampicillin detection limit.

Kirchmann et al. [79] describe the detection of penicillin antibiotics in milk. For this work, a reversed phase C-18 chromatography column was used with the organic modifier gradient the same as described above. An on-column preconcentration step was developed to improve the detection limits when working with real samples. The preconcentration step employed a dual-pump scheme in which the sample was pumped onto the head of the column with a 100% acetate buffer mobile phase and then the valve was switched and a second pump delivered the analytical mobile phase which consisted of an organic modifier/acetate buffer mixture. Separation of eight penicillins was achieved with detection limits ranging from 2 to 0.7 ppm for direct injections and an order of magnitude lower using the preconcentration scheme. Figure 19 shows the separation of eight penicillins within a milk extract matrix. Indirect detection was employed in all cases in this work.

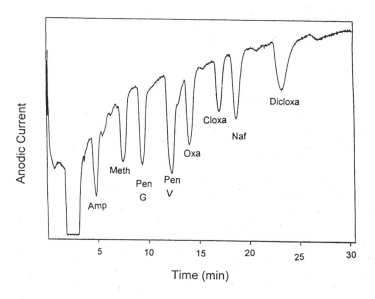

Fig. 19 The separation of eight penicillins in a milk sample following concentration for 8 min at 3 mL/min. All penicillins present at 3–4 ppm. (Reprinted by permission from Ref. 79.)

Altunata et al. [80] attempted to improve upon earlier work with penicillin through the investigation of various three- and four-step waveforms in both direct and indirect detection modes. They investigated the effect of the ionic strength of the mobile phase on the response for penicillin and determined that the response increases with increased ionic strength for direct detection modes and the response decreases for increased ionic strength for indirect detection. This increase in signal is attributed to the slowed oxidation kinetics which occurs with an increase in i-R drop. Slowed oxidation kinetics is the basis for the indirect response. In this work, 6 different waveforms are compared. The best detection limits of 0.2 ppm and 0.3 ppm are accomplished with three-step indirect PAD and four-step indirect PAD, respectively.

LaCourse and Owens [65] report on the use of PED following microbore separation of biologically important compounds. The analysis of the sulfur-containing antibiotics, lincomycin, penicillin G, cephalexin using IPAD following reversed-phase liquid chromatography is shown in Fig. 20. This work shows that PED detection is applicable to micro-

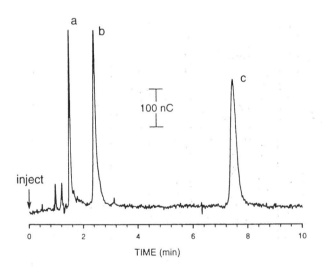

Fig. 20 Application of HPLC-IPAD to sulfur-containing antibiotics. Conditions were 82% 100 mM acetate buffer (pH 4.75) and 18% acetonitrile. Peaks: (a) cephalexin, 25 pmole; (b) lincomycin, 50 pmole; (c) penicillin-G, 150 pmole. (Reprinted from Ref. 65 with kind permission of Elsevier Science.)

Table 2 Summary of the Analytical Figures of Merit Achievable with HPLC-IPAD[a]

Compound	Target Level (ppb)	LOD[b] (ppb)	LOL[b] (ppb)	CV[b] at 50 (ppb)
Amoxicillin	10	1	50	1.4%
Ampicillin	10	2	500	1.9%
Cephapirin	20	1	200	1.7%
Penicillin G	5	5	500	3.6%
Cloxacillin	10	5	500	2.4%

[a]These data were collected by injecting each compound separately with a mobile phase chosen to yield similar k'-values for each compound.
[b]LOD = limit of detection (S/N = 3); LOL = limit of linearity; CV = coefficient of variation.

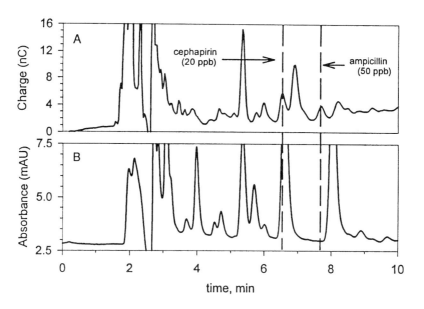

Fig. 21 Detection of ampicillin (50 ppb) and cephapirin (20 ppb) within a milk matrix for IPAD (A) and UV detection at 254 nm (B).

bore separations and that good mass detection is achievable with this technique.

Recent studies by Dasenbrock, Zook, and LaCourse [81] have focused on the RPLC-IPAD analysis of five sulfur-containing antibiotics for their determination within milk extracts. The compounds are separated on a C-8 column with a mobile phase of 20% 500 mM acetate buffer (pH 3.75)/5% acetonitrile/75% water. Table 2 shows the detection limits achievable with single injections of each compound. All of these detection limits are at or below the target levels for antibiotic resudues in milk set by the FDA. The IPAD analysis was also compared with a UV detection scheme by adding a UV detector in line. In all cases, the IPAD detection was more sensitive and selective for the compounds of interest than UV detection. Work was also done on milk extracts spiked with selected target compounds (Fig. 21). The analytes of interest were detectable within the matrix at the 20 to 50 ppb level for cephapirin and ampicillin, respectively. Ongoing work has focused on improving the extraction/concentration of the target compounds from milk.

IV. FUTURE RESEARCH

Numerous variations of PED waveforms have been tested, and some offer unique advantages for PED. For example, Polta et al. [62] advocated the used of "reversed"-PAD (RPAD) to achieve significant decreases in baseline currents for Mode II detections. The RPAD waveform simply reverses the order of application of E_{oxd} and E_{red}. The theory proposed is that oxide formation, which is reflected in the background signal, ceases with the negative step from E_{oxd} to E_{det}, and the preformed oxide generated at E_{oxd} functions to effectively catalyze the oxide-catalyzed detection.

A more complex waveform combines the effect of activating the oxide, as in RPAD, and of maintaining the cleaning and adsorption characteristics of the PAD waveform. The waveform is known as either "activated"-PAD (APAD) [82] or four-step PAD. In APAD, a fourth potential step is placed immediately after the application of E_{red} in the PAD waveform. This positive potential pulse initiates the formation of surface oxide, and the rate of oxide formation is arrested upon stepping back to a less positive detection potential. Hence, lower background currents are achieved for Mode II detection. The Welch group has been the most prominent in testing advanced waveforms for Mode II detections [77–80].

A. Capillary Electrophoresis—PED

Capillary electrophoresis (CE) is a relatively new and powerful technique for the separation of compounds of interest in complex sample matrices. As a consequence of the separation mechanism in a capillary tube (i.e., <100 μm ID), CE produces very narrow electrophoretic bands and high separation efficiencies. In general, the small pathlength of the capillary tube is deleterious to optical detection methods, especially for the detection of polar aliphatic compounds (e.g., aminoglycosides) with weak extinction coefficients.

Since electrochemical detection is based upon a reaction at an electrode surface, the electrode can be made very small with no adverse affect on sensitivity [83–85]. CE-EC has been used extensively for the determination of neurotransmitters [86] and other easily oxidized compounds at carbon electrodes. PAD has been applied following CE to the determination of carbohydrates under highly alkaline conditions [87–89]. Recently, LaCourse and Owens have extended the application of IPAD to the direct detection of many polar aliphatic

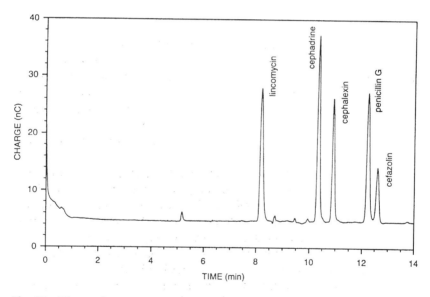

Fig. 22 Electropherogram of a mixture of sulfur-containing antibiotics. Conditions: 10 mM phosphate buffer, pH 7.0, 10 kV, 60 cm, 50 μm i.d.; waveform: IPAD.

compounds over a wide range of pH conditions [90]. Figure 22 shows an electropherogram for the detection of (a) lincomycin, (b) cephadrine, (c) cephalexin, (d) penicillin G, and (e) cefazolin using CE-IPAD. Limits of detection are typically 1 to 10 fmole injected. This electropherogram highlights the sensitivity which can be obtained with PED following CE. In addition, fast-cycle PED waveforms can be used to maintain the narrow bands of CE, and the integrity of the high-efficiency separation can be maintained. The authors feel that CE-IPAD will have a major impact on the analysis of complex biological samples for antibiotics.

B. Future Applications

Even though the majority of applications have focused on a limited number of aminoglycosides, penicillins, and cephalosporins, numerous antibiotics are amenable to PED. The requirement for PED activity is that the analyte possess one or more alcohol, amine, or sulfur-containing group. For instance, Figures 23A and 23B shows pulsed voltammograms for bacitracin, a peptide-based antibiotic, and tetracycline, respectively. From the PVs, it is obvious that both these compounds are well-suited to PED using either PAD or IPAD. Table 3 summarizes all the published antibiotic applications and test antibiotics, which have been shown to be appropriate for PED.

V. CONCLUSIONS

The significance of PED within chemical and biochemical analysis can be best appreciated in view of the commonly-held impression, based on attempted detections at constant (dc) applied potential, that polar aliphatic compounds are generally not electroactive. Furthermore, these compounds often do not possess chromophoric or fluorophoric groups and, as a consequence, direct and sensitive detection by photometric techniques is not possible. PED has been applied to the *direct*, *sensitive*, and *reproducible* detection of a large variety of aliphatic compounds of biological consequence including bioactive carbohydrates, peptides, and thiocompounds. This technique is *simple*, in that it requires no derivatization.

Although relatively untapped, PED is ideally suited to the detection of virtually all aminoglycosides, penicillins, and cephalosporins. In addition, the optimized waveform for any one compound with a class is

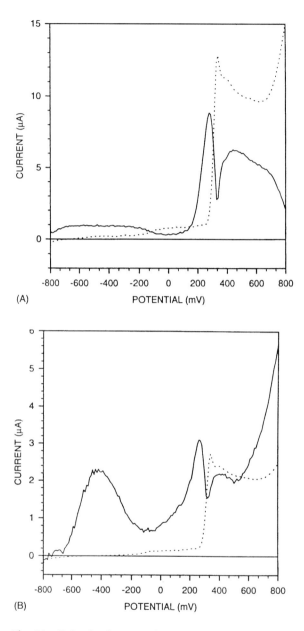

Fig. 23 Pulsed voltammetric response of (A) bacitracin and (B) tetracycline at a Au RDE in 100 mM NaOH. These plots are "background-corrected." Conditions/waveform: see Fig. 5. Solutions: (—), 100 µM each compound; (·····) residual, de-aerated.

Table 3 Summary of all Antibiotic Applications by PED

Compound(s)	Comments	Reference
2-deoxystreptamine, nebramycin, nebramycin F5, apramycin, tobramycin, nebramycin F4	Ion Chromatography, Mode I, Pt electrode AELC, Mode I, Au electrode, post column base	71 72
spectinomycin	RPLC, Mode I, Au electrode, post column base	73, 74
hygromycin B	RPLC, Mode I, Au electrode, post column base	73
streptomycin	RPLC, Mode I, Au electrode, post column base	73, 93
dihydrostreptomycin	RPLC, Mode I, Au electrode, post column base	73
gentamycin	AELC, Mode I, Au electrode, post column base	75
neomycin B, C	AELC, Mode I, Au electrode, post column base	76
amoxicillin	RPLC, Mode II/III, Au electrode	77–79, 81
ampicillin	RPLC, Mode II/III, Au electrode	77–79, 81
cloxacillin	RPLC, Mode II/III, Au electrode	77–79, 81
dicloxacillin	RPLC, Mode II/III, Au electrode	77–79
nafcillin	RPLC, Mode II/III, Au electrode	77–79
oxacillin	RPLC, Mode II/III, Au electrode	77–79
penicillin G	RPLC, Mode II/III, Au electrode	65, 77–81
penicillin V	RPLC, Mode II/III, Au electrode	77–79
methicillin	RPLC, Mode II/III, Au electrode	78–79
cephalexin	RPLC, Mode II, Au electrode	65
lincomycin	RPLC, Mode II, Au electrode	65
clindamycin	Mode I	92
kanamycin	Mode 1	93
amikacin	Mode I	93
ribavirin	Mode I	93
foscarnet	Mode I	93
cephapirin	RPLC, Mode II, Au electrode	81

applicable to the all others in that class. IPAD facilitates Mode II detections and increases gradient compatibility. PED is compatible with all aqueous-based separations, and sulfur-based compounds are selectively detected under typical reversed-phase conditions.

PED offers many advantages over alternate detection schemes for liquid chromatography and capillary electrophoresis. Because electrochemical detection relies on reaction at the electrode surface, detector cells can be miniaturized without sacrificing sensitivity. This advantage makes them especially suited for microbore techniques. As microbore techniques gain prominence, low-volume-detection schemes will become more important. Pulsed potential cleaning eliminates the need for daily polishing of the electrode which renders PED more convenient experimentally than dc amperometry. PED is sensitive and selective for specific functional groups on the analyte, excluding components which are do not contain those functional groups. In this manner, some potential interferents are not detected and the data interpretation is simplified. When analyzing complex sample matrices, such as biological fluids or fermentation broths, this advantage becomes quite significant.

ACKNOWLEDGMENTS

The authors would like to acknowledge the assistance of Christine M. Zook and Ann Burquist. The authors gratefully acknowlege the support of USFDA Grant No. FD-R-000903.

REFERENCES

1. D. M. Radzk and S. M. Lunte, *CRC Rev. Anal. Chem.*, *20*:317 (1989).
2. A. Marzo, N. Monti, M. Ripamonti, E. A. Martelli, and M. Picari, *J. Chromatogr.*, *507*:235 (1990).
3. W. A. Moats, *J. Chromatogr.*, *507*:177 (1990).
4. T. D. Schlabach and R. Weinberger, in *Reaction Detection in Liquid Chromatography*, I. S. Krull, ed., Marcel Dekker, New York, 1986.
5. D. L. Mays, R. J. Van Apeldorn, and R. G. Laubauk, *J. Chromatogr.*, *120*:93 (1976).
6. M. C. Cartula, E. Cusido, and D. Westerlund, *J. Chromatogr.*, *593*:69 (1992).

7. B. Shaikh, J. Jackson, G. Guyer, and W. R. Ravis *J. Chromatogr.* *571*:189 (1991).
8. S. D. Burton, J. E. Hutchins, T. L. Friedericksen, C. Ricks, and J. K. Tyczkowski *J. Chromatogr.*, *571*:209 (1991).
9. P. T. Kissinger, *Anal. Chem.*, *49*:447A (1977).
10. S. Abou-Khalil, W. H. Abou-Khalil, A. N. Masoud, and A. A. Yunis, *J. Chromatogr.*, *417*:111 (1987).
11. A. K. Ghone, R. P. Mehendre, and H. P. Tipnis, *Indian Drugs*, *32*:65 (1995).
12. R. M. Shepard, G. S. Duthu, R. A. Ferraina, and M. A. Mullins, *J. Chromatogr.*, *565*:321 (1991).
13. D. Croteau, F. Vallee, M. G. Bergeson, and M. LeBel, *J. Chromatogr.*, *419*:205 (1987).
14. G. S. Duthu, *J. Liq. Chromatogr.*, *7*:1023 (1984).
15. M. L. Chen and W. L. Chiou, *J. Chromatogr.*, *278*:91 (1983).
16. S-Y CHu, L. E. Sennello, and R. C. Sonders, *J. Chromatogr.*, *571*:199 (1991).
17. M. Hedenmo and B. M. Eriksson, *J. Chromatogr.*, *692*:161 (1995).
18. M. A. Brooks, M. R. Hackman, and D. J. Mazzo, *J. Chromatogr.*, *210*:531 (1981).
19. T. A. Getek, A. C. Haneke, and G. B. Selzer, *JAOAC*, *66*:172 (1983).
20. B. Ogorevc and S. Gomiscek, *J. Pharm. Biomed. Anal.*, *9*:225 (1991).
21. D. Leech, J. Wang, and M. R. Smyth, *Analyst*, *115*:1447 (1990).
22. D. C. Johnson and W. R. LaCourse, *Anal. Chem.*, *62*:589A (1990).
23. P. T. Kissinger and W. R. Heineman, in *Laboratory Techniques in Electroanalytical Chemistry*, Marcel Dekker, New York, 1984.
24. S. A. McClintock and M. L. Cotton, *J. Liq. Chromatogr.*, *12*:2961 (1989).
25. C. M. Selavka, I. S. Krull, and K. Bratin, *J. Pharm. Biomed. Anal.*, *4*:83 (1986).
26. H. Fabre and W. T. Kok, *Anal. Chem.*, *60*:136 (1988).
27. S. Hughes, P. L. Meschi, and D. C. Johnson, *Anal. Chim. Acta*, *132*:11 (1981).
28. L. P. Hammett, *J. Am. Chem. Soc.*, *46*:7 (1924).
29. G. Armstrong, F. R. Himsworth, and J. A. V. Butler, *Proc. Roy. Soc. London (A)*, *143*:89 (1934).
30. S. Gilman, *J. Phys. Chem.*, *67*:78 (1963).
31. M. W. Breiter, *Electrochim. Acta*, *8*:973 (1963).
32. J. Giner, *Electrochim. Acta*, *9*:63 (1964).

33. S. Gilman, in *Electroanalytical Chemistry*, A. J. Bard, ed., Vol. 2, Marcel Dekker, New York, 1967.
34. D. Clark, M. Fleishman, and D. Fletcher, *J. Electroanal. Chem.*, *36*:137 (1972).
35. A. MacDonald and P. D. Duke, *J. Chromatogr.*, *83*:331 (1973).
36. W. Stulik and V. Hora, *J. Electroanal. Chem.*, *70*:253 (1976).
37. S. Hughes and D. C. Johnson, *Anal. Chim. Acta*, *132*:11 (1981).
38. P. Edwards and K. K. Haak, *Am. Lab.*, *April*:78 (1983).
39. R. D. Rocklin and C. A. Pohl, *J. Liq. Chromatogr.*, *6*:1577 (1983).
40. S. Hughes and D. C. Johnson, *J. Agric. Food Chem.*, *30*:712 (1982).
41. S. Hughes and D. C. Johnson, *Anal. Chim. Acta*, *149*:1 (1983).
42. G. G. Neuberger and D. C. Johnson, *Anal. Chim. Acta*, *192*:205 (1990).
43. G. G. Neuberger and D. C. Johnson, *Anal. Chem.*, *60*:2288 (1988).
44. L. E. Welch, W. R. LaCourse, D. A. Mead Jr., and D. C. Johnson, *Anal. Chem.*, *61*:555 (1989).
45. W. R. LaCourse and D. C. Johnson, in *Advances in Ion Chromatography*, P. Jandik and R. M. Cassidy, eds., Vol. 2, Century International, Medfield, MA, 1990.
46. D. C. Johnson and W. R. LaCourse, *Electroanalysis*, *4*:367 (1992).
47. W. R. LaCourse, *Analysis*, *21*:181 (1993).
48. B. Fleet and C. J. Little, *J. Chromatogr. Sci.*, *12*:747 (1974).
49. H. W. Van Rooijan and H. Poppe, *Anal. Chim. Acta*, *130*:91 (1991).
50. A. G. Ewing, M. A. Dayton, and R. M. Wightman, *Anal. Chem.*, *53*:1842 (1981).
51. T. A. Berger (Hewlett-Packard Corp., Avondale, PA, U.S.A.), U.S. Pat. No. 4,496,454; Jan 29, 1985.
52. J. Tengyl, in *Electrochemical Detectors*, T. H. Ryan, ed., Plenum Press, New York and London, 1984.
53. R. Woods, in *Electroanalytical Chemistry*, A. J. Bard, ed., Vol. 9, Marcel Dekker, New York, 1976.
54. G. G. Neuberger and D. C. Johnson, *Anal. Chem.*, *59*:203 (1987).
55. D. C. Johnson, *Nature*, *321*:451 (1986).
56. W. T. Edwards, C. A. Pohl, and R. Rubin, *Tappi J.*, *70*:138 (1987).
57. J. Olechno, S. R. Carter, W. T. Edwards, and D. G. Gillen, *Am. Biotech. Lab.*, *5*:38 (1987).
58. D. S. Austin, J. A. Polta, T. Z. Polta, A. P. C. Tang, T. D. Cabelka, and D. C. Johnson, *J. Electroanal. Chem.*, *108*:227 (1984).
59. J. A. Polta and D. C. Johnson, *J. Liq. Chromatogr.*, *6*:1727 (1983).
60. D. C. Johnson, J. A. Polta, T. Z. Polta, G. G. Neuberger, J. Johnson,

A. P. C. Tang, I. H. Yeo and J. Baur, *J. Chem. Soc. Faraday Trans. 1*, *82*:1081 (1986).
61. T. Z. Polta and D. C. Johnson, *J. Electroanal. Chem. 209*:159 (1986).
62. T. Z. Polta, D. C. Johnson, and G. R. Luecke, *J. Electroanal. Chem. 209*:171 (1986).
63. D. C. Johnson and T. Z. Polta, *Chromatogr. Forum*, *1*:37 (1986).
64. A. Ngoviatchai and D. C. Johnson, *Anal. Chim. Acta*, *215*:1 (1988).
65. W. R. LaCourse and G. S. Owens, *Anal. Chim. Acta*, *307*:301 (1995).
66. W. R. LaCourse and D. C. Johnson, *Anal. Chem.*, *65*:50 (1993).
67. L. A. Larew and D. C. Johnson, *J. Electroanal. Chem.*, *262*:167 (1989).
68. L. A. Larew and D. C. Johnson, unpublished data, Iowa State University, Ames, IA, 1989.
69. W. R. LaCourse, D. C. Johnson, M. A. Rey, and R. W. Slingsby, *Anal. Chem.*, *63*:134 (1991).
70. W. R. LaCourse, D. A. Mead Jr., and D. C. Johnson, *Anal. Chem.*, *62*:220 (1990).
71. J. A. Polta and D. C. Johnson, *J. Chromatogr.*, *324*:407 (1985).
72. J. A. Statler, *J. Chromatogr.*, *527*:244 (1990).
73. L. G. McLaughlin and J. D. Henion, *J. Chromatogr.*, *591*:195 (1992).
74. J. G. Phillips, and C. Simmonds, *J. Chromatogr.*, *675*:123 (1994).
75. L. A. Kaine and K. A. Wolnik, *J. Chromatogr.*, *674*:255 (1994).
76. Application Note 66R, *Neomycin in Topical Lotions*, Dionex, Sunnyvale, CA (1991).
77. L. Koprowski, E. Kirchmann, and L. E. Welch, *Electroanalysis*, *5*:473 (1993).
78. E. Kirchmann and L. E. Welch, *J. Chromatogr.*, *633*:111 (1993).
79. E. Kirchmann, R. L. Earley, and L. E. Welch, *J. Liq. Chromatogr.*, *17*:1755 (1994).
80. S. Altunata, R. L. Earley, D. M. Mossman, and L. E. Welch, *Talanta*, *42*:17 (1995).
81. C. O. Dasenbrock, C. M. Zook, and W. R. LaCourse, presented at Ohio Valley Chromatography Symposium, June 1996.
82. D. G. Williams and D. C. Johnson, *Anal. Chem.*, *64*:1785 (1992).
83. R. A. Walligford and A. G. Ewing, *Anal. Chem.*, *59*:1762 (1987).
84. P. D. Curry, C. E. Engston-Silverman, and A. G. Ewing, *Electroanalysis*, *3*:587 (1991).

85. T. J. O'Shea, R. D. Greehagen, S. M. Lunte, C. E. Lunte, M. R. Smyth D. M. Radzik, and N. Watanabe, *J. Chromatogr.*, *593*:305 (1992).
86. M. A. Hayes, S. D. Gilman, and A. G. Ewing, in *Capillary Electrophoresis Technology*, N. A. Guzman, ed., Marcel Dekker, New York, 1993.
87. T. J. O'Shea, S. M. Lunte, and W. R. LaCourse, *Anal. Chem.*, *65*:2878 (1993).
88. W. Lu and R. M. Cassidy, *Anal. Chem.*, *65*:2878 (1993).
89. P. L. Weber, T. Kornfelt, N. K. Klausen, and S. M. Lunte, *Anal. Biochem.*, *225*:135 (1995).
90. William R. LaCourse and G. S. Owens, *Electrophoresis*, *17*:1 (1996).
91. W. R. LaCourse and D. C. Johnson, *Carbohydr. Res.*, *215*:159 (1991).
92. J. A. Statler and P. Williams, presented at Pittsburgh Conference, #480, March 1990.
93. J. A. Statler and R. B. Rubin, presented at Pittsburgh Conference, March 1989.

6
Theory of Capillary Zone Electrophoresis

H. Poppe *Amsterdam Institute for Molecular Studies (AIMS), University of Amsterdam, Amsterdam, The Netherlands*

I.	IN MEMORY OF J. CALVIN GIDDINGS	234
II.	INTRODUCTION	236
III.	MOBILITIES	237
	A. Definitions	237
	B. Models Describing the Mobility of Ions	239
IV.	IONS IN EQUILIBRIUM	243
	A. Reactions and Equilibrium	243
	B. Protolysis, Dependence of Mobility on pH	244
	C. Mobility Not Proportional to Charge	246
	D. Influence of Ionic Strength on Equilibria	247
	E. Correlation of Mobility with Molecular Structure	248
V.	ELECTROPHORESIS IN CAPILLARIES	252
	A. Experimental Setup	252
	B. Efficiency and Speed	254
	C. Limitations to Field Strength	256
VI.	ELECTROOSMOTIC FLOW	256

	A. Charge Separation	256
	B. Electroosmotic Mobility and Zeta Potential	256
VII.	DISPERSION	258
	A. Introduction, Injection, and Detection Effects	258
	B. Broadening During Electrophoresis	260
	C. Alternative Geometry	263
	D. Resolution	264
VIII.	INTERACTION OF IONS	266
	A. Nonlinear Transport	266
	B. Steep, Self-Sharpening Boundary	267
	C. Non-Self-Sharpening Boundary	269
	D. Triangular Zones	271
	E. Mutual Interaction of Ions	274
	F. Moving-Boundary Equations	275
	G. Kohlrausch Regulating Function (KRF)	276
	H. Numerical Approaches	282
	I. Intensity of System Peaks	287
IX.	EFFECT OF ELECTROLYSIS ON BUFFER COMPOSITION	288
X.	PEAK INTEGRALS	290
	REFERENCES	296

I. IN MEMORY OF J. CALVIN GIDDINGS

From the time that I became acquainted with modern separation science, about 1970, the name "Giddings" was associated for me with his wonderful book, *Dynamics of Chromatography, Part I*. I had long discussions on the approach Giddings followed in his book with my teacher, Josef Huber, and my first colleague in this field, Henk Lauer. It was the intuitive side rather than the formal mathematical approach that Josef Huber preferred.

After trying for more than 25 years to contribute to the science of separation, I marvel at the depth and comprehensiveness of *Dynamics of Chromatography*. There is not much, that cannot be found in this book. I advise everyone presenting new ideas on chromatographic dispersion:

The first thing to do in such an endeavor is look in *Dynamics*, to see if the idea has been worked out there. If so, in all likelihood one cannot do a better job, even after so many years. If not, there is a good chance that the idea is mentioned but for some reason or other the author left it to others to work it out. In this case, the researcher is lucky, and may benefit by checking to see whether his starting ideas have any merit.

The foreword to the book says the following about theory and practice in chromatography: "Yet, if some attempt is not made by the field's active workers to correlate the two, the study of chromatography will be in danger of becoming, on one side, an unrelated array of tens of thousands of separate facts and observations, and, on the other side, a meaningless set of mathematics." At this time, we may have hundreds of thousands of facts, but thanks to Cal Giddings' activities as an editor and author, such a dichotomy has not occurred.

The announced parts II and III on GC and LC never appeared. Apparently, Cal was so excited by the new horizons opened up by his invention, field flow fractionation (FFF), and saw so many theoretical activities in GC and LC around him, that he did not have time to write about these. While regrettable, he made up for it by writing another marvelous book, *Unified Separation Science*, which, by its larger scope, is again a "fundgrube" of inspiration.

Since the first announcement of field flow fractionation in 1966, Cal has been the godfather of activities on this exciting method, in the Utah FFF Research Center but also for the many scientific workers all over the world that have been exploring FFF and its possibilities. Many of them spent some time in Utah to learn the trade of "one-phase chromatography".

One such scientist was Arian van Asten, the first Ph.D. student working on FFF we had in our institute. It was a remarkable experience, not only for Arian, but also for me. Arian came home with so many stories about Cal, that I learned a lot more about him. Of course, I had met Cal at symposia, and was already impressed by his personality, his science, and his unique insight; however Arian's stories gave them much more flavor.

We had wondered about the amount of activity that can be compacted into one life. We learned how Cal managed the Research Center and still found time for all sorts of social, nature, and sports activities, and especially, environmental problems.

A great man has passed away.

II. INTRODUCTION

In this chapter some theoretical aspects of capillary zone electrophoresis (CZE) will be discussed. The treatment will be mainly limited to simple free-solution electrophoresis (FSCE). The reason is that more complicated varieties of capillary electrophoresis (CE), such as micellar electrokinetic chromatography (MEKC), capillary gel electrophoresis (CGE), capillary isoelectric focusing (CIEF), and affinity capillary electrophoresis (ACE) contain a number of theoretical aspects far too great to be treated in one chapter of a reasonable size. Also, many of these aspects have been treated extensively in literature predating the invention of CE, or were developed in recent years out of the context of CE. Thus, e.g., the modeling of the migration behaviour of DNA fragments and similar polyelectrolytes in sieving media such as polyacrylamide gels [1] or polymer [2] solutions will not be discussed; neither will the development of pH-gradient and of the focused zones of analytes in CIEF. For these aspects, the reader is referred to other publications, such as Refs. 3–7. This chapter focuses on the special theoretical questions that became important shortly after the introduction of CE.

A special note has to be made. All equations and variables (with the exception of concentration, given in mol/L) are given in the SI system, the modernized version of what was once known as the meter-kilogram-second-ampere (MKSA) system. This has to be mentioned because most of the literature explored to "dig out" useful physical laws and relations stems from before 1960, and the system used by the earlier authors often is the electrostatic centimeter, gram, second (cgs) system. When electrical phenomena are involved, conversion is often not easy. In the first place the electrostatic cgs system is not "rationalized"; a factor 4π occurs where one would not expect them. In the second place there is often a factor of 300 (not exact) in the equations, stemming from the conversion of pure electrostatic cgs unit for electrical potential into the "practical" unit volt.

These two conversions in themselves can generate considerable confusion. What makes things even worse is that in many equations the factors are combined with numerical factors (e.g., stemming from calculus on more complicated functions, to yield 2π, 1200π, 600π, 200π, etc.

One source of SI-formulated equations is Ref. 8. For older sources it is nearly always possible to find the SI equivalent of the equations by substituting a dielectric constant (often D) for $4\pi\varepsilon_r\varepsilon_0$, where ε_r is the relative permittivity (not "permissivity" as we saw in articles on protein electrophoresis), equivalent to the dielectric constant (78 for water at 25°C), and ε_0 is the "permittivity of the vacuum," equal to 8.85×10^{-12} (This horrible number is in reality $10^7/4\pi c^2$, c being the velocity of light in vacuum. In the electrostatic cgs system the parameter does not occur; i.e., it is set to unity). Further, if any factors 300, 600, 1200, etc., occur, it is probably necessary to replace all voltages and field strengths by 300 times the value. For example, voltage U becomes $(300\,U)$, electric field X becomes $300X$. It is likely that some factor 300 or similar cancels then in the end result.

More trivial is the conversion of concentrations in mol/liter to mol/m^3, required for consistent equations in the SI system.

III. MOBILITIES

A. Definitions

Before going into the details of CZE, we need to discuss electrophoresis as such, so that the advantages and peculiarities of the experimental variety that CZE constitutes can be appreciated.

An electrically charged particle, i, e.g., an ion or a charged sol particle, in an aqueous electrolyte solution moves under the influence of an electric field. The acquired velocity u_j is generally found to be proportional to the electric field strength, E, (V/m), according to

$$u_j = \mu_j E \tag{1}$$

The proportionality factor, μ_j (m^2/V/s), is known as the electrophoretic mobility of the species j in the medium considered. Values for common ions such as sodium and nitrate are about 50×10^{-9} m^2/V/s. The mobilities turn out to be practically independent of the field strength. They depend, however, on the temperature and slightly on the ionic composition of the medium.

Before discussing the factors that determine the magnitude of the mobilities, it is useful to point out their relation to two other variables: the equivalent conductance, λ_j and the transference numbers, T_j.

In the early decades of this century ionic motion under influence of electric fields was extensively studied via the overall electrical conductivity of electrolyte solutions, κ, the reciprocal of the resistance of a cube of 1 meter filled with the solution. It was found that this is, under conditions, proportional to the concentration, c, of the electrolyte, and thus the molar conductivity, Λ, was derived:

$$\kappa = \Lambda c \tag{2}$$

Next, the value of λ was successfully considered to be composed of the contributions of individual ions, j, with charge z_j:

$$\kappa = \sum_{\text{all ions } j} \lambda_j z_j c_j \tag{3}$$

where c_j is the concentration of an individual ion j.

The unit of the λ's in the SI system should be Ω^{-1} m^{-1}/(mol/m^3) = Ω^{-1} m^2 mol^{-1}; quite often people want to stick to the mol/L scale and write them in units Ω^{-1} m^{-1}/(mol/L). In the old system, the unit often used is Ω^{-1} cm^{-1}/(mol/L).

When considering the movement of all ions j, velocity u_j, under the influence of the field, adding up their contributions to the electrical current, each equal to $F\, u_j z_j c_j$, with F being the Faraday (9.65 × 10^4 Coulomb/mol), using Eq. (1) one obtains directly an expression for the current density, I, and with that for the conductivity, $\kappa = I/E$:

$$\kappa = \frac{I}{E} = F \frac{\sum_{\text{all ions } j} E \mu_j z_j c_j}{E} = F \sum_{\text{all ions } j} \mu_j z_j c_j \tag{4}$$

Comparison of Eq. (3) and Eq. (4) shows that

$$\mu_j = \frac{\lambda_j}{F} \tag{5}$$

Equation (5) is practical because it allows us to derive mobilities for a multitude of common ions, under various conditions, from tabulated [9] values of equivalent conductivities, λ, in case μ's cannot be found in electrophoretic literature [10–13]. The only thing we have to take care of is the powers of 10 involved in the different units; the tabulated (cgs) values of λ have to be divided not only by F but also by 10^2.

Transference numbers for ions are defined as the fraction of the current carried by that ion. The formula is

$$T_k = \frac{z_k \mu_k c_k}{\sum_{\text{all ions } j} z_j \mu_j c_j}, \quad \text{leading to} \quad \sum_{\text{all ions } k} T_k = 1 \tag{6}$$

As far as we can see, the transference number does not add much to our insight. It was useful in the early days as a means to understand the interaction between electrode reactions and electrolytic conduction. Presently we can describe the behavior of ions while using the mobilities, charges, and concentration of each ion. We mention it only because some derivations in the older electrophoresis literature used this concept.

B. Models Describing the Mobility of Ions

Many models exist for the description and prediction of mobilities in (mostly aqueous) solutions, with a great variety in the degree of sophistication. In the simplest model, the ion undergoes an electric force $Ez_j q_0$, q_0 being the elementary charge. This force is counteracted by a friction force due to the movement of the sphere with velocity u_j, given by $\rho_j u_j$. The friction factor ρ_j in a medium of viscosity, η, for a rigid spherical unit with radius R, is given by the Stokes equation: $\rho_j = 6\pi\eta R_j$. Equating these two forces, one obtains

$$u_j = \frac{Ez_j q_0}{6\pi\eta R_j} \quad \text{or} \quad \mu_j = \frac{z_j q_0}{6\pi\eta R_j} \tag{7}$$

Despite the primitive model on which it is based, this equation gives good insight. It gives about the right magnitude, and explains why in general larger ions with the same charge have smaller mobilities. It also explains the influence of the viscosity of the medium, and with that, most of the temperature dependence of the mobilities (about +2% per °C, which holds approximately for 1/viscosity). An apparent anomaly, observed when considering a series of Li^+ ($\approx 40 \times 10^{-9}$ m^2/V/s), through Na^+ and K^+, to Cs^+ ($\approx 80 \times 10^{-9}$ m^2/V/s), showing the "wrong" order with respect to ion size, is easily resolved. It is assumed that the ions of smaller atomic number such as Li^+ in solution have a larger radius because they are more strongly solvated by solvent molecules that become part of the moving sphere.

Unfortunately, this simple model also leaves much unexplained. The most important is the observed dependence of mobility on the ionic concentration. Equation (7) gives consistent results for very dilute solutions. However, even at (total) ionic concentrations as low as 10 mmol/L, deviations of about 5% are observed. Thus, for use in CZE Eq. (7) is too primitive.

Theories describing these effects have been the subject of a remarkable amount of physicochemical research during the first half of the twentieth century. A cornerstone of all this is the Debye-Hückel concept of an electrical charge cloud around an ion under consideration, formed by other ions. Of the latter, those of the same charge are repelled, while those of opposite charge are attracted by the central ion. A good description of this is to be found in Refs. 8 and 14. We skip the derivations and just state the results and the limitations.

The system is analyzed by using a combination of the electrostatic Poisson equation and the Boltzmann distribution of the "satellite" ions in the electric field of the central ion. To obtain an explicit solution we need to linearize the latter, something that is only admissible when the concentrations are low and the charges are small. Under these conditions the charge density, $\rho(r)$, and the potential, $\phi(r)$, around the ion are

$$\rho(r) = -z_j q_0 \kappa^2 \frac{1}{1+\kappa a} \frac{e^{-\kappa r}}{4\pi r}$$

$$\psi(r) = z_j q_0 \frac{1}{\varepsilon_0 \varepsilon_r} \frac{1}{1+\kappa a} \frac{e^{-\kappa r}}{4\pi r} \qquad (8)$$

where r is the distance from the center of j, a is the radius of the ion (more precisely, the distance of a satellite ion on closest approach, thus in fact the sum of two radii), and the ε's are the permittivity of the vacuum and the relative permittivity of the solvent (78 for water at 25°C).

The parameter κ (m^{-1}) plays a key role in the treatment. Its inverse is known as the Debye length, and gives a measure of the thickness of the charge cloud. It depends on the solvent, the concentration of ions (but not their nature), and the temperature, according to

$$\kappa^2 = \frac{2ISFq_0}{k_B T \varepsilon_0 \varepsilon_r} \qquad (9)$$

where IS is the ionic strength ($1/2 \sum c_k z_k^2$) in mol/m^3 of the solution. k_B is Boltzmann's constant, and T is the temperature in kelvins.

Insertion of the physical constants and conversion of IS into units of mol/L gives

$$\kappa = 3.3 \times 10^9 \sqrt{\frac{IS}{(\text{mol/L})}} \ \text{m}^{-1} \qquad (10)$$

For a buffer solution with IS = 10 mmol/l, a reasonable average for CZE experiments, this leads to a cloud thickness, $1/\kappa$, of 3.0 nm, in ultrapure water, $IS=10^{-7}$ mol/L, the value would be 0.96 µm.

The concept of the charge cloud explains the dependence of activity coefficients of electrolytes as a function of ionic strength. More important, in the present context, is its bearing on the electrophoretic mobilities, leading to corrections to Eq. (7). There are two effects involved.

A. The cataphoretic effect: This is also called, to our surprise, the electrophoretic effect. In the days these theories were developed, electrophoresis was mostly understood as applying to entities much larger than the small ions considered in the treatment of electrolyte conductivity. The electrophoresis meant here was the migration of the (large) charge cloud. The charge cloud, having a charge opposite and equal to that of the considered ion, moves in the "wrong" direction. The ion therefore experiences a stronger friction force. This leads [14,15] to a correction:*

$$\Delta\mu_{j.\text{cata}} = -\frac{q_0}{6\pi\eta}z_j\frac{\kappa}{1+\kappa a} \qquad (11)$$

which at 25°C in water is equivalent to

$$\Delta\mu_{j.\text{cata}} = -31.4\times 10^{-9}z_j\frac{\sqrt{\frac{IS}{(\text{mol/L})}}}{1+1.0\sqrt{\frac{IS}{(\text{mol/L})}}}\ \text{m}^2/\text{V/s} \qquad (12)$$

An example to demonstrate the magnitude: For Na$^+$, µ changes from 52×10^{-9} m^2/V/s in pure water to 47.6×10^{-9} m^2/V/s in solution of ionic strength of 20 mmol/L.

---

*As in Ref. 15, we followed here the suggestion by Robinson and Stokes [14] to retain the $1/(1+\kappa a)$ factor in the expressions for both the cataphoretic and the relaxation effects, by which the equation becomes much more suitable than the classical Debye-Hückel-Onager equation for correlating data under the relatively high ionic strengths occurring in CE.

B. The relaxation effect: The charge cloud becomes deformed when the ion migrates; it must be built up ahead of the ion and must decay in its wake. A delay occurs in these processes. The resulting asymmetry in charge distribution retards the ion; it is pushed back. An expression for the correction is

$$\Delta\mu_{j.\text{relax}} = -\frac{q_0^2}{24\pi\varepsilon_r\varepsilon_0 k_B T}\frac{\kappa}{1+\kappa a}\mu^0\omega \tag{13}$$

where μ^0 is the mobility at zero ionic strength. At 25°C in water this is equivalent to

$$\Delta\mu_{j.\text{relax}} = -0.395\frac{\sqrt{\frac{IS}{(\text{mol/L})}}}{1+1.0\sqrt{\frac{IS}{(\text{mol/L})}}}\mu^0\omega \tag{14}$$

where ω for an electrolyte of the type $A^+\ B^-$ equals 0.586 (2×0.29289). This correction is usually of equal importance. For the change from $IS = 0$ to $IS = 20$ mmol/L, assumed above, the correction amounts to some 5%.

Unfortunately, the application of Eq. (14) in CZE is problematic in several ways. One point is that the factor ω depends on the mobilities and charges of all ions in solution. A calculation of ω for a trace ion in a buffer composed of other ions, required in CZE, leads to a fair amount of matrix algebra [16]. Figure 1 shows the dependence of ω, the cataphoretic correction, and the relaxation correction for a singly charged anion in a sodium acetate–acetic acid buffer of ionic strength 20 mmol/L as a function of the mobility of the anion. As can be seen, both corrections are a nonlinear function of the mobility; in other words, they cannot be expressed as a constant percentage.

Even when one would be prepared to carry out the complex calculations in practical cases, the results would be often unreliable for several reasons. In the derivation of these equations the linearized Boltzmann distribution has been used. This is hardly admissible in most cases; only for a 1:1 electrolyte a fortuitous canceling of terms leads to reliable results for up to 10 mmol/L ionic strength. Also, the (partial) treatment of ions as undeformable spheres introduces uncertainties. In view of these difficulties it comes as no surprise that appli-

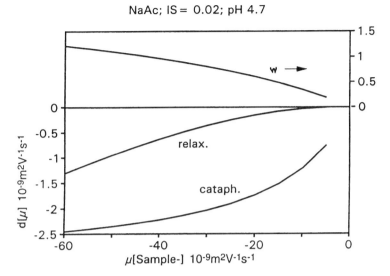

Fig. 1 Dependence of the cataphoretic and relaxation effects on the mobility, as a function of the mobility for an singly charged ion in a sodium acetate buffer of ionic strength 20 mmol/L.

cations of these theories in CZE have been limited to singly charged ions in simple buffers [15,17,18], where good results have been obtained. For multiply charged ions and/or large ions, the corrections become much more important, but unfortunately, as indicated, less predictable.

IV. IONS IN EQUILIBRIUM

A. Reactions and Equilibrium

Most sample and buffer components that undergo electrophoresis occur in various forms. The best-known case of this is of course in the acid-base reactions, which can change the charge of many organic analytes and to which most of this section will be devoted. However, other equilibria may occur. In classical texts one finds the example of cadmium(II), which can migrate as Cd^{2+}, but in chloride-containing buffers it may also migrate in the opposite direction as negatively charged species such as $CdCl_3^-$.

One can give this an even wider twist. For ions that we normally consider as simple and nonreacting, such as H^+, Na^+, NO_3^-, it is known that they are strongly solvated. That is, they are in continuous equilibrium between various forms. What we observe as the (macroscopic) mobility is just an average over numerous molecular forms. The most convincing example is the H^+ ion, often now written as H_3O^+, as if that would be the exclusive form it occurs in. There is much evidence that clusters with more water molecules also occur. Thus, what we observe, e.g., in terms of mobility, is in fact an average.

It is not so useful, though, to take these solvation reactions into account explicitly in equations. The reason is that the activity of the solvent is virtually constant in experiments, with the result that the distribution over various forms is always the same. Except when sophisticated spectroscopic techniques are applied, it is impossible even to notice the presence of these different species. Thus, identifying an averaged property, such as mobility with a (conventionally assumed) form such as H^+, does not pose any problems; a system which contained H^+ only in this form would behave exactly the same.

Equilibria are only interesting when their position depends on a varying concentration of one or more reactants. The pH dependence of mobilities is treated in the following, as the example of overwhelming importance, but with remarks referring to other equilibria.

B. Protolysis, Dependence of Mobility on pH

This issue will be treated with the concept of formation constants, now widely accepted in the treatment of successive complexation equilibria, of which acid-base equilibria form a special instance. This leads to a complication that at first sight appears annoying: Customary K_a values have to be converted to a formation constant for an acid to be formed in a reaction from a base and a proton. This, however, is well worth the effort, in view of the elegance and generality of the resulting formalism.

Imagine a base B, of charge z_0 (z_0 can be negative (acetate), zero (ammoniac), or even positive (ethylene diamine with one amine group present as ammonium group)). As a base it can accept a proton, and in the general case more protons in successive steps:

$$B^z + H^+ \Leftrightarrow BH^{z+1} \qquad K_1 = \frac{[BH^{z+1}]}{[B^z][H^+]}$$

$$BH^{z+1} + H^+ \Leftrightarrow BH_2^{z+2} \qquad K_2 = \frac{[BH_2^{z+2}]}{[BH^{z+1}][H^+]}$$

$$BH_2^{z+2} + H^+ \Leftrightarrow BH_3^{z+3} \qquad K_3 = \frac{[BH_3^{z+3}]}{[BH_2^{z+2}][H^+]}$$

etc., where [X] denotes the actual concentration of species X; in the following c_X will be used for the total concentration of X, whatever its form. That is, $c_B = [B^z] + [BH^{z+1}] + [BH_2^{z+2}]\ldots$.

We note that the same formalism applies to, e.g., the successive reactions of Cd^{2+} with Cl, with replacement in the equations of B by Cd and H^+ by Cl, or for borate ion complexing with a sugar, amine ions with cyclodextrin, etc.

The number of protons (in general ligands) acquired is denoted by n. Thus, here the charge of the corresponding species n is z_0+n. The fraction of B that occurs in the form n (i.e., B^{z_0+n}), α_n, can be calculated from:

$$\alpha_n = \frac{\beta_n x^n}{1 + \beta_1 x + \beta_2 x^2 + \beta_3 x^3 + \ldots} \tag{15}$$

$$= \frac{\beta_n}{\sum_{j=0,\ldots,\max} \beta_j x^j}$$

where $\beta_0 = 1$

$$\beta_1 = K_1$$
$$\beta_2 = K_2 \beta_1$$
$$\beta_3 = K_3 \beta_2 \quad (= K_3 K_2 K_1)$$
$$\vdots$$
$$\beta_{nmax} = K_{nmax} \beta_{nmax-1} \quad (= K_{nmax} K_{nmax-1} \ldots K_1)$$

x = activity of the hydrogen ion, a_{H+}, i.e., 10^{-pH}

The K-values are the formation constants. Thus K_1 is the reciprocal of the acid dissociation constant of BH+, K_a,BH+. Fortunately, the conversion presents little difficulties because of the conventional logarithmic scale: $\log(K) = pK_a$. It should be noted, though, that the K_j's are numbered in reversed order. Thus, for phosphate one has $\log(K_1) = pK_{a,3} = 12.0$; $\log(K_2) = pK_{a,2} = 7.1$; $\log(K_3) = pK_{a,1} = 2.0$.

These acid-base equilibria are established within a time of about 1 μs (see Ref. 19 for other time scales in CE). That is, for an electrophoretic experiment, usually of several minutes duration, it can be safely assumed that the components are always in equilibrium. Therefore the migration of B, irrespective of its form, can be described [20] with an average electrophoretic mobility, $^a\mu_B$:

$$^a\mu_B = \alpha_0\mu_{B,0} + \alpha_1\mu_{B,1} + \alpha_2\mu_{B,2} + \alpha_3\mu_{B,3}$$
$$= \sum_{n=0..n_{max}} \alpha_n \mu_{B,n} \tag{16}$$

Figure 2 shows the dependence of the average mobility and the α-values on the pH for glutamic acid. Note that for this case the starting charge on the base is –2, corresponding to the dissociated form of both carboxyl groups and the neutral form of the amino group. The maximum charge acquired is +1. At the isoelectric point, indicated by IP, at pH = (4.376 + 2.155)/2 = 3.26, the form with two protons predominates, whereas the forms with one and three protons (having opposite charge) are present in equal amounts.

The advantage of this formalism is that there is no distinction between acids and bases as solute or buffer components. Indeed, every acid has a corresponding base. With one equation one can handle all cases. Especially in numerical simulation work this simplifies the algorithms.

In the sequel subscript j will usually refer to an individual ionic species, whereas other subscripts, such as i and k, will refer to a "component," that is, a property irrespective of the equilibrium form (e.g., the total concentration of glutamic acid/glutamates, or the average mobility $^a\mu$).

C. Mobility Not Proportional to Charge

The data for the mobilities used in Fig. 2 reveal an annoying aspect. It is seen that $\mu_{G,0}$, belonging to the doubly charged anion, 49.6, is not twice as large as the values for singly charged forms, 28.9 (all in units $10^{-9}/m^2/V/s$). The phosphate ions constitute an even more drastic exam-

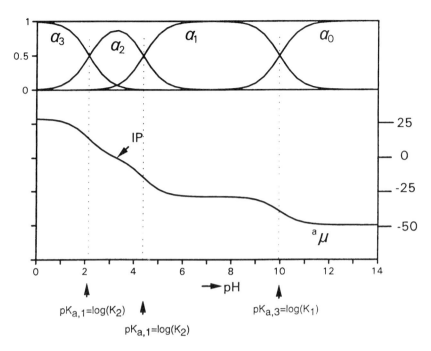

Fig. 2 α-values and $^a\mu$-values for glutamic acid, as a function of pH. Curves were calculated with the data of Hirokawa [10,11]: $\log(K_1) = pK_{a,3} = 9.96$; $\log(K_2) = pK_{a,2} = 4.376$; $\log(K_3) = pK_{a,1} = 2.155$; $\mu_0 = -49.6$; $\mu_1 = -28.9$; $\mu_2 = 0$; $\mu_3 = +28.9$;

ple: the μ-values are −71.5, −61.5, and −35.1, not at all in proportion to the charges −3, −2, and −1, respectively. It may be clear that this can be rationalized again as being the result of different degrees of hydratation of the ions. However, it has an unfortunate consequence in the treatment of multiply charged ions (e.g., peptides and proteins). The average mobility $^a\mu$ cannot be calculated from the average charge (see Section IV.E); one needs to insert specific (experimental or estimated) values for the individual mobilities in Eq. (16).

D. Influence of Ionic Strength on Equilibria

The equilibrium constants K_n referred are dependent on the ionic strength of the solution. The well-known Debye-Hückel expression for the activity coefficient of an ion j with charge z_j is

$$\mathrm{LN}(f_j) = \frac{\kappa a z_j^2 q_0^2}{\varepsilon_r \varepsilon_0 k_B T (1 + \kappa a) 4\pi}$$

In water at 25°C this corresponds to

$$\mathrm{Log}(f_j) = \frac{0.514 z_j^2 \sqrt{\frac{IS}{(\mathrm{mol/L})}}}{1 + \frac{1.01 a}{(10^{-9}\,\mathrm{m})} \sqrt{\frac{IS}{(\mathrm{mol/L})}}} \tag{17}$$

where a is the distance of closed approach of the counterions. For a 20-mmol/L solution, the activity coefficient for a singly charged ion is 0.86.

The importance of such corrections on the position of acid-base equilibria is comparable to the mobility corrections discussed above. As an example we take the case of phenylacetate ion, with μ_0 = 31.7×10^{-9} m²/V/s, and for which $pK_a = \log(K_1) = 4.70$, in a buffer of pH = pK_a = 4.70, with an ionic strength of 50 mmol/L.

Without corrections we would obtain $\alpha_0 = 0.5$, $\alpha_1 = 0.5$; the average mobility would be 0.5×31.7 = 15.85×10^{-9} m²/V/s. With corrections, the effective K-value changes by a factor of 0.86, and the average mobility would be 17.01×10^{-9} m²/V/s (apart from any corrections on the mobility μ_0, as discussed above). Such changes are large enough to cause peak inversions in an electropherogram. The situation is further complicated by the change in pK_a of the buffer. For instance, if the buffer is of the acetic acid–acetate type (neutral acid, negative base) and the stoichiometric proportions (rather than the pH) are held constant, the IS exerts an influence on the pH by the same mechanism, with the end result that IS has no effect. Each situation therefore requires careful examination.

For well-studied solutes, such as amino acids, many dipeptides, some oligopeptides, and very few larger peptides and proteins, the approach given above may have some use. However, in many cases the data are simply not known, and the only thing one can do is perform the experiment, wait and see, or try to make estimates on the basis of molecular structure.

E. Correlation of Mobility with Molecular Structure

It would be extremely useful when average mobilities of ions could be predicted from the chemical structure. A starting point for such predictions is Eq. (7). Most efforts have been devoted to following this approach for peptides (and proteins). This material has been reviewed recently [21]. A brief summary is given here.

For the use of Eq. (7), or similar equations, one needs the charge z and some measure of the hydrodynamic size, R_j (Eq. (7)). For simple peptides only one stage of ionization may occur, giving an integer value for z. For example, all peptides having no side acid or base groups have a charge of +1 at pH <2 or of –1 at pH >10. But, e.g., each histidine group increases the charge to +2 at low pH. Other cases, however, are not as clear-cut. A second carboxylic group (cf. the log(K) values given for glutamic acid, given above) changes the average charge at low pH by a fractional amount, as in these cases Log(K_3) is usually around 2; it would require a pH impractically low for CZE to suppress its dissociation entirely. Indeed, Fig. 2 shows for glutamic acid that only near pH = 0 the charge is fully developed to +1.

For many practical cases, therefore, we have to do with equilibrium mixtures; several species bearing different charges being present. In principle, it would be appropriate to use Eq. (16), after estimating the mobility of each species of importance. Although this approach has been taken once [22], most authors have worked with the average charge, $^a z$, from

$$^a z = \alpha_0(z_0 + 0) + \alpha_1(z_0 + 1) + \alpha_2(z_0 + 2) + \alpha_3(z_0 + 3) \ldots$$

$$= \sum_{n=0,\ldots,n_{max}} \alpha_n (z_0 + n)$$

(18)

which is then inserted in Eq. (7). Effectively this means that it is assumed that the friction factor, represented in Eq. (7) as $6 \pi \eta R_j$, is taken to be equal for all charged forms (cf. Section IV.C).

For the K-values needed to apply Eq. (18) (while using Eq. (15)) usually a set of "standard" pK_a-values are used (with the exception of Ref. 22). That is, each terminal –COO$^-$ and –NH2 group is assigned a fixed pK_a-value, as is each side group occurring in histidine, aspartic acid, etc. These sets [23–30] have been shown to be quite useful [25,26] in the prediction of isoelectric points of peptides, but their use for the calculation of charge has been criticized [12,22,23,28,31–34]. Still, this is often the only practical way, and the results are not too disappointing, as will be shown.

Table 1 summarizes the various forms for the correlation function that have been proposed. Note that two of the four expressions have a direct proportionality with the charge. In expressions 2 and 4 a propor-

Table 1 Semiempirical Equations Relating Peptide Structure and Mobility

Equation No.		pK_a data from Ref.	Applied in Ref.
1	$\mu = \dfrac{A_z}{M^{2/3}}$	23–25	17,22–24,36,106–116
2	$\mu = \dfrac{A\log(1+q)}{N^B}$	23,25,26	24,113,22,117–119
3	$\mu = \dfrac{A_q}{M^m}$	26,27	27,38
4	$\mu = \dfrac{A\log(1\_B_q)}{M^c}$	22	21,22

tionality with some logarithmic function is assumed. For example, the second one would predict a ratio of mobilities for charges 1, 2 and 3 being in the proportion 1. : 1.58 : 2.00. For the three forms of phosphate ions one finds experimentally that $\mu_2 : \mu_1 : \mu_0$ = 1 : 1.75 : 2.04, which is not exactly the same but it makes some sense. Citrate ions give a series with proportions –28.7 : –54.7 : –74.4, of similar approximate agreement.

The physical meaning of such an expression remains obscure though, unless one assumes that by some special freak of nature the hydration of multiply charged ions goes in proportion to $(1/z)\mathrm{Log}(1+z)$. Similar criticism applies to Eq. 4 in Table 1.

Considering now the denominators of the four equations, it can be seen that they all are some fractional power of the molecular mass (MM) or, what is roughly equivalent for peptides of average composition, the number of amino acid residues. The exponents, either predetermined or fitted to the data, vary between 0.41 and 2/3.

Equation 1, coined by Offord in 1966, has been used by many workers. It is interesting to discuss the value 2/3 of the exponent from the theoretical point of view. For a globular, tightly coiled form of the peptide one would expect the volume to be proportional to the molecular mass (MM), the radius, and with that the friction factor according to Stokes, would go up with $MM^{1/3}$, half the exponent value successfully used by Offord.

When the peptide behaves as a random coil, the gyration radius is ≈$MM^{1/2}$ [35]. From diffusion measurements on polymers it is well known that friction factors of such molecules indeed go up with this, or a slightly larger, exponent. Still, the exponent is seldom as high as 2/3.

A third molecular model is the rigid rod. For not too large peptides the flexibility in the peptide bonds could be insufficient to allow anything like a random coil; a stretched molecule would be more likely as a model. For such a situation indeed an exponent close to 1 would not be surprising.

However, before jumping to the conclusion that the pertinent peptides behave as rodlike molecules, the ionic strength effects on mobility should be considered. In none of the correlations referred to in Table 1 has an extrapolation to zero ionic strength been made; so the ionic strength effects are indeed "absorbed" by the fitting procedures.

As pointed out, the ionic strength has quite an influence on mobilities. Therefore the exponent found may be strongly dependent on ionic strength. In this respect, it is interesting to recall Offord's [36] reasoning leading to a 2/3 exponent: frictional force was assumed to be proportional to the area of the spherical ion, as "the solvated ionic atmosphere, which surrounds the peptide, causes a backward flow of solvent in the immediate vicinity of the migrating peptide." This is, in effect, nothing else than a description of the cataphoretic effect, when the ion is large compared to the Debye length. In terms of the Debye-Hückel-Onsager description (Eq. (15)) (leaving the relaxation out, as this leads to a similar argument):

$$\mu = \frac{zq_0}{6\pi\eta R} - \frac{zq_0\kappa}{6\pi\eta}\frac{1}{1+\kappa a} \tag{19}$$

With R large, the radius of closest approach, $a \approx R$, and $\kappa a \gg 1$, so this can be approximated by

$$\mu = \frac{zq_0}{6\pi\eta\kappa R^2} = \frac{zq_0}{6\pi\eta\kappa \text{ Const } MM^{2/3}} \tag{20}$$

Thus, there is no problem in finding a rationalization of the value of 2/3 for the exponent; as shown [37,38] it is even possible to explain a transition from this value to other ones, depending on size and ionic strength. From this point of view it is not surprising that with other exponents satisfactory results have been obtained using other data sets. Summarizing: On the basis of theoretical models that are at least reasonable for a subset of peptides, and taking the ionic strength effects into account, it is possible to expect virtually any exponent (within reasonable limits of course).

A better approach to the problem would be to use zero-ionic-strength extrapolated values in the fitting procedure; prediction for practical CZE then of course would necessitate applying the reverse correction on the predicted data. The possibility of realizing this concept appears to be remote; only a few workers have gone through the tedious procedure to arrive at ionic-strength-corrected mobilities. Meanwhile one has to deal with the more empirical correlations published, as summarized in Table 1. It seems wise to use these only for peptides and conditions that do not differ too much from the "learning set" used.

All these objections do not take away that the availability of these correlations can be very helpful in understanding electropherograms of real samples and identifying unknown peaks. They are certainly not precise enough to predict resolutions; however, they are sufficiently accurate to discard incorrect peak assignments in many cases. Examples can be found in Refs. [22 and 39].

V. ELECTROPHORESIS IN CAPILLARIES

A. Experimental Setup

Capillary electrophoresis was introduced nearly at the same time by Mikkers et al. [40,41] and Jorgenson and Lukacs [42,43]. The essential experimental feature of it is in the small cross section of the separation channel, which allows an increase in the electric field strength to above 40 kV/m. Such intense fields are quite inaccessible in other experimental versions (with the exception of the micromachined channels [44–47], which have similar or even larger advantages in this respect), because of the prohibitively strong heat dissipation. This small cross section was and is realized by using fused silica (Teflon in the case of Ref. 40) capillaries, with inner diameters around 75 µm. The admissible high fields improve the analytical performance of the capillary format in two ways: the separations are faster, and they are more efficient. This section discusses some of the pertaining relations semiquantitatively.

The basic elements of a CE instrument are shown in Fig. 3. Although this is not the place to go into the details of the experimental setup, it is necessary to stipulate a few points of importance for the present "theoretical" discussions.

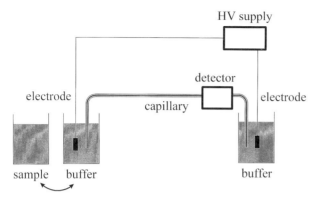

Fig. 3 Capillary electrophoresis scheme.

The capillary is inserted at both ends in a buffer vial, normally filled with the buffer solution present in the capillary. In these vials electrodes, often of platinum, are inserted, on which a high voltage (10 to 40 kV) is applied. The resulting electrical current brings about reactions at both electrodes; the consequences [48] of these reactions are briefly discussed in Section VIII.

On injection, one buffer vial is replaced by a vial containing the sample. Injection can be done by the "electromigration" method; that is, the sample ions are simply electrophoresed into the capillary by applying a metered voltage during a metered time. Alternatively, and preferably, a controlled pressure difference across the two vials can be used to press a certain amount of sample solution into the capillary. Depending on the origin of this pressure difference, one speaks about "hydrostatic" or "hydrodynamic" injection. In the former case the levels of the two vials are made different, in the latter case pressure or vacuum from a gas manostat is applied.

Thermostating is nearly always present. In most instruments a circulating air bath surrounds the capillary, in one or a few designs a liquid sheath is present around the capillary. However, none of the experimental versions are perfect in this respect. The coupling to the vials and to the on-column detector does not allow perfect thermostating of the whole capillary. Many artefacts and limitations can be traced to these imperfections [49].

B. Efficiency and Speed

Efficiency is measured by the plate height, H, or the plate number N. H is best defined as

$$H_i = \frac{d\sigma_{i,z}^2}{dz_i} \qquad (21)$$

where $\sigma_{i,z}^2$ is the variance (square of standard deviation) of the distribution of the component over the length coordinate z and z_i is the traveled distance. The plate number, N_i, is derived from it as

$$N_i = \frac{l}{H_i} \qquad (22)$$

where l is the distance from injection to detection point. Experimentally, N_i is determined from

$$N_i = \left(\frac{t_{i,M}}{\sigma_{i,t}}\right)^2 \qquad (23)$$

where $t_{i,M}$ is the migration time and $\sigma_{i,M}$ is the time standard deviation of the observed peak.

The relations [41,42] with the properties of i are, under ideal conditions (see Section VI),

$$\sigma_{i,z}^2 = 2D_i t, \quad t_{i,M} = \frac{1}{u_i}, \quad u_i = \mu_i E, \quad E = \frac{V}{L} \qquad (24)$$

where the first relation is known as the Einstein relation; further, E is the electric field strength, L is the length of the entire capillary, and V is the applied voltage.

By manipulating Eq. (24), one arrives at a more direct expression for the efficiency, characterized by the plate number:

$$N_i = \frac{V\mu_i}{2D_i}\frac{1}{L} \qquad (25)$$

and for the migration time, $t_{i,M}$:

$$t_{i,M} = \frac{L1}{\mu_i V} \qquad (26)$$

As is immediately clear from Eqs. (25) and (26), the application of high voltages is beneficial for both efficiency and separation time. Also, decreasing the length would only bring the benefit of speed. This fact, puzzling for chromatographers that are used to a trade-off between speed and efficiency, forms one of the attractive features of capillary electrophoresis.

At this point it is instructive to work out Eq. (25) further, along the lines stipulated by i.a. Giddings [50] and Kenndler [51]. Both μ_i and D_i are related to the friction factor, ρ_i, for a spherical particle given by the Stokes relation:

$$\rho_i = 6\pi\eta R_i \tag{27}$$

where R_i is the hydrodynamic radius of the particle (or ion).

The diffusion coefficient is, according to the Nernst-Einstein equation,

$$D_i = \frac{k_B T}{\rho_i} \tag{28}$$

Likewise, the electrophoretic mobility can be expressed as

$$\mu_i = \frac{q_0 z_i}{\rho_i} \tag{29}$$

That is, the ratio of mobility and diffusion coefficient, as in Eq. (25), is

$$\frac{\mu_i}{D_i} = \frac{q_0 z_i}{k_B T} \quad \left(= \frac{z_i}{0.025 \text{ Volt at } 25°C} \right) \tag{30}$$

and the expression for the efficiency assumes the simple form

$$N_i = \frac{V q_0 z_i}{k_B T} \frac{1}{L} \quad \left(= \frac{V z_i}{0.025 \text{ Volt}} \frac{1}{L} \text{ at } 25°C \right) \tag{31}$$

According to this relation, the efficiency obtainable is determined only by the available voltage and the valency of the ion. The physical fact behind this is that migration and dispersion are governed by the same friction factor. Thus, change in temperature would only have a minor effect (0.3% per Kelvin rather than 1.5–2%): change in viscosity of the medium, or even changes in size of the ion (e.g., by complexation with a neutral entity), have no effect at all on N_i. Only a, possibly strong, effect on $t_{i,M}$ will be observed. Kenndler [51] pointed out that the role of z_i in Eq. (31) helps us to understand the excessive plate numbers obtained in DNA-sizing CE experiments. It is interesting to note [50] that the predicted plate number is equal (apart from a factor of 2) to the potential energy difference for the ion between the beginning and the end of the separation track, when measured on the scale of the thermal energy, k_B T. Such equations should also apply to, e.g., ultracentrifugation.

C. Limitations to Field Strength

Equation (31) suggests that there is an unconstrained advantage in the use of higher fields and shorter lengths. The following practical limitations exist, though. In the first place the maximum voltage that can be delivered by reasonably priced sources is about 30–40 kV. Also, it turns out that higher voltages are difficult to work with, because of discharges in the air surrounding the equipment parts.

A high field of course could also be obtained by shortening the tubes. This would not increase resolving power (cf. Eq. (31)), but from Eq. (26), considering that $l \approx L$, a gain in separation speed proportional to the square of the length reduction would be anticipated. Although this has been demonstrated [52], the approach fails to be practical because of the extremely small diameters needed to keep the heat dissipation under control (see Section VI).

VI. ELECTROOSMOTIC FLOW

A. Charge Separation

In fused-silica capillaries electroosmotic flow is often present. This flow is the result of the charge separation at the interface between the liquid and a solid material of the wall. The charge separation is a universal phenomenon. However, in CE the solid is nearly always silica, and the charge separation can be interpreted as being the result of the ionization of silanol groups; the leaving hydrogen ions charge the liquid, leaving the charged surface with siloxy groups (=SIO⁻) behind. These groups have a pK_a of about 3, but depends very much on the conditions. Indeed at pH's lower than 3 the effect largely disappears.

Detailed discussions of electroosmosis in CE can be found in Refs. 53–55. Here a brief synopsis is presented.

B. Electroosmotic Mobility and Zeta Potential

The electroosmotic flow is characterized by a mobility, μ_{EO}, defined in a similar way as the mobility of ions. It has a theoretically sound relation with the zeta potential, ζ. This is the electrical potential in the liquid, close to the surface (δ 0.5 to 1.0 nm), at the boundary between stagnant (i.e., not behaving as a liquid) and mobile (normal) liquid. The relation for a flat surface is

$$\mu_{EO} = \frac{\zeta \varepsilon_w \varepsilon_0}{\eta} \tag{32}$$

Equation (32) is not very practical, as the only experimental way to find values for ζ for a wall material is to measure μ_{EO}, so ζ is not much more than another format for the data. (We note in passing that the ζ-potential is also often used in the discussion of ion mobility. However, the reasons that make this a useless exercise are even stronger in the case of small ions. Only when discussing very large multiply charged ions or particles, where the ζ-potential can be believed to be independent of size, may this be useful in the interpretation.)

The relation between the ζ-potential and the surface charge, σ, is complicated and discussed in Refs. 55 and 56. A few essential points emerging from these theory have to be mentioned. At low σ-values the relation is linear:

$$\zeta = \frac{\sigma}{\varepsilon_w \varepsilon_0 \kappa} e^{-\kappa \delta} \quad \text{or} \quad \mu_{EO} = \frac{\sigma}{\eta \kappa} e^{-\kappa \delta} \tag{33}$$

where δ is the distance of stagnant-fluid boundary to the surface. The second form [56] in Eq. (33) is more appealing, because the awkward permittivity is removed, and one directly sees the relation between the charge density, determining the electric force, and the thickness of the layer, Debye length, $1/\kappa$, determining the viscous force. As δ is reasonably constant, this shows clearly why, as observed in practice, EO flow decreases with ionic strength, IS, since κ is proportional to $IS^{1/2}$.

The other point is the behavior at high σ. When σ is above about 0.1 C m^{-2}, the ζ-potential (and with it the EO mobility) increases much less than proportionally, reaching a plateau value about 100 mV at a ionic strength of 10 mmol/L. That corresponds to a maximum value in μ_{EO} of 100×10^{-9} m^2/V/s. Indeed, higher values are seldom observed, also when an external radial field, as discussed in Ref. 55, is applied for controlling the EO flow.

A key element in the theory is that the flow profile of EO flow is virtually rectangular; one speaks about plug flow. The propulsion takes place exclusively in the diffuse double layer, of thickness $1/\kappa$, the inner part of the liquid is passively carried forward, as if it is held in a cylindrical container. Electrical and viscous forces (velocity gradients) occur only in the diffuse layer. Under nearly all practical conditions, its thickness (10 nm for $IS = 0.01$ mol/L) is very small compared to the channel diameter; thus, only in a small part of a channel with only 1 μm diameter is the flow nonuniform.

Noteworthy is that, according to this treatment, any viscosity change in the core of the liquid is immaterial. Thus, a temperature profile within it would not lead to a parabolic flow pattern nor to the associated dispersion (of course, the temperature profile does lead to dispersion when a separation selectivity of any kind, e.g., mobility differences for ions, or micellar distribution, is present).

The counterpart is that it is the viscosity in the diffuse layer that determines EO flow, not the one in the core of the liquid. This may explain why certain physical and chemical modifications, with no ionic groups present and for which an effect on the charge density is unlikely, sometimes dramatically affect the EO flow; these agents apparently work via the change of the local viscosity in the diffuse layer.

When EO flow is present, the migration of ions is effectively described by the sum of the ion mobility μi and that of the EO flow, μ_{EO}. This sum will be indicated in the sequel by the superscript t for total: $^t\mu_i = \mu_i + \mu_{EO}$, where c.q. it is understood that μ_i is an average mobility, $^a\mu_i$.

The uniformity of the flow profile is of great importance for the efficiency of electrophoretic separations. It means that dispersion effects due to flow uniformity, as present in all forms of pressure-driven chromatography, are absent. Thus, as discussed in the next section, the ultimate limit to efficiency is determined only by the longitudinal diffusion of analytes.

VII. DISPERSION

A. Introduction, Injection, and Detection Effects

Peak width in CZE is often determined to a great extent by nonlinearity, i.e., dependence of migration rate on concentration, as well as by injection effects. The former, known as electromigration dispersion (EMD), will be treated in Section VII; in the present section some attention is given to the injection and detection effects, but it is mainly devoted to the dispersion occurring during the electrophoretic transport.

In CE it is often more convenient to use longitudinal positions rather than retention times or volumes, which are more appropriate in chromatography. One reason is that injection and detection effects on peak width are nearly always known in terms of position (injection plug length, detection optical window) rather than in time or volume. Another, pointed out by Rathore and Horvath [57], is that position and velocity increments and not time increments due to various processes are additive in CZE. Therefore in the sequel the use of time expressions will be avoided as much as possible.

The total variance, σ_z^2 (second moment, see Ref. 58) in length units, of the zone, on arrival at the detection position, l, can be expressed as (neglecting EMD, and often dropping the index i, as everything applies to one analyte ion)

$$\sigma_z^2 = \sigma_{z,\text{inj}}^2 + \sigma_{z,\text{tra}}^2 \tag{34}$$

where $\sigma_{z,\text{inj}}^2$ and $\sigma_{z,\text{tra}}^2$ are the contributions generated by the injection and the electrophoretic transport, respectively.

The first term is determined by amount of sample solution injected, but the relation depends very much on the conditions. Only under "easy" academic conditions can the length of the injection plug, L_{inj}, be found from basic considerations: L_{inj} calculated from the Poiseuille law in the case of hydrodynamic injection (applying a factor between 1 and 2 to allow for the parabolic profile, distorting and diluting the plug) or from the product of injection voltage, time, and mobility of the analyte in the case of electromigration injection. In real-life situations usually the difference in conductivity and pH between injection solution and BGE distorts this simple picture. Whatever the injection mode, these differences will lead to strongly nonlinear electrophoretic migration during the first stages of the electrophoresis or during the injection itself. These can be used to advantage ("stacking") for concentrating the analyte, i.e., inject more volume while keeping the effect on zone width small. The ultimate form of this is doing an isotachophoretic step before the CZE.

It was decided not to dwell on this subject here, as the pertinent theory on such transition stages, depending very strongly on the particular conditions, is not well developed yet. Thus, for the moment we take it that there are no such phenomena, and find [58], for an idealized condition of plug-shaped injection,

$$\sigma_{z,\text{inj}}^2 = \frac{L_{\text{inj}}^2}{12} \tag{35}$$

Another contribution to peak width stems from the detection. Although it does not affect the properties of the zone itself, it broadens the observed peaks. With the frequently used on-column UV absorbance or fluorescence detection the illuminated part of the tube may not be infinitely short. When L_{det} represents the length of the illuminated part, the peak broadening can be expressed, formally, in position units by

$$\sigma^2_{z,det} = \frac{L^2_{det}}{12} \tag{36}$$

Although for the actual zone itself there never is such a spatial broadening, Eq. (36) corresponds to the width increase as observed in the signal trace (i.e., a variance in time units equal to $\sigma^2_{z,det}/U^2$). In pertinent cases that term should be added to Eq. (34).

B. Broadening During Electrophoresis

One main contributor to the dispersion during the transport was already discussed—the straightforward, longitudinal, molecular diffusion. It leads to a variance, $\sigma^2_{z,diff}$:

$$\sigma^2_{z,diff} = 2Dt = 2D\frac{l}{\mu} = \frac{2Dl}{\mu E} = \frac{2DlL}{\mu V} \tag{37}$$

When expressed as a plate height ($H = d\sigma_z^2/dz$),

$$H_{diff} = \frac{2D}{\mu E} \tag{38}$$

Now we need to discuss three other important sources of zone broadening:

- A. Parabolic migration profile, brought about by the temperature profile in the capillary, indicated by subscript PaT
- B. Parabolic migration profile, brought about by pressure gradients across the capillary, indicated by subscript PaP
- C. Adsorption on the capillary wall, indicated by subscript Ads

Each of these processes lead to additional contributions to the total variance $\sigma^2_{z,tot}$. This addition can be similarly expressed in terms of plate height:

$$H_{tot} = H_{diff} + H_{PaT} + H_{PaP} + H_{Ads} \tag{39}$$

In some texts H-contributions due to injection and detection effects are included in equations such as Eq. (39). This is not preferred, as H should be a parameter describing the dispersion during the transport and should not have anything to do with the external conditions. These effects should be accounted for by using the additivity of variances (Eq. (34)).

The source under A has been the subject of much theoretical investigation [59–62]. All authors arrive at the same result (i.e., known since 1974):

$$H_{\text{PaT}} = \frac{1}{384} \frac{d_c^2 \Delta u^2}{Du} \tag{40}$$

where d_c is the capillary diameter, and Δ_u is the difference in migration velocities at the wall and in the center of the capillary. The value of Δ_u is determined by the temperature difference ΔT and the relative change in migration rate per kelvin, δ (about 0.02 K^{-1}). Note that this applies only to the mobility of the ion. The EO flow is assumed to be nonparabolic, so for strong cooperative EO flow δ must be smaller. The value of ΔT is found [53,63] from the power dissipated per volume unit, equal to the product of the field, E, and current density, I, and the thermal conductivity of the buffer solution, k_β, as

$$\Delta T = \frac{1}{16} \frac{d_c^2 EI}{k_\beta} \tag{41}$$

The value of Δ_u is therefore

$$\Delta\mu = \frac{1}{16} \frac{d_c^2 EI\delta u}{k_\beta} \tag{42}$$

Substituting $u = \mu E$ and $I = \kappa E$ into Eq. (40), one obtains

$$H_{\text{PaT}} = \frac{1}{98304} \frac{\kappa^2 \delta^2 \mu d_c^6 E^5}{Dk_\beta^2} \tag{43}$$

Calculations of this term demonstrate [53] that the limits to which d_c and E can be increased are very tight, because of the high exponents occurring. Later, however, workers realized (cf. Refs. 64–67), that conditions of such large d_c and E-values are not accessible, at least in the case of air cooling, for a much simpler reason: The heat transfer to the thermostating air is so slow that the buffer liquid will boil before H_{PaT} becomes really significant. That is, any increase in plate height observed before this happens has to be interpreted with much care, because of the overall increase in temperature and diffusion coefficient. Also for liquid cooling this may apply, as some centimeters of the tube are, at the inlet and outlet vials, surrounded by still air.

The source mentioned under B has been studied less. An expression can be derived [65], using similar reasoning that led to Eq. (43):

$$H_{\text{PaP}} = \frac{1}{98,304} \frac{d_c^6 \Delta P^2}{D\eta^2 L^2 E\mu} \tag{44}$$

As the effect is less intensely discussed, it is worthwhile to devote a figure to it (Fig. 4). The figure conveys the same message: There is a rather sharp upper limit to the capillary diameter. Three undesirable increases (H_{PaT}, H_{PaP}, and temperature) coincidentally impose roughly the same limitation. This has an important bearing on the sample capacity of CZE, a point discussed in Refs. 65 and 93.

Adsorption of analyte (point C) on the capillary wall can have a disastrous effect on the efficiency in CZE. This has been treated in an early contribution by Martin and Guiochon [69]. Their analysis is detailed, but for perfect plug flow and a linear adsorption isotherm the result is

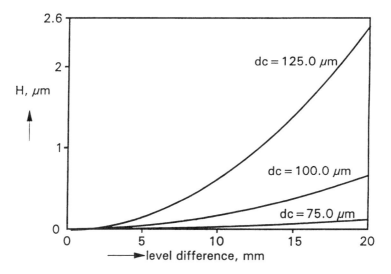

Fig. 4 Dependence of the plate height contribution due to hydrostatically induced Poiseuille flow in the capillary, H_{PaP}, for three inner diameters of the capillary, on the level difference of the buffer vials. Calculated according to Eq. (44), with $\eta = 0.00089$ Pa s, $\rho = 1000$ kg m^{-3}, $L = 0.75$ m; $D = 1 \times 10^{-9}$ m^2 V^{-1} s^{-1}.

$$H_{\text{Ads}} = \frac{1}{16}\left(\frac{k'}{1+k'}\right)^2 \frac{d_c^2 u}{D}$$

where k' is the retention factor, mass adsorbed over mass in solution.

An examples may serve to demonstrate the importance of the effect. With d_c = 100 µm, u = 2 mm/s, D = 10^{-9} m²/V/s, and k' as small as 0.03 (3% increase in migration time due to adsorption), the result is H_{Ads} = 1.1 µm, whereas the value under ideal conditions is $H = H_{\text{diff}} = 2D/u$ = 1 µ. Thus, with this minor extent of adsorption already half of the efficiency is lost. For an analyte of higher molecular mass, with a smaller diffusion coefficient, the H_{Ads}-value goes up in proportion, whereas H_{diff} goes down. For such analytes the effect can indeed be catastrophic.

Therefore, attempts to improve selectivity by means of wall modifications (that can only work via adsorption, as genuine electrophoretic selectivity is determined by the mobilities in solution) have to be considered with great suspicion. Unless the capillary diameters are very small, the resulting loss in efficiency may nullify the gain in peak spacing, and the peak capacity of the system will be degraded. In fact, doing such experiments is equivalent to carrying out open tubular liquid chromatography, a technique attractive only when column diameters can be kept below 10–15 µm.

C. Alternative Geometry

The capillary diameters occurring in contemporary CE are close to the value originally used by Jorgenson and Lukacs. Smaller ones would allow much higher field strength and concurrent higher speed [52], but the small sample capacity would seriously degrade detection possibilities. Larger ones are often desirable for the separation larger samples, both in terms of volume and in terms of amounts. This is the case, e.g., when detection is problematic or it is desired to collect sizable amounts of the separated compounds. However, larger diameters degrade separation performance, as may be clear from Section VI.B. Thus, there is clearly a conflict between speed of separation and sample capacity/detectability.

As with cylindrical geometry further improvement appears to be impossible, workers have tried other solutions. One is to use a bundle of many capillaries in parallel. This idea has been pursued in (open channel) LC as well [68]. One problem with the approach is in the synchronization of the migration in the set of parallel tubes. In CZE, this is, on the one hand, easier than in chromatography; the EO flow does not depend on the tube diameter, so the tolerance for spreading in diameters during fabrication of the set is not so tight. Also, stationary phase coating, which is likely to result in serious synchronization defects in LC, is not needed. On the other hand, distribution of material at the injection and the collection of exiting material for detection or collection seems more difficult. As far as we know the idea has not been pursued intensely in CZE.

Another approach is to use rectangular geometry, something that has been tried in LC [70]. The advantage can be explained as follows: Limitations to the efficiency and the speed of separation stem (see above section) from insufficiently fast transfer of either heat or mass across the cross section of the channel (note that in Eqs. (40), (43), and (44) a transport coefficient is in the denominator and the diameter is in the numerator). A rectangular system having one dimension (say height) small and the other (width) large could enhance the speed of heat/mass transfer while maintaining a given area for the cross section, which is required to obtain a large sample capacity. For CZE the advantages have been analyzed in Ref. 71. Similar arguments apply to the uses of micromachined [44–47] channels; virtually all of them have a height much smaller than 75 µm and a width much larger than that: in other words, they combine relatively large sample capacity and fast equilibration, as argued in Ref. 72, in the same manner.

D. Resolution

The question of the resolution in CZE can be solved in a straightforward manner. Resolution is defined here (following Huber [73]) as the ratio of the peak spacing Δ_w over the peak standard deviation σ_w, the latter taken for the peak with the smallest coordinate (i.e., the fastest component of the two when reckoning in the time domain, the slowest one when the position is used, as done here). This definition,

$$R_{k,i} = \frac{\Delta_W}{\sigma_W} \qquad (45)$$

where W is either time or position, differs from the one in most publications in separation science in that it omits a factor 4, inserted apparently in order to have $R_{k,i} = 1$ for reasonably well separated peaks, and in the fact that σ_w is not taken as the average for the two peaks. The resolution required may be between 1, or even less, and 6, or even more, depending on the analytical problem and the method of data handling. Thus, 1 is by no means a magic number in this context. Taking the average may describe the situation under some extreme conditions more accurately, but in general it makes little difference: The concept of resolution is only relevant for those cases where the components are difficult to separate, and thus their peak width is virtually the same. In this, and other texts, we therefore stick to Eq. (45).

The position difference of i and k after a time lapse t and the standard deviation at that moment can be found by applying basic equations:

$$\Delta z_{k,i} = (\mu_k - \mu_i)Et \tag{46}$$

$$\sigma_{1,2} = (H_{tot}{}^t\mu_i Et)^{1/2} \tag{47}$$

The resolution is therefore

$$R_{k,i} = (\mu_k - \mu_i)E^{1/2}t^{1/2}\,{}^t\mu_i^{-1/2}H_{tot}^{-1/2} \tag{48}$$

The value of t is set equal to $t_{i,M}$, according to Eq. (26). Substitution yields

$$R_{k,i} = (\mu_k - \mu_i)\,{}^t\mu_i^{-1}\left(\frac{1}{H_{tot}}\right)^{-1/2} = (\mu_k - \mu_i)\,{}^t\mu_i^{-1}N^{1/2} \tag{49}$$

While using Eq. (49) it should be kept in mind that H_{tot} and thus N depend strongly on the values of the mobilities of the EO and of i and k. For the case where only diffusion is of importance, the full equation, expressing everything in mobilities (${}^t\mu_i = \mu_i + \mu_{EO}$), diffusion coefficients, and external conditions is

$$R_{k,i} = (\mu_k - \mu_i)\left(\frac{V}{2D(\mu_i + \mu_{E0})}\right)^{1/2}\left(\frac{l}{L}\right)^{1/2} \tag{50}$$

This equation demonstrates clearly the effect of the ratio l/L, as well as that of the EO flow. The latter is two-fold. Assuming positive μ_{EO}-values (silica negatively charged), as is most common, the analysis of cations is speeded up. However, the equation shows that the resolution goes down. The peak spacing (in position) goes down in proportion to the residence time, the peak width goes down only in proportion to the square root of the time. Thus, their ratio, being the resolution, is degraded. For negative ions or, in general, for ions with the same charge sign to the wall charge, the effects are the opposite: speed is degraded, but resolution is enhanced. The extreme is when μ_i approaches $-\mu_{EO}$, when both resolution and migration time approach infinity.

VIII. INTERACTION OF IONS

A. Nonlinear Transport

In the equations thus far it has been assumed that the velocities of the components during the time cours of the CE run are constant. For that case, linear behavior, the concepts of position of a zone, width of a zone, migration time, peak width, etc., are simple and unambiguous. However, when allowance is made for the variation of velocities, one enters the field of nonlinear transport, of importance in (preparative) chromatography as well as for electrophoresis.

In this section this topic, as well as the related topic of system peaks, will be explored in some depth. In order to avoid an (as yet) unmanageable mathematical complexity, in such a treatment it is necessary to make a very drastic simplification: All dispersive mechanisms are ignored (with the exclusion of electromigration dispersion, EMD, a central issue in this chapter). That is, it is assumed that diffusion and dispersion due to, say, parabolic profiles, are absent. In chromatography this model is indicated as "nonlinear ideal chromatography," whereas the model underlying the previous chapters corresponds to "linear, nonideal chromatography." It appears logical to extend this nomenclature and to speak about "linear nonideal electrophoresis" for the material of Sections I–IV; this section then is about "nonlinear, ideal electrophoresis."

This ignoring of all dispersion effect may seem too drastic, as the very power of CZE resides in the effective control of dispersion. However, as said, a theory of nonlinear, nonideal electrophoresis is not available. That field is presently only accessible via numerical methods (as it is for the major part in its chromatographic analogue).

Nonideality as well as nonlinearity lead to zone broadening. Which model is the most appropriate depends on the conditions: When nonlinearity produces zone widths much larger than nonideality does, the predictions of the nonlinear ideal model are closer to the truth, and vice versa. For instance, it will be seen that the nonlinear ideal model may predict triangular peaks; when dispersion is small relatively, these are indeed observed. However, with increasing dispersion the triangles will first appear in a slightly deformed version, rounded off at the edges, and then gradually transform into the familiar Gaussian shape predicted by by the linear, nonideal model.

Movement of zones in (ideal) electrophoresis and their deformations have nearly always been described in terms of the moving boundary equations [74,75]. It appears that the theory has been developed independently from the theory of nonlinear and multicomponent chromatography, described, e.g., in detail by Helfferich and Klein [76]. In the following an attempt will be made to use concepts from both fields. In doing this, the idea that boundaries can be self-sharpening or "diffuse" (the indication non-self-sharpening is preferred) will be taken for granted from the outset, as direct theoretical derivation of these leads to too much mathematical complexity.

The running buffer, or background electrolyte, BGE, in a CZE experiment is required to obtain constant values for the pH as well as for the conductivity. Both are required for the velocity of an analyte ion within its zone to be constant. When this condition is fulfilled ("linear"), the expression for the velocity of a zone of i, $u_{z,i}$ is

$$u_{z,i} = \frac{{}^a\mu_i V}{L} = {}^a\mu_i E \tag{51}$$

where it is again understood that c.q. ${}^a\mu_i$ is the average over various forms i that may occur in.

In this case the zone velocity, $u_{z,i}$, is equal to component velocity of i, u_i, also being the product of mobility and field strength (here "component velocity" is used rather than "species velocity" as in Ref. 76, since the word "species" is used here for the individual charged forms of a compound). The two velocities can be different, as discussed below.

If we had means to follow one ion, ${}^a u_i$ would appear as the average velocity for such an ion over a time long enough to allow a sufficient number of equilibrium reactions between the various charged forms as described in Eqs. (15) and (16).

What can only be observed macroscopically are zone velocities, or better even, boundary velocities, i.e., velocities of transitions from one concentration to another. They may be different from the component velocities. The difference can be best explained [76] by analogy with the highway traffic. Although cars, the analogues of the ions, may have velocity around 100 km/h, a pile-up of cars, the analogue of an increased concentration, may move with quite another speed or even backwards.

B. Steep, Self-Sharpening Boundary

Assume that Fig. 5A, B represents the migration of a zone, indicated by α, where the concentration of i is ${}^\alpha c_i$, in a medium, β, (buffer) where its

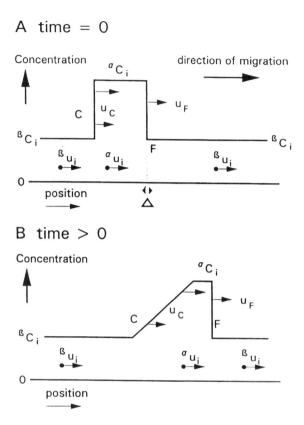

Fig. 5 Migration and deformation of a rectangular plug of concentration $^{\alpha}c_i$ injected into a medium of concentration $^{\beta}c_i$. It is assumed that the plug is "inserted" into the medium rather than fed into the system at one end. Figure does not apply directly to electrophoresis, as only one component is considered, and is shown only to demonstrate the deformation of the zone when the velocity u depends on the concentration c_i.

concentration is $^{\beta}c_i$, $^{\beta}c_i$ not necessarily being zero, but $^{\alpha}c_i \neq {^{\beta}c_i}$. Consider the location indicated by F (front) in Fig. 5A. Assume that the transition from $^{\alpha}c_i$ to $^{\beta}c_i$ is sharp and remains so, and that in this part the zone is not deformed during transport. This situation is known as a self-sharpening boundary, and for the moment it is taken for granted that it exists. Use a coordinate system that moves to the right with velocity u_F. In this coordinate system the boundary is not only of constant shape but is stagnant. Therefore the amount of i in the segment Δ in Fig. 5 has to be constant and the moving-boundary equation can be found by considering that the amounts that enter and leave this small segment per second from left and right (fluxes) must cancel. That is,

$$^\alpha c_i(^\alpha u_i - u_F) = {}^\beta c_1(^\beta u_1 - u_F)$$

or

$$^\alpha c_i{}^\alpha u_i - {}^\beta c_1{}^\beta u_1 = (^\alpha c_i - {}^\beta c_1)u_F \tag{52}$$

Note that the u_F terms appear because of the use of the moving-coordinate system. Note also that diffusion terms are missing, demonstrating that the ideal model is treated.

If the u_i's are equal, Eq. (52) can be satisfied only when $u_F = {}^\alpha u_i = {}^\beta u_i$, that is, the simple result given above is recovered; boundary velocity equals component velocity. However, the µ's may differ in α and β zones. For Fig. 5, where only one compound is considered, it could be that the velocity changes as a result of self-association of i, this being more severe at higher concentrations. For real CZE systems, with the effect of other ions present, the change of field strength and pH come into play. But before treating such topics in detail, it is instructive to look at the consequences of a change in u_i as a function of c_i in general terms.

With given values for $^\alpha c_i$ and $^\beta c_i$, and with that for $^\alpha u_i$ and $^\beta u_i$, u_F is found from Eq. (52) as

$$u_F = \frac{^\alpha c_i{}^\alpha u_i - {}^\beta c_i{}^\beta u_i}{^\alpha c_i - {}^\beta c_i} \tag{53}$$

With $^\beta C_i = 0$ it follows again $u_F = {}^\alpha u_i$; that is, when the boundary is self-sharpening and the concentration in one zone is zero, its velocity is equal to the component velocity in the non-zero-concentration zone.

C. Non-Self-Sharpening Boundary

Equation (53) is not of much use when the boundary is not self-sharpening, in which case the velocity of the boundary is hardly defined. In that case one has to resort to the "concentration velocity," u_c, as defined by Helfferich and Klein [76], i.e., the speed of migration of a point with constant concentration, which indeed one can again follow experimentally. For that case the integral form Eq. (53) has to be transformed into a differential one. A change from one concentration to another differing only infinitesimally from it, is considered, by taking $^\beta c_i$ equal to the constant c_i, and $^\alpha c_i$ equal to $c_i + \Delta c_i$. When $\Delta c_i \to 0$,

$$u_c = \frac{d(c_i u_i)}{dc_i} \tag{54}$$

Expanding the derivative yields

$$u_c = u_i(c_i) + c_i \frac{du_i(c_i)}{dc_i} \tag{55}$$

Thus, for a given concentration c_i it is possible to find "its" velocity (i.e., answering the question: where is this concentration to be found after a time lapse?) by evaluating Eq. (55).

It is instructive to discuss the history of a zone such as given in Fig. 5, when u_i is a linear function of the concentration, something that is quite often a most effective approximation:

$$u_i(c_i) = {}^0 u_i + s_{u,i}\, c_i \tag{56}$$

with $s_{u,i}$ being a constant.

When we take, for simplicity, $^\beta c_i = 0$, the result for the two boundaries F and C is

$$u_F = {}^0 u_i + s_{u,i}\, {}^\alpha c_i \tag{57}$$

$$u_C = {}^0 u_i + s_{u,i}\, {}^\alpha c_i + s_{u,1}\, {}^\alpha c_i = {}^0 u_i + 2 s_{u,i}\, {}^\alpha c_i \tag{58}$$

It is seen that the deviation in u from the starting value $^0 u_i$ in the nonsharpening boundary (u_C) is twice as large as that in the sharpening boundary (u_F). Thus, as shown in Fig. 5, the horizontal roof of the originally rectangular zone becomes smaller and smaller, until the nonsharpening trailing boundary, C, overtakes the leading boundary, F. From that moment on the zone will be triangular in shape and will continuously decrease in height.

It is also to be noted that the faster migration of the top part of the nonsharpening boundary is required to preserve the mass balance for the entire zone. This mass balance requires that the area of the trapezium-shaped zone in Fig. 5B equals that of Fig. 5A. The average movement of the trailing edge must therefore be as large as the movement of the leading edge; the average can only be so large when its maximum value (in the top) is larger.

It is now appropriate to discuss the issue of self-sharpening. Rather difficult mathematics has been devoted to this subject. Here the matter is approached in a more intuitive manner.

With the conditions as in Fig. 5 and in case of a negative value for u'_i (or $s_{u,i}$), the application of Eq. (55) (or Eq. (57)) would have an unacceptable consequence: The top part of the trailing boundary would move slower that the lower part, and we would have two or three concentration values at one position. In that case, the trailing boundary becomes the self-sharpening one (Eq. (53) should be applied) and the leading edge the non-self-sharpening one (Eq. (55)).

Yet more general: The condition for self-sharpening is

$$\text{self-sharpening if } (^\rho c_i - ^\sigma c_i)\frac{du_i(c_i)}{dc_i} > 0 \qquad (59)$$

where σ and ρ refer to the subsequent regions separated by the boundary, indexed according to the direction of positive velocity. Anticipating on the topic of mutual interaction, we mention that Eq. (59) also holds for multicomponent systems, provided one takes for the c_i in the derivative the same one as in the other factor.

D. Triangular Zones

As indicated above, the zones, although injected as rectangular plugs, eventually, and often before they reach the detection, deform into triangles. The triangular shape is often quite well discernable in the signal trace of the detector. It is therefore important to have some insight in their movement.

The history course of a triangular zone is approached most conveniently by first focusing on the non-self-sharpening boundary. It behaves exactly the same as when it were part of a trapezoidal zone. In other words, the position, z_c, "of a concentration", of which the value is c_i, relative to the starting position of the corresponding boundary at $t = 0$, can be given as

$$z_c(c_i) = u_c(c_i) t \qquad (60)$$

Note that for a given choice of c_i in Eq. (60) the corresponding point in the boundary may not exist any more because the zone has eroded away too much.

Nevertheless, for judicious choices of c_i, meaningful values for $u_c(c_i)$ can be explicitly found from Eq. (55) and a curve $z_c(c_i)$ versus c_i—that is, one boundary of the zone—can be constructed. At that stage it is unknown where it ends: i.e., it is unknown where the zone crest is. For that purpose a calculation of the self-sharpening boundary would lie at hand; from the intersection of two lines the peak crest would be found. Unfortunately, this leads to complicated calculations, resulting from the fact that the velocity of this steep boundary keeps changing during the history of the triangle.

Therefore, it is easier to use the integral mass balance for the zone. The area of the triangle must always be equal to that of the area (say Q) of the original, rectangular, injected plug, related via the cross section of the capillary to the amount injected. Knowing where the hypotenuse of the rectangular triangle is, and knowing its area, the triangle is determined (Fig. 6).

The effect is very important as the peak width in the electropherogram, and with that the attainable resolving power, is determined by it. It is therefore appropriate to derive a quantitative relation.

Assume again that the dependence of u_i on c_i is linear. The shape of the zone (plotted versus position) is then exactly triangular. One then has

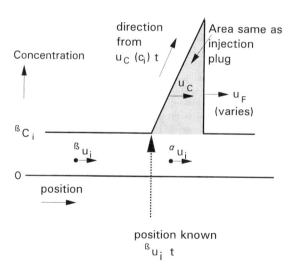

Fig. 6 Calculation of the profile of a zone that has degenerated into a triangle. With the known position at the rear end of the triangle, the known slope of the rear boundary, and the known area, the curve can be constructed.

$$q = \text{Area triangle} = \frac{1}{2} w c_{crest}$$

where q is a measure of the amount injected: in fact $q = Q_i/\phi$, where Q_i is the number of moles injected and ϕ is the cross section of the capillary, but it can also be expressed as $q = c_{inj} L_{inj}$, where c_{inj} is the concentration in the injected solution; w is the width at baseline of the triangular zone (in length); and c_{crest} is the concentration in the maximum. However, the relation between height (concentration) and base width is fixed by the nonlinearity parameter $s_{u,i}$, describing the change of velocity with concentration. Equation (58) gives, when multiplied by time to obtain position rather than velocity,

$$\Delta\text{position} = c_i \, 2 \, s_{u,i} \, t$$

Thus $w = c_{crest} 2 s_{u,i} t$ or $c_{crest} = w/2s_{u,i} t$, and the area of the triangle is

$$q = \frac{w^2}{4 s_{u,i} t} \tag{61}$$

Reversing Eq. (61) the expression of the baseline zone width becomes

$$w = 2\sqrt{q s_{u,i} t} = 2\sqrt{L_{inj} c_{inj} s_{u,i} t} \tag{62}$$

(Note: q is in mol/L · m, $s_{u,i}$ is in m/s/(mol/L), t is in seconds; thus under the root sign the unit is m^2, as should be the case.)

The width, w, appears to be in proportion to the square root of the time. The same has been noted in the case of overloaded chromatography [77,78]. It means that peak broadening due to the dependence of migration rate on concentration behaves in the same way as the familiar dispersion effects, described, e.g., by the plate height; the peak width goes up with the square root of the time or of the traveled distance. The proportion of the two effects remains the same during the history of the triangular zone. It is therefore useful but slightly inaccurate (cf. Ref. 79), to express the width in a dispersion-like form and add it to the other dispersion effects. For this purpose first the variance, σ^2_{EMD}, of the triangle is calculated, using its definition [58]:

$$\sigma^2_{EMD} = \frac{1}{18} w^2 = \frac{2}{9} q s_{u,i} t \tag{62a}$$

Casting this as a plate height contribution gives

$$\sigma^2_{EMD} = H_{EMD} z = H u_i t$$

so that

$$H_{EMD} = \frac{2}{9}\frac{qs_{u,i}}{u_i} = \frac{2}{9}\frac{L_{inj}c_{inj}s_{u,i}}{u_i} \tag{63}$$

which then can be added to other plate height contributions.

In the electropherogram this leads to zone width, w_t, that can be found from Eq. (62) and the substitution $t = L/u_i$:

$$w_t = \frac{w}{u_i} = \frac{2\sqrt{qs_{u,i}L/u_i}}{u_i}$$

$$= 2\sqrt{qs_{u,i}Lu_i^{-3}} \tag{64}$$

E. Mutual Interaction of Ions

The preceding sections were nonspecifically devoted to electrophoresis. It is now needed to describe the electrophoretic factors that cause variation in u_i. The velocity is given by Eq. (1):

$$u_i = {}^a\mu_i E \tag{65}$$

where the superscript a has been added to allow for changes in the average mobility brought about by equilibria that are shifted within the zone with respect to the buffer. For strong ions ${}^a\mu = \mu_i$ = constant, so for those cases the variation of E is the only thing that matters. However, for the general case of weak ions, we must also allow for a variation in ${}^a\mu$. First the variation in E is discussed, and for strong ions this suffices.

What is constant across the tube is the current density, I (at constant diameter, which is assumed). The latter is, according to Ohm's law, equal to $V = V/R/\phi \approx V/\beta\kappa$, where R is the resistance of the liquid in capillary, ϕ is the cross section, and V is the voltage. It may not be constant in time because R may change due to the presence of analyte zones, but we neglect this effect here, arguing that in high-resolution electrophoresis the zones occupy only a small part of the tube. One could circumvent this small difficulty by working with constant current rather than constant voltage.

At each position, z, E is found from the local application of Ohm's law:

$$E = \frac{I}{\kappa} \tag{66}$$

where κ is the conductivity defined in Eq. (4).

From here on the units of I and κ will be changed by a factor equal to the Faraday, F, in order to avoid repetitious occurrences of F that always cancel. In this system (current expressed as mol*elementary charge/second rather than Coulomb/second) the conductivity, κ, becomes

$$\kappa = \sum_{\text{all ions } j} \mu_j z_j c_j \tag{67}$$

It may be clear that the conductivity within a zone, α, will in general not be the same as that in the buffer, β, when the analyte ion i has a nonzero concentration, as i counts in the sum of Eq. (67). Also, other ion concentrations will not be undisturbed in the α-zone. The effects on κ cancel when one ion in the buffer has the same mobility as i, but this can only be the case for one or a few sample ions.

Thus, it is needed to figure out how the buffer ion concentrations change in a zone containing i. There are several, not independent, ways to approach this problem.

F. Moving-Boundary Equations

The first way is to use the set of moving-boundary equations. For n ions (n-1 being buffer components, $i = n$), and a self-sharpening boundary, there are n such equations:

$$^\alpha c_1{}^\alpha u_1 - {}^\beta c_1{}^\beta u_1 = ({}^\alpha c_1 - {}^\beta c_1) u_F$$

$$^\alpha c_2{}^\alpha u_2 - {}^\beta c_2{}^\beta u_2 = ({}^\alpha c_2 - {}^\beta c_2) u_F$$

.
.
.

$$^\alpha c_{n-1}{}^\alpha u_{n-1} - {}^\beta c_{n-1}{}^\beta u_{n-1} = ({}^\alpha c_{n-1} - {}^\beta c_{n-1}) u_F \tag{68}$$

$$^\alpha c_{n=i}{}^\alpha u_{n=1} - {}^\beta c_{n=i}{}^\beta u_{n=i} = ({}^\alpha c_{n=i} - {}^\beta c_{n=i}) u_F$$

For each u the value $\mu E = \mu I/\kappa$ has to be substituted.

Equation (68) demonstrates a peculiar aspect: The u's are functions only of the composition. Therefore a "random" combination of $^\alpha c$'s (with given $^\beta c$'s) in general makes the set of equations inconsistent. The $^\alpha c$'s must obey certain relations. If not, the boundary cannot migrate as an entity, it will be resolved in two or more boundaries where the set of Eq. (68) is consistent, i.e., can be solved. The latter state was called "coherent" by Helfferich and Klein [76], a word that we did not encounter in electrophoretic literature. There one usually describes this situation by "satisfying the Kohlrausch function." However, as will be seen, this is a necessary but, for a case with more than three ions, not a sufficient condition for coherence, i.e., it does not guarantee that Eq. (68) has a solution.

Because of the required coherence one should regard Eq. (68) as equations with, e.g., a set of equations with unknowns $^{\alpha}c_1, ^{\alpha}c_2, \ldots, ^{\alpha}c_{n-1}$ and u_F and all μ's, all $^{\beta}c$'s and $^{\alpha}c_{n=i}$ supposed to be known. The value $^{\alpha}c_{n=i}$, the analyte concentration in the zone, can be used as a starting point in the evaluation, i.e., assumed to be still equal to the injected concentration.

The brute-force approach with the set of boundary equations is useful only for demonstrating the complexity of the problem, and possibly when numerical techniques would be applied. For practical purposes simplifications are needed, which are discussed in the sequel.

G. Kohlrausch Regulating Function (KRF)

Much insight can be gained by using the function introduced by Kohlrausch [80] around the turn of the century, dubbed by him the *beharrliche Funktion*. The commonly used translation is "regulating function," although the word "regulating" does not convey very well the meaning of the German work "beharrlich", reluctant to change. In the symbols used here it reads

$$KRF = \sum_{\text{all ions } j} \frac{z_j c_j}{\mu_j} \qquad (69)$$

with

$$\frac{d(KRF)}{dt} = 0 \qquad (70)$$

Note that all terms are positive, since z_j and μ_j always have the same sign.

The KRF is of great usefulness just because it is "beharrlich": it is independent of time, when observed at a particular point that does not move relative to the liquid. A full treatment is given in Refs. 81–83. Here the proof that it is constant in time is given for strong ions and constant μ's:

$$\frac{d(KRF)}{dt} = \sum_{\text{all ions } j} \frac{z_j}{\mu_j} \frac{dc_j}{dt} \qquad (71)$$

The derivative of each c_j to time follows from the differential mass balance:

$$\frac{dc_j}{dt} = -\frac{d(E\mu_j c_j)}{dz} \tag{72}$$

Thus,

$$\frac{d(KRF)}{dt} = \sum_{\text{all ions } j} \frac{z_j}{\mu_j} \frac{d(E\mu_j c_j)}{dz}$$

$$= -\sum_{\text{all ions } j} z_j \frac{d(Ec_j)}{dz}$$

$$= -E \sum_{\text{all ions } j} z_j \frac{dc_j}{dz} - \frac{dE}{dz} \sum_{\text{all ions } j} z_j c_j \tag{73}$$

The sums in Eq. (73) describe the charge density and the derivative of the charge density, respectively. Because of the electroneutrality, both sums are zero, and with that the time derivative of KRF.

The "regulating" action of KRF can be described as follows. When we inject a sample plug, its KRF usually will be different from that of the BGE. The plug resolves in zones, each traveling with its own velocity through the liquid. The passage of a zone through a given point cannot lead to any change in KRF. Thus, each zone, provided it migrates through the liquid, must have a KRF equal to that of the BGE.

Obviously, there is a problem here when all zones would move, and nothing except BGE would ultimately be left at the starting position. All zones then should have a KRF equal to that of the buffer, so that the sum of their intensities—which should represent the injection composition—cannot reconstitute the injection. So, if the injected solution differs in KRF from the buffer, there must be one zone (at least) accounting for this difference. As a consequence this particular, *stagnant*, zone cannot move. It stays at the starting position (unless it does not exist when by chance of because of deliberate manipulation, the KRF of the injected plug was equal to that of the BGE). The composition of this stagnant zone equals that of the BGE, except for a dilution or concentration factor. In other words, all concentrations in it differ from the BGE ones by the same factor.

This situation forms an extreme and illustrative example of the difference between concentration and component velocity. Imagine a NaCl "buffer," in which a solution of NaCl of another (for the sake of the argument, smaller) concentration is injected. According to the regulating action of KRF, the zone is stagnant, concentration velocities are zero. Meanwhile, Na⁺ and Cl⁻ ions do have their regular electrophoretic velocity; they pass through the stagnant zone; i.e., the identity of the ions in the zone is changing all the time. The concentration in the zone can be kept lower indefinitely (except for diffusion) only because the ions are much faster within it (which allows the product $c_j u_j$ to be the same throughout). They are faster because the conductivity is smaller in the more diluted zone. Analogously to a triathlon, when the athletes are on the running track, they are much farther apart, i.e., of smaller concentration, than when swimming.

The repetitious "through the liquid" in the foregoing text refers to the role of EO flow. When the latter is present, the stagnant zone travels with it. Often with UV detection, and always with any sort of indirect detection, the "stagnant" zone is visible in the electropherogram, sometimes as a distorted peak resulting from refractive index effects.*

The KRF is further very useful for the calculation of zone shape in simple cases. Consider, as an example, an ion I^{i+}, the analyte of charge $i+$ during electrophoresis in a "buffer," β, consisting of P^{p+} and M^{m-}, in concentration $^\beta[P^{p+}] = c$, thus $^\beta[M^{m-}] = cp/m$, because of the electroneutrality condition, EN (note $z_p = p$, $z_M = -m$). The first task is to calculate the concentrations $^\alpha[P^{p+}]$ and $^\alpha[M^{m-}]$ in the zone, α, where a given concentration of I^{i+} prevails.

For the buffer (β) KRF equals:

$$^\beta KRF = \frac{pc}{\mu_P} + \frac{(-m)cp/m}{\mu_M} = pc\left(\frac{1}{\mu_P} + \frac{-1}{\mu_M}\right) \tag{74}$$

For the zone, α, of i we have thus

---

*As it is often practice to express mobilities such that they include the EO flow, $'\mu_j = \mu_j + \mu_{EO}$, the question lies at hand whether Eq. (70) can be derived with $'\mu$'s instead of μ's. This would invalidate all what has been said above about the stagnant peak, because the new KRF could neither change in time. However, Eq. (70) cannot be derived with $'\mu$'s, since Eq. (73) fails: The E acting on the μ_{EO} is not the local one, as intended in Eq. (73), but an average over the capillary.

$$^{\alpha}KRF = \frac{p^{\alpha}[P^{p+}]}{\mu_P} + \frac{(-m)^{\alpha}[M^{m-}]}{\mu_M} + \frac{i^{\alpha}[I^{i+}]}{\mu_I} = {}^{\beta}KRF \qquad (75)$$

This gives us one equation for $^{\alpha}[P^{p+}]$ and $^{\alpha}[M^{m-}]$. Another equation is given by EN:

$$EN = 0 = p\ {}^{\alpha}[P^{p+}] + m\ {}^{\alpha}[M^{m-}] + i\ {}^{\alpha}[I^{i+}] \qquad (76)$$

so that it is possible to solve this case with awkward, but basic algebra:

$$^{\alpha}[P^{p+}] = {}^{\beta}[P^{p+}] - \frac{i\mu_p(\mu_I - \mu_M)}{p\mu_I(\mu_p - \mu_M)}\ {}^{\alpha}[I^{i+}] \qquad (77)$$

$$^{\alpha}[M^{m+}] = {}^{\beta}[M^{m+}] + \frac{i\mu_M(\mu_I - \mu_p)}{m\mu_I(\mu_p - \mu_M)}\ {}^{\alpha}[I^{i+}] \qquad (78)$$

NB: This exercise has been performed by Mikkers et al. [41].

Equation (77) and (78) describe what is called the displacement effect, of importance to arrive to an estimate of the change of migration rate with concentration, but also for indirect detection. The fraction in Eqs. (77) and (78) have been dubbed transfer ratios or displacement ratios. They are often for convenience taken as $-i/p$, the charge ratio for the buffer ion of the same sign (co-ion), and as zero for the ion of opposite sign.

Inspection of Eqs. (77) and (78) shows this to quite a crude approximation, correct only when $\mu_p = \mu_I$. Only in that case the ion I^{i+} displaces it electrochemical equivalent of P^{p+}, while the concentration of M^{m-} remains unaffected.

The values found for the buffer concentrations in the α-zone can be used to find the value of $^{\alpha}\kappa$, by insertion in Eq. (67). This, in turn can be used [41] to find how the component velocity, being $I\mu_I/\kappa$, changes with concentration. Some algebra yields

$$\frac{\Delta\kappa}{^{\beta}\kappa} = \frac{{}^{\alpha}[I^{i+}]}{c}\frac{i(\mu_i - \mu_M)(\mu_I - \mu_P)}{p\mu_I(\mu_P - \mu_M)} \qquad (79)$$

where both the change in κ and the concentration of the analyte have been expressed relative to the buffer values. The RHS of Eq. (79) also describes (for small changes, and with a minus sign) the relative change in the field E.

The change in component velocity, as a function of $^{\alpha}[I^{i+}]/c$, the relative concentration, then follows by insertion of Eq. (79) in Eq. (66):

$$\frac{{}^{\alpha}u_I}{{}^{0}u_I} = -\frac{{}^{\alpha}[I^{i+}]}{c}\frac{i(\mu_I - \mu_M)(\mu_I - \mu_P)}{p\mu_I(\mu_I - \mu_M)} \tag{80}$$

where the index 0 (rather than β) indicates behavior at infinite dilution in the buffer.

The second part of the RHS of Eq. (80) has been indicated as an EMD constant, β_{EMD} [65] (or as $k_{EMD}/{}^0\mu_I$ in Ref. 84). It is related to the slope in velocity used above: $s_{u,i} = E\,{}^0u/c\beta_{EMD}$. Note that it is zero when the mobility of the analyte ion equals that of the co-ion. For triangular zones it follows from the combination with Eq. (63):

$$H_{EMD} = \frac{2}{9}\beta_{EMD}\frac{c_{i,inj}}{c}L_{inj} \tag{81}$$

where c_{inj} and L_{inj} refer to the injection plug.

The KRF can also be used to derive the position of system peaks in three-ion buffers. Consider, as an example, a "buffer." β, consisting of P⁺, M⁻, and N⁻, in concentration ${}^{\beta}[P^+] = c$, ${}^{\beta}[M^-] = \theta c$, and ${}^{\beta}[N^-] = (1-\theta)c$. Imagine that a minor disturbance in the concentration of P, of magnitude Δ_p, travels through the capillary, as zone α, i.e., ${}^{\alpha}[P^+] = {}^{\beta}[P^+] + \Delta_P$. The task is to calculate the concentrations ${}^{\alpha}[M^-]$ and ${}^{\alpha}[N^-]$ in the zone, α. Also these concentrations are expressed relative to the buffer concentrations: ${}^{\alpha}[M^-] = {}^{\beta}[M^-] + \Delta_M$; ${}^{\alpha}[N^-] = {}^{\beta}[N^-] + \Delta_N$.

KRF is a linear addition, so saying that it is the same in α and β is equivalent to saying that the KRF definition expression, Eq. (69), with the Δ's instead to the c's, should be zero:

$${}^{\alpha}KRF - {}^{\beta}KRF = 0 = \frac{\Delta_P}{\mu_P} + \frac{(-1)\Delta_M}{\mu_M} + \frac{(-1)\Delta_N}{\mu_N} \tag{82}$$

In addition, the EN yields

$$0 = \Delta P + \Delta M + \Delta N \tag{83}$$

The solution of this pair of equations for ΔM and ΔN is

$$\Delta_M = +\frac{(\mu_N + \mu_P)\mu_M}{(\mu_N - \mu_M)\mu_P}\Delta_P \tag{84}$$

$$\Delta_N = \frac{(\mu_M + \mu_P)\mu_N}{(\mu_N - \mu_M)\mu_P}\Delta_P \tag{85}$$

essentially the same result as Eqs. (77) and (78).

Application of one moving boundary equation suffices to find the velocity of the zone. Taking component P for that purpose:

$$u_c = I \frac{d\left(\dfrac{c_P \mu_P}{\kappa}\right)}{dc_P} \tag{86}$$

The derivative requires quite some algebra (for which a symbolic manipulator is convenient) as the κ-expression, Eq. (67), not only involves the change of $[P^+]$; also those of $[M^-]$ and $[N^-]$ given by Eqs. (84) and (85) have to be taken into account. The result is

$$u_c = \frac{\mu_M \mu_N - (1-\theta)\mu_M \mu_P - \theta \mu_N \mu_P}{\theta \mu_M + (1-\theta)\mu_N - \mu_P} \tag{87}$$

It is instructive to analyze Eqs. (85) and (87) for some special conditions:

1. When $\mu_M = \mu_N$, Eqs. (84) and (85) degenerate, giving infinity, e.g., for Δ_M/Δ_P. This only means that the disturbance in the zone only involves the exchange of M^- and N^-, while P^+ remains unaffected, i.e., $\Delta_P = 0$. By taking the limit, one finds $\Delta_M = -\Delta_N$. This is a stoichiometric exchange of the two ions with the same charge sign, occurring when the mobilities are the same. Of course the speed of the zone correspond to that common mobility value, as can be ascertained by inspection of Eq. (87). The degeneration of the equations could have been avoided by using Δ_P instead of Δ_M or of Δ_N as one of the unknowns, because μ_P can never be equal to one of the µ's of the negatively charged ions. However, this was not done here, in order to demonstrate that it is immaterial in the treatment which choice for the unknowns is made.

2. When θ approaches 0 (i.e., the fraction of M^- goes to zero), the mobility of the zone approaches that of M^-, corresponding to the behavior of M^- as an analyte. Likewise, when θ approaches 1, μ_C goes to μ_N, in that case N^- is at infinite dilution.

3. With θ intermediate between 0 and 1, and differing μ_M and μ_N, the zone velocity corresponds to a µ between those values.

The above illustrates the possibilities of the use of the KRF for the analysis of strong ion systems. For weak ions, the conductivity effect also plays a role, but in addition the variation of $^a\mu_i$ in Eq. (65) has to be accounted for. For monovalent weak ions, buffer as well as analytes, this can be done by complementing the KRF with another function, ω_2 [81–83], that also can be derived to be constant in time. With this, and the moving boundary equations, cases with weak electrolytes have been studied theoretically by Beckers [85,86] and Xu et al. [84]. However, explicit analytical expressions, when derived, become rather complicated. Moreover, there are a few important limitations. In the first place, with all these approaches it as (as yet?) impossible to account for the contribution of H^+ and OH^- ions to the conductivity and the buffer capacity. This implies a serious drawback as in many CZE analyses pH-values are used (i.e.,< about 3 or > about 11) where such an approximation is quite crude. In the second place the approach offers no solution when there are three or more ions present in the buffer. For example, if there is one analyte, of which the concentration is taken as known, two equations (KRF and EN) are simply not sufficient to solve for the three unknowns.

For all these reasons, in our opinion it is more effective to handle the moving-boundary equations with a more general approach, in which the step to numerical procedures is made earlier. Therefore the variation of $^a\mu_i$ with c_i will be discussed in the next section.

H. Numerical Approaches

Much numerical work has been devoted to electrophoresis. Here, the simulation in the strict sense, where the electrophoretic transport is approached directly by numerically solving the partial differential transport equations, will not be discussed, as it is quite a subject on its own (cf. Ref. 87). What is left to be discussed are the numerical methods to solve the moving-boundary equations, Eq. (68).

Also in this context it is most effective to focus first on the non-sharpening boundary. Then Eq. (68) has to be transformed into a differential form. The step is the same as the one that, in the one-ion case, Eq. (53) has led to Eq. (54), the latter describing the concentration velocity.

Before doing so, it is to be decided which variables are to be considered as "unknowns," and which are to be considered as "derived." The ion concentrations in a solution are not independent, there are relations such as equilibrium constants (including the water equilibrium, and the electroneutrality). Putting them all in the equations leads to a mathematically intractable set of equations. The choice of variables we have used can be described as follows:

For each component, the total concentration c_x is taken as an unknown variable. Concentrations [H$^+$] and [OH$^-$] are not considered as componentns, they are "derived"; i.e., they can be calculated with the equilibrium constants and the electroneutrality. Thus, a sodium phosphate buffer is characterized by two numbers: c_{PO_4} and c_{Na}. For pair emerging in a solution of the equations, c_{PO_4} (= [H$_3$PO$_4$] + [H$_2$PO$_4^-$] + [HPO$_4^{2-}$] + [PO$_4^{3-}$]) and c_{Na} (= [Na$^+$]), the values of [H$^+$] and [OH$^-$] and pH can be found afterwards. The transport of each component is described by its $^a\mu$, which can be found by Eq. (16); the α-values follow via Eq. (15) from the pH.

Of course, alternative choices would be possible, e.g., taking [H$^+$] as a variable, and an ion such as Na$^+$ as "derived," via EN. However, such a scheme is not convenient, as an asymmetry in many equations is introduced. The results should nevertheless be the same. That means the following: When a solution is found, the profiles for a "derived" ion (H$^+$ in our case) can be calculated. These should satisfy the moving-boundary equations for that ion. We have checked that, with positive result, in a few cases. Although we are not able to analyze this mathematically in full depth, it is apparently so that the inclusion of the conductivity contribution of H$^+$ and OH$^-$ (in our scheme) in the calculation of κ, suffices to have the solution obey the moving-boundary equations for these ions.

Next we need to generalize u_c, the concentration velocity. One cannot be sure that the u_c's are the same for all components. However, as indicated above and discussed at length in Ref. 76, the situation where this is the case is nearly always the "natural" condition to which the system evolves, corresponding to separated zones or boundaries. This is indeed the desirable end result in CZE as it is in moving-boundary electrophoresis and isotachophoresis (ITP). In the following this situation, indicated [76] as "coherence" will be assumed from the outset; the phenomena occurring before this coherent state is reached presently go beyond our (author's) mathematical possibilities.

Coherence thus means that for a given point in a non-self-sharpening boundary there is one u_c, being the same, irrespective of which concentration is used to define it. Another way to express this is to state that a given concentration of one component keeps being accompanied by the same set of other concentrations. For self-sharpening boundaries a u_F exists such that the boundary can move without being split.

The result for the multicomponent case is further complicated by the influence exerted by all ions on each u_j, via their effect on conductivity and pH. For the first component, 1, Eq. (53) can be written as:

$$\Delta(c_1 u_1) = u_c \Delta c_1 \tag{88}$$

The LHS is a function of all concentrations (via conductivity and pH). For small Δc_i's, as relevant for the non-self-sharpening boundary, it can be expanded to first order:

$$\Delta(c_1 u_1) = u_1 \Delta c_1 + c_1 \sum_{\text{all ions } k} \frac{\partial u_1}{\partial c_k} \Delta c_k \tag{89}$$

Inserting the average mobilities, and $E = I/\kappa$, this becomes

$$\Delta(c_1 u_1) = I \frac{{}^a\mu_1}{\kappa} \Delta c_1 + c_1 \sum_{\text{all ions } k} \frac{\partial\left(\frac{{}^a\mu_1}{\kappa}\right)}{\partial c_k} \Delta c_k \tag{90}$$

Combining Eqs. (88) and (90) for all ions, i.e., $1 \to i$, the result is, in matrix notation:

$$|\mathbf{A}| \, \Delta \mathbf{c} = u_c \, \Delta \mathbf{c} \tag{91}$$

where $|\mathbf{A}|$ is an $N \times N$ matrix, with elements

$$A_{i,k} = I \left(\frac{\delta(i,k) \, {}^a\mu_k}{\kappa} + c_i \frac{\partial\left(\frac{{}^a\mu_i}{\kappa}\right)}{\partial c_k} \right) \tag{92}$$

and $\Delta \mathbf{c}$ is an N-vector with elements Δc_i.

Equations such as Eq. (91) are a standard problem in linear algebra. There are (apart from pathological behavior) N solutions, with unequal values for u_c. In more physical terms: Only for these values for the concentration velocity the boundary can be coherent, stable, i.e., not resolve in more than one boundary. Note that the same approach was used in chromatography[88–90].

This gives the solution for u_c and the proportions of the Δc_i values as the eigenvalues and eigenvectors, respectively for $|\mathbf{A}|$.

The derivatives of $\partial^a\mu_i/\partial c_k$ and $\partial k/\partial c_k$ can be found, when the species mobilities, $\mu_{i,j}$, and the pk_a values are known, from relations such as Eqs. (15), (16), and (67). Deriving expressions for these derivatives, i.e., the matrix elements, is a laborious exercise, but it has to be done only once, as the result is frozen in computer source code [91]. With the obtained numerical values for $|A|$ the vectors and eigenvalues can be found by standard mathematical packages.

As mentioned, the eigenvalues are concentration velocities. When normal CZE operation is studied, one (or more) of the ions, the analyte(s), would have concentration zero. Inspection of Eq. (92) shows that for that case ($c_i = 0$) one row of the matrix contains zero's, except for $A_{i,i}$ that is equal to $I/\kappa_\beta \, ^a\mu_i$. Then one (or more) eigenvalue has the value $u_i = I/\kappa_\beta/^a\mu_i \, (= \mu_i E)$. This is the velocity of i, in very small concentration, according to Eq. (65).

The eigenvectors satisfy the coherence condition of Ref. 76. They describe the proportions in which the concentrations change in concentration (relative to the starting BGE composition). They are equivalent to the transfer ratios mentioned above in the context of indirect detection, and expressions such as Eqs. (78) and (85). When, in the example of sample I in buffer $P + M$ (Section VII.G), the transfer ratio toward P would be T_P, and that toward M T_M, the eigenvector is $(1, T_P, T_M)$. Note that the vectors are mathematically not defined in size, only the proportions of the elements matter, so $(0.1, 0.1t_P, 0.1t_M)$ would be as good.

The other eigenvalues "belong to the buffer." One of them has a value zero, or very close to that. It corresponds to the stagnant zone, mentioned under the discussion of the KRF. When ions of the same charge sign occur in the BGE, there is always one eigenvalue between the component velocities of the two. Other rules can be found in Refs. 91 and 92.

These "buffer eigenvalues" emerge experimentally, especially when indirect detection is used, as "ghost" or "system" peaks. They are further discussed in Ref. 92. Some recent work on their intensity is reviewed in Section VII.I.

The prediction of electromigration dispersion necessitates a slight extension of the procedure. It is analogous to what has been described in Section VII.D, "triangular zones": the nonsharpening boundary is followed by calculation of u_i for successive values c_i. Here this works quite the same, except that the concentrations of the other ions vary with c_i, according to their element in the eigenvector (transfer ratio) proportions. Thus, for each new c_i-value (starting from zero), all the other c's are changed in proportion, according to the eigenvector. At this new composition the procedure is repeated, yielding a new value for the velocity, etc. Also here, the integral mass balance is used to calculate the position of the sharp boundary.

This procedure yields peak positions as well as displacement ratios and peak asymmetries. It works with complicated buffer systems, where the approach with the KRF fails: With more than two buffer components, two equations (KRF and EN) are not enough to arrive at a solution.

Also, with this method polybasic acids as a buffer present no problem. Most important, probably, is that the contribution of H^+ and OH^- ions to the conductivity and their part in exchange processes are taken into account. In that respect, this procedure is unique; in most other analytical approaches to the problem these influence have to be neglected. High, respectively low pH buffer for which this is of concern are i.a. used in the CZE analyses of sugars and peptides.

The schemes described by Beckers[85,86] and Xu et al. [84] are less general, but follow the same route. The difference is i.a. in the mathematics, i.e., where the breakpoint between algebraic and numerical operations is.

In view of the complexity of the calculations in this type of work, it is reassuring that the calculations described in Refs. 84–86, as well as those explained above in Section VII.G are in agreement with the more general scheme used in Ref. 91. First of all, the eigenvectors given by the general scheme always satisfy (except when $u_c = 0$) the Kohlrausch equation, on which the other treatments are based. Further comparison applies of course only when the conditions for the less general treatments are fulfilled, i.e., intermediate pH values and at most one ionization stage for each constituent. In those cases the results, in terms of peak positions (system peaks included) and peak asymmetry, agree quantitatively, at least in all cases that we had the opportunity to compare. Furthermore, as demonstrated [84–86,91], experimental results follow closely the predictions of these calculations.

pH Shifts in Zone. From these activities an important conclusion emerges: For protolytic analytes and buffer components the change in pH in the zone forms an independent source of EMD; it can enhance or counteract the effect of changing conductivity. Thus, where for strong ions a mobility match rule applies (same values for analyte and co-ion prevent EMD), no such rule can be derived for "weak" ions. However, the equations given in Ref. 84, as well as the program described in Ref. 91 allow a quick assessment of this effect.

In a number of articles [22,39,91,94,95] this approach was taken to predict the shape of electropherograms. In combination with prediction of pK_a and µ-values from peptide structure [22,39], this was even possible without resource to any literature data on the solutes, just the AA sequence and the CZE instrumental parameters sufficed. In all but a few cases quite a reasonable agreement with the experiment was obtained.

I. Intensity of System Peaks

It may be clear from the above that system peaks, often degrading the utility of indirect detection schemes, can be predicted well in terms of position (mobility) and of the eigenvector ("displacement ratios") either by the use of KRF (Section VII.G) or by numerical work (Section VII.H). However, the intensity of the system peaks escaped analysis and prediction thus far.

Recently we have found [96] that such prediction is possible according to the following approach. The composition of the injection solution is considered relative to that of the BGE. The deviations can be taken as an N-vector, the "injection vector," **i**, where N is the number of ions involved in both buffer and sample. Thus, injecting LiCl 7 mmol/L in water, into a "buffer" of NaCl 20 mmol/L gives for the components of the injection vector (in the order Li, Na. Cl): (7, –20, –13). The injected solution is eventually resolved in zones, each having a specific eigenvector, that is, proportions in the concentration deviations relative to the BGE. However, the integral mass balance for each ion requires that the injection vector can be written as a linear sum of zone intensities (k_z), each multiplied by the corresponding eigenvector. Mathematically speaking, that is projection of **i** on the set of eigenvectors. The pertinent equation is

$$\mathbf{i} = \mathbf{k} \cdot |\mathbf{E}| \tag{93}$$

where $|\mathbf{E}|$ is the matrix formed by the N eigenvectors of the system with all the ions involved, and the **k**-vector has the zone intensities k_z as the elements. Equation (93) expresses the fact that all deviations in the injection solution much be accounted for by the sum of the zones.

Thus, finding the **k**-vector when i is known comes down to inverting $|\mathbf{E}|$. The result $\mathbf{k} = \mathbf{i}|\mathbf{E}|^{-1}$ gives the intensities of the system zones as well as those of the analyte zones. The latter could have been found in a simpler way by considering separately the mass balance for each analyte ion involved, as each analyte occurs only on one zone. However, for buffer components the mass balance is usually "distributed" over system zones as well as analyte zones, and the full inversion of **E** is required to obtain the intensities of the system zones. Satisfactory agreement with experimental results was found.

IX. EFFECT OF ELECTROLYSIS ON BUFFER COMPOSITION

The change of pH brought about by the electrode reactions has been the subject of several recent papers [97–99,48]. In the following a brief account is given for a simple buffer and conditions, from which the situation in more complicated situations can be derived easily.

As an example, it is assumed that originally the buffer consists of sodium, Na^+, in a concentration $^0[Na^+]$ and a total amount of acetic acid-acetate of $^0c_{HA}$. Neglecting H^+ and OH^-, we have $^0[HA] = {}^0c_{HA} - {}^0[Na^+]$ and $^0[A^-] = {}^0[Na^+]$, with index 0 for "original."

Usually inert electrodes such as platinum are used that do not take part in the reaction, so that reactants and products of the electrode reaction are either part of the buffer system or gases. The most common reactions are

$$2H_2O + 2e \rightarrow 2OH^- + H_2 \quad \text{at the cathode}$$
$$2H_2O \rightarrow 4H^+ + O_2 + 4e \quad \text{at the anode}$$

It is seen that for each electron passed one H^+ or one OH^- ion is produced. With more complicated electrode reactions this is not much different, except for some more complicated electrochemical reactions.

The amount of charge that each electrode has to handle is determined by the transport of ions through the capillary. For the anode vial: Because the electroneutrality has to be maintained, each A^- arriving and each Na^+ leaving, necessitates that the electrode reaction produces one H^+ ion.

We assume for simplicity that the mobilities of both buffer ions are equal in absolute magnitude and are also equal to the mobility of the sample. Also, it is assumed that the detection is at the end of the capillary, $l = L$.

During the time of one CZE run the sample travels over the length L, so that is also what the buffer ions do: the buffer ion content of the capillary is exactly refreshed during one run. In other words, considering the cathode vial: the Na^+ content of the capillary is discharged into the vial, the A^- content is refreshed from the vial. Thus, the total amount of ions delivered to the vial, or removed from it, are respectively (in mol)

$$q_{Na} = V_{cap}[Na^+], \qquad q_A = V_{cap}[A^-] \qquad (94)$$

and the amount of H⁺ that has to be released by the electrode during the run is equal to the sum of these.

It seems to lie at hand to obtain the resulting change in pH from this amount of H⁺ released, dividing by the volume and by the buffer capacity of the buffer. This gives a reasonable estimate, but is inaccurate since the buffer equation is not accurate in a situation, as here, where the total buffer concentration, c_{HA}, is changing (by the amount q_A/V_{vial}). Therefore it is necessary to go back to the underlying equilibrium calculation.

At any time the relation

$$x = \frac{[HA]}{K[A]} = \frac{C_{HA} - [Na^+]}{K[Na^+]} \qquad (95)$$

holds, where $x = [H^+]$.

Indicating $^0[A^-]/^0c_{HA} = {}^0[Na^+]/^0c_{HA}$ by α (dropping the subscript 0 (α_0) of the nomenclature used in Section III), this becomes

$$x = \frac{1}{K}\frac{1-\alpha}{\alpha} \qquad (96)$$

Before the run we have

$$^0x = \frac{1}{K}\frac{1-\alpha}{\alpha}, \qquad {}^0pH = Log(K) - Log\frac{1-\alpha}{\alpha} \qquad (97)$$

Indicating variables after the run by superscripts R, and indicating V_{cap}/V_{vial} by v, we have

$$^R[Na^+] = {}^0[Na^+] + \frac{V_{cap}\ {}^0[Na^+]}{V_{vial}} = {}^0[Na^+](1+v) \qquad (98)$$

$$^Rc_{HA} = {}^0c_{HA} + \frac{V_{cap}\ {}^0[A^-]}{V_{vial}} = {}^0c_{HA}(1-v\alpha) \qquad (99)$$

Inserting these new values in Eq. (95) one obtains

$$^Rx = {}^0x\ \frac{1 - 2v\frac{\alpha}{(1-\alpha)}}{1+v} \qquad (100)$$

$$^R\text{pH} = {}^0\text{pH} - \text{Log}\frac{1 - 2v\dfrac{\alpha}{(1-\alpha)}}{1+v} \qquad (101)$$

As $v \ll 1$, this can be simplified to

$$^R\text{pH} = {}^0\text{pH} - 0.4343v\frac{1+\alpha}{1-\alpha} \qquad (102)$$

Note: By just applying the buffer equation $\Delta\text{pH} = c_{added}/\beta$, where $\beta = c_{HA}\alpha(1-\alpha)/0.4343$, one obtains

$$\Delta\text{pH} = -0.4343v\frac{2}{1-\alpha} \qquad (103)$$

Notable in Eq. (102) is that hardly any physical parameters such as mobilities, conductivity, or Faraday's constant occur. In part we enforced that by the foregoing assumptions. Still, a more general treatment is easily derived from the above, by introducing dimensionless quantities, e.g., ratio of mobilities, ratio of L/l, etc., α-values different from 0.5, etc. Especially noteworthy is the nonimportance of increasing the buffer concentration: What one gains in buffer capacity is nullified by the increased current. Electroosmotic flow would ameliorate the effect for positive ions, but enhance it for negative ones (because the run times are different). Under all conditions the dominant role is played by the ratio v of capillary and vial volumes.

Common values for V_{cap} and V_{vial} are 2 µL and 500 µL, while α is often, for reasons of good buffer capacity, close to 0.5. Equation (102) then would predict, in this favorable case, a pH change of 0.007 during one run. Assuming that 20 runs can be made during a working day, one would extrapolate to a change of 0.14 pH unit. This would certainly mix up many eletropherograms; before that happens, replacement of the solution is indicated.

It should be noted that the need to change the end buffer vial depends on the presence of osmotic flow; with it the change in pH in that vial remains without effect, as no ions can enter the capillary from that side.

X. PEAK INTEGRALS

The value of integrated signals in CZE need some consideration, as the relations with the amount injected are slightly different from those pertinent to chromatography. This is the more important as quantitation on the basis of integrated signals (rather than of peak heights) is often desirable in CZE for two reasons. First, due to the nearly ubiquitous influence of electromigration dispersion there is often no proportionality between peak height and injected amount. Second, poor signal-to-noise ratio is slightly improved by using areas.

The difference with, e.g., HPLC is a result of the fact that analyte velocity in the detection region is not a constant, as it is in chromatography.

Assume that the detection has zero offset and perfect linear calibration, with sensitivity (slope of calibration) τ. The latter value is thus defined as the ratio of signal S_i (e.g., volts or AU units) to concentration c_i (mol/L).

For HPLC the integrated peak area A_i (e.g., V s) equals

$$A_i = \frac{TQ_i}{F_c} \tag{104}$$

where F_c is the flow rate in the detector (L/s). This seldomly used, but useful equation can be derived by considering the integral mass balance of i:

$$A_i = \int S_i \, dt = \int \tau c_i \frac{1}{F_c} \, dV$$

$$= \frac{\tau}{F_c} \int c_i \, dV = \frac{\tau}{F_c} Q_i \tag{105}$$

where V is the volume flow through.

The substitution $dt = 1/F_c \, dV$, leading to Eq. (104), cannot be made that easily in CZE. With on-column detection the cubic flow rate is irrelevant; it is the linear velocity of the analyte that matters. More importantly, this velocity is not a system constant as F_c is in HPLC. It varies between analytes, it may even vary (electromigration dispersion!) within the zone of one analyte.

Returning again to the use of position rather than time coordinate, we have for CZE

$$\frac{Q_i}{\phi} = c_{i,\text{inj}} L_{\text{inj}} = \int c_i \, dz \tag{106}$$

where ϕ is the area of the cross section of the tube and $c_{i,\text{inj}}$ and L_{inj} are the concentration and the length of the injection plug.

The "easy but inaccurate" way to proceed now is to set $dz = u_i\, dt$ (forgetting that u_i is not constant):

$$\int c_i\, dz = \int \frac{S_i}{\tau} u_i\, dt = \frac{u_i}{\tau} A_i \qquad (107)$$

or

$$A_i = \frac{\tau}{u_i}\frac{Q_i}{\phi} = \tau \frac{L}{t_{i,M}}\frac{Q_i}{\phi} \qquad (108)$$

Equation (104) forms the rationale for multiplying [100–104] the areas given by the data handling system with the migration time. This practice, advisable with hydrodynamic/static injection, eliminates the influence of varying migration times; especially troublesome in this respect are variations in electroosmotic flow. Also, ratios of areas, of importance with the use of internal standards, are no longer dependent on variations in u_i ($t_{i,M}$).

With electrokinetic injection there is a caveat [105]: A larger u_i, e.g., as a result of increased EO flow, increases the injected amount, but it decreases the area according to Eq. (107). The net result is a constant value for the area. Thus, in this case areas and their ratios are better used without any multiplication with $t_{i,M}$.

When one does take the variation of u_i within the zone into account an intractable expression for the time-integrated signal, A_i, results. However, numerical work (not shown) indicates the following. The relative error made by neglecting the variation in u_i depends only on the degree of EMD. For this purpose a value Δ_{EMD} is defined:

$$\Delta_{\text{EMD}} = 2\beta_{\text{EMD}} \frac{c_{i,\text{crest}}}{c_{\text{buffer}}} \quad (\text{cf. Section VII.G}) \qquad (109)$$

Δ_{EMD} is a convenient measure in practice. It is the total peak width in time units divided by the migration time, i.e., the fraction of the time scale from 0 to $t_{i,M}$ occupied by the peak of i. It follows from the numerical exercise:

$$\text{Error} = \frac{2}{3}\Delta_{\text{EMD}} \qquad (110)$$

Thus, in a case with $\Delta_{\text{EMD}} = 0.1$ (an excessive value that nevertheless is observed in a few published electropherograms), one would expect a deviation of 6.7% when comparing the peak area with one that is obtained using a much smaller amount injected.

SYMBOLS

Symbol	Description	Unit	C.q. value
ε_0	permittivity of the vacuum	F m^{-1}	8.85×10^{-12}
ε_r	relative permittivity (dielectric constant)	—	
$\alpha, \beta, \sigma, \rho$	as prefixes: electrophoretic zones		
α_n	fraction of component in form with n protons		
β_{EMD}	EMD constant	—	
β_n	overall formation constant	varies	
δ	relative change in mobility per kelvin	K^{-1}	
δ	thickness of immobile liquid layer on capillary wall	m	
Δ	as prefix: difference, often difference with BGE		
Δc	vector of concentration differences		
Δ_{EMD}	part of electropherogram occupied ed by triangular peak	—	
ΔP	pressure difference across capillary	Pa	
ζ	zeta potential	V	
η	viscosity of buffer solution	Pa s	
κ	Debye constant, 1/Debye length	m^{-1}	
κ	conductivity of solution in Section VIII mol q_0 V^{-1} m^{-1}	Ω^{-1} m^{-1}	
Λ	equivalent conductivity of salt	Ω^{-1} m^2 mol^{-1}	
λ_j	equivalent conductivity of ion j	Ω^{-1} m^2 mol^{-1}	
μ^0	electrophoretic mobility at zero ionic strength	m^2 V^{-1} s^{-1}	
μ_{EO}	mobility of electroosmotic flow	m^2 V^{-1} S^{-1}	
μ_j	electrophoretic mobility of ion j	m^2 V^{-1} s^{-1}	
ρ	charge density	C m^{-3}	
σ	surface charge density	C m^{-2}	
$\sigma_{i,t}$	time standard deviation of zone of component i	s	

$\sigma_{i,z}$	position standard deviation of zone of component i	m
τ	detection sensitivity (response/concentration)	e.g., AU mol^{-1} L
ϕ	area of cross section of capillary	m^2
ψ	electrical potential	V
ω	parameter in relaxation effect	—
A_i	peak area of peak of component i	e.g., AU s
a	distance of closest approach of two ions	m
$^a\mu_i$	average electrophoretic mobility of component i	m^2 V^{-1} s^{-1}
c	concentration	mol L^{-1}
$D\,(D_i)$	diffusion coefficient (of i)	m^2 s^{-1}
d_c	internal diameter of capillary	m
E	electric field strength	V m^{-1}
EMD	electromigration dispersion	—
EN	electroneutrality condition	—
F	Faraday C	96,500
F_c	flow rate	m^3 s^{-1}
f_j	activity coefficient of ion j	—
$H\,(H_i)$	plate height (for component i)	m
$\mathbf{i}$	vector of injection differences	
I	electrical current density in Section VIII	A m^{-2} mol q_0 m^{-2}
IS	ionic strength	mol m^{-3}
$\mathbf{k}$	vector of zone intensities	
k'	retention factor, mass adsorbed over mass in solution	—
K	equilibrium constant	mol^{-1} L (here)
k_β	heat conductivity of buffer	W m^{-1} K^{-1}
k_B	Boltzmann's constant	J/K
KRF	Kohlrausch regulating function (Eq. 68)	mol L^{-1} m^{-2} V s
l	length to detection position	m
L	total length of capillary	m
L_{det}	length of detection area	m
L_{inj}	length of injected sample plug	m
MM	molecular mass	Dalton

n	number of amino acid residues in peptide	—
N_i	theoretical plate number for component i	—
$Q\,(Q_i)$	injected amount (of component i)	mol
q_o	elementary charge 1.6×10^{-19}	C
R_j	radius of ion j	m
$R_{k,i}$	resolution between components k and i	—
S_i	signal in detector from component i	e.g., absorbance unit
$s_{u,i}$	derivative of velocity with respect to concentration	m s^{-1} mol^{-1} L
t	time	s
T	transference number	
$^t\mu_i$	total averaged electrophoretic mobility of component i, $\mu_{EO} + {}^a\mu_i$	m^2 V^{-1} s^{-1}
$t_{i,M}$	migration time of zone of component i	s
u_c	concentration velocity in a non-self-sharpening boundary	m s^{-1}
u_F	velocity of self-sharpening boundary	m s^{-1}
u_i	velocity of component i	m s^{-1}
v	volume of capillary/volume buffer in vial	—
V	applied voltage in Section IX	V volume
V_{cap}	volume of capillary	m^3
V_{vial}	volume of buffer in vial	m^3
w	width at base of triangular zone	m
x	concentration or activity of H$^+$	mol L^{-1}
z_j	formal charge of ion j	
[X]	concentration of species X	mol m^{-3}
|**A**|	matrix describing transport of components	
|**E**|	matrix consisting of eigenvectors	

REFERENCES

1. P. Righetti, in E. Heftmann, Ed., *Chromatography, Part A*, Elsevier, Amsterdam, 1992.
2. E. Kenndler and H. Poppe, *J. Capillary Electrophoresis 1*, 144 (1994).
3. J. Noolandi, *Ann. Rev. Phys. Chem. 43*, 237 (1992).
4. Ch. Schwer, *Chromatographia 78*, 1 (1995).
5. Y-H. Chu, L. Z. Avilla, H. A. Biebuck, and G. M. Whitesides, *J. Org. Chem. 57*, 3524 (1992).
6. M. H. A. Busch, L. B. Carels, H. F. M. Boelens, J. C. Kraak, and H. Poppe, *J. Chromatogr. A*, in press.
7. J. Vindevogel and P. Sandra, *Introduction to Micellar Electrokinetic Chromatography*, Hüthig Buch Verlag GmbH, Heidelberg, 1992, pp. 179–180.
8. R. J. Hunter, *Foundations of Colloid Science*, Clarendon Press, Oxford, 1987.
9. J. Bartels et al. (eds.), *Landolt-Börnstein Zahlenwerte und Funktionen*, 7. Teil, Springer-Verlag, Berlin, 1960; see also: B. E. Conway, *Electrochemical Data*, Elsevier, Amsterdam, 1952.
10. T. Hirokawa, M. Nishino, N. Aoki, Y. Kiso, Y. Sawamoto, T. Yagi, and J-I. Akiyama, *J. Chromatogr. 271*, D1–D106 (1983).
11. T. Hirokawa, S. Kobayashi, and Y. Kiso, *J. Chromatogr. 318*, 195 (1995).
12. T. Hirokawa, Y. Kiso, B. Gas, I. Zuskova, and J. Vacik, *J. Chromatogr. 628*, 283 (1993).
13. J. P. Pospichal, P. Gebauer, and P. Bocek, *Chem. Rev. 89*, 419 (1989).
14. R. A. Robinson and R. H. Stokes, *Electrolyte Solutions*, Butterworths, London, 1959.
15. M. A. Survay, D. M. Goodall, S. A. C. Wren, and R. C. Rowe, *J. Chromatogr. A 741*, 99 (1993).
16. H. Falkenhagen, *Theorie der Elektrolyte*, S. Hirzel Verlag, Leipzig, 1971.
17. T. A. A. M van de Goor, P. S. L. Janssen, J. W. van Nispen, M. J. M. van Zeeland, and F. M. Everaerts, *J. Chromatogr. 545*, 379 (1990).
18. J. L. Beckers, F. M. Everaerts, and M. T. Ackermans, *J. Chromatogr. 537*, 407 (1991).
19. E. V. Dose and G. Guiochon, *J. Chromatogr. A 652*, 263 (1993).
20. A. Tiselius, *Nova Acta Regiae Soc. Sci. Uppsaliensis 7*, 1 (1930).

21. A. Cifuentes and H. Poppe, *Electrophoresis 16*, 2051 (1995).
22. A. Cifuentes and H. Poppe, *J. Chromatogr. A 680*, 321 (1994).
23. E. C. Rickard, M. M. Strohl, and R. G. Nielsen, *Anal. Biochem. 197*, 197 (1991).
24. N. Adamson, P. F. Riley, and E. C. Reynolds, *J. Chromatogr. 646*, 391 (1993).
25. B. Skoog and A. Wichman, *Trends Anal. Chem. 5*, 82 (1986).
26. A. Sillero and J. M. Ribeiro, *Anal. Biochem. 179*, 319 (1989).
27. B. J. Compton, *J. Chromatogr, 559*, 357 (1991).
28. A. E. Martell and R. M. Smith, in *Critical Stability Constants: Amino Acids*, Vol. 1, Plenum Press, London, 1974.
29. D. D. Perrin, in *Dissociation Constants of Organic Bases in Aqueous Solution*, Butterworths, London, 1972.
30. E. P. Serjeant and B. Dempsey, in *Ionization Constants of Organic Acids in Aqueous Solution*. IUPAC Chemical Data Series No. 23, Pergamon Press, Oxford, 1979.
31. N. Bjerrum, *Z. Phys. Chem. 106*, 219 (1923).
32. C. Tanford and J. G. Kirkwood, *J. Am. Chem. Soc. 79*, 5333 (1957).
33. J. B. Matthew, *Ann. Rev. Biophys. Biophys. Chem. 14*, 387 (1985).
34. M. K. Gilson, A. Rashin, R. Fine, and B. Honing, *J. Mol. Biol. 184*, 503 (1985).
35. R. Tijssen and J. Bos, in F. Dondi and G. Guiochon, Eds., *Theoretical Advancement in Chromatography and Related Separation Techniques*, Kluwer, Dordrecht, 1992, pp. 397–441.
36. R. E. Offord, *Nature 211*, 591 (1966).
37. B. J. Compton, *J. Chromatogr. 559*, 357 (1991).
38. B. J. Compton and E. A. O'Grady, *Anal. Chem. 63*, 2597 (1991).
39. A. Cifuentes and H. Poppe, *Electrophoresis 16*, 516 (1995).
40. F. E. P. Mikkers, F. M. Everaerts, and Th. P. E. M. Verbruggen, *J. Chromatogr. 169*, 1 (1979).
41. F. E. P. Mikkers, F. M. Everaerts, and Th. P. E. M. Verbruggen, *J. Chromatogr. 169*, 11 (1979).
42. J. W. Jorgenson and K. D. Lukacs, *J. Chromatogr. 218*, 209 (1981).
43. J. W. Jorgenson and K. D. Lukacs, *Anal. Chem. 53*, 1298 (1981).
44. A. Manz, D. J. Harrison, E. M. J. Verpoorte, J. C. Fetinger, A. Paulus, and H. M. Widmer, *J. Chromatogr. 593*, 253 (1992).
45. A. Manz, C. S. Effenhauser, N. Burggraf, E. Verpoorte, D. E. Raymond, and H. M. Widmer, *Analusis 22*, 25 (1994).
46. S. C. Jacobson, R. Hergenöder, L. B. Koutny, R. J. Warmack, and J. M. Ramsey, *Anal. Chem. 66*, 1107 (1994).

47. A. T. Wooley and R. A. Mathies, *Anal. Chem. 67*, 3676 (1995).
48. H. Corstjens, to be published.
49. X. Xu, W. Th. Kok, and H. Poppe, *J. Chromatogr. A*, in press.
50. J. C. Giddings, *Unified Separation Science*, Wiley, New York, 1991.
51. E. Kenndler and Ch. Schwer, *J. Chromatogr. 595*, 313 (1992).
52. A. W. Moore and J. W. Jorgenson, *Anal. Chem. 65*, 3550 (1993).
53. J. H. Knox and I. H. Grant, *Chromatographia. 24*, 135 (1987).
54. J. H. Knox, *Chromatographia 26*, 329 (1988).
55. H. Poppe, A. Cifuentes, and W. Th. Kok, *Anal. Chem. 68*, 888 (1996).
56. M. M. Dittman and G. Rozing, *J. Chromatogr. A 744*, 63 (1996).
57. A. S. Rathore and Cs. Horvath, *J. Chromatogr. A 743*, 231 (1996).
58. J. C. Sternberg, in J. Calvin Giddings, Ed., *Advances in Chromatography*, Vol. 2, Marcel Dekker, New York, 1966.
59. R. Virtanen, *Acta Polytechn. Scand. 123*, (1974).
60. J. H. Knox and I. H. Grant, *Chromatographia 24*, 135 (1987).
61. G. D. Roberts, P. H. Rhodes, and R. S. Snyder, *J. Chromatogr. 480*, 35 (1989).
62. E. Grushka, R. M. McCormick, and J. J. Kirkland, *Anal. Chem. 61*, 241 (1989).
63. H. Poppe and J. C. Kraak, *J. Chromatogr. 282*, 399 (1983).
64. M. S. Bello, E. I. Levin, and P. G. Righetti, *J. Chromatogr. 652*, 329–336 (1993).
65. A. Cifuentes, X. Xu. W. Th. Kok, and H. Poppe, *J. Chromatogr. A 716*, 141–156 (1995).
66. A. Cifuentes, W.Th. Kok, and H. Poppe, *J. Microcol. Separations 7*, 365 (1995).
67. S. V. Ermakov, M. Y. Zhukov, L. Capelli, and P. G. Righetti, *Anal. Chem. 66*, 4034 (1994).
68. R. F. Meyer, P. B. Champlin, and R. A. Hartwick, *J. Chromatogr. 211*, 433 (1983).
69. M. Martin and G. Guiochon, *Anal. Chem. 57*, 561 (1985).
70. J. C. Giddings, J. P. Chang, M. N. Myers, J. M. Davis, and K. D. Caldwell, *J. Chromatogr. 211*, 433 (1983).
71. A. Cifuentes and H. Poppe, *Chromatographia 39*, 391 (1994).
72. H. Poppe, *Analusis 22*, 22 (1994).
73. J. F. K. Huber and A. I. M. Keulemans, *Z. Anal. Chem. 205*, 263 (1964).
74. H. Svensson, *Acta Chim. Scand. 2*, 841 (1948).

75. R. A. Alberty, *J. Am. Chem. Soc. 72*, 2361 (1950).
76. F. Helfferich and G. Klein, *Multicomponent Chromatography*, Marcel Dekker, New York, 1970.
77. H. Poppe and J. C. Kraak, *J. Chromatogr. 255*, 395 (1983).
78. J. H. Knox and H. M. Pyper, *J. Chromatogr. 363*, 1 (1986).
79. S. Golshan-Shirazi and G. Guiochon, *J. Chromatogr. 506*, 495 (1990).
80. F. Kohlrausch, *Ann. Phys. 62*, 209 1897.
81. V. P. Dole, *J. Am. Chem. Soc. 67*, 1119 (1945).
82. L. G. Longworth, *J. Phys. Chem. 51*, 171 (1947).
83. L. M. Hjelmeland and A. Chrambach, *Electrophoresis 3*, 9 (1982).
84. X. Xu, W. Th. Kok and H. Poppe, *J. Chromatogr. A 742*, 211 (1996).
85. J. L. Beckers, *J. Chromatogr. A 693*, 347 (1995).
86. J. L. Beckers, *J. Chromatogr. A 696*, 285 (1995).
87. R. A. Mosher, D. A. Saville, W. Thormann, *The Dynamics of Electrophoresis*, VCH. New York, 1992.
88. P. C. Mangelsdorff, Jr.; *Anal. Chem. 38*, 1540 (1966).
89. F. Riedo and E. Sz. Kovats, *J. Chromatogr. 239*, 1 (1982).
90. J. Crommen, G. Schill, and P. Herné, *Chromatographia 25*, 2642 (1988).
91. H. Poppe, *Anal. Chem. 64*, 1908 (1992).
92. H. Poppe, in M. Khaledi, ed., *Capillary Electrophoresis*, Wiley, New York, 1997.
93. H. F. Yin, C. K. Templin. and D. McManigill, *J. Chromatogr. A 744*, 45 (1996).
94. G. J. M. Bruin, A. C. van Asten, X. Xu, and H. Poppe, *J. Chromatogr. 608*, 97 (1992).
95. S. J. Williams, E. T. Bergström, D. M. Goodall, H. Kawazumi, and K. P. Evans, *J. Chromatogr. A 636*, 39 (1993).
96. H. Sellmeijer and H. Poppe, paper presented at the 21st International Symposium on High Performance Liquid Phase Separations, Birmingham, UK, June 22–27, 1997.
97. M. A. Strege and A. L. Lagu, *J. Liq. Chromatogr. 16*, 51 (1993).
98. A. Vinther and H. Soeberg, *J. Chromatogr. 589*, 315 (1992).
99. J. Vindevogel and P. Sandra, *J. High Resolut. Chromatogr. 14*, 795 (1991).
100. S. Hjerten, K. Elenbring, F. Kilar, J-L. Liao, A. J. C. Chen, C. J. Siebert, and M-D. Zhu, *J. Chromatogr. 403*, 47 (1987).
101. M. Zhu, D. L. Hansen, S. Burd, and F. Gannon, *J. Chromatogr. 480*, 311 (1989).

102. M. W. F. Nielen, *J. Chromatogr.* 588, 3212 (1991
103. D. M. Goodall, S. J. Williams, and D. K. Lloyd, *Trends Anal. Chem.* 10, 272 (1991).
104. K. D. Altria. *Chromatographia* 35, 177 (1993).
105. J. H. van der Moolen, H. F. M. Boelens, H. Poppe, and H. C. Smit, *J. Chromatogr.*, in print.
106. J. Frenz, S. L. Wu, and W.S. Hancock, *J. Chromatogr.* 480, 379 (1989).
107. J. Bongers, T. Lambros, A. M. Felix, and E. P. Heimer, *J. Liq. Chromatogr.* 15, 111 (1992).
108. H. J. Issaq, G. M. Janini, I. Z. Atamna, G. M. Muschik, and J. Lukszo, *J. Liq. Chromatogr.* 15, 1129 (1992).
109. M. H. J. M. Langenhuizen and P. S. L. Janssen, *J. Chromatogr.* 638, 311 (1993).
110. J. R. Florance, Z. D. Konteatis, M. J. Macielag, R. A. Lessor, and A. Galdes, *J. Chromatogr.* 559, 391 (1991).
111. Z. Deyl, V. Rohlicek, and M. Adam, *J. Chromatogr.* 480, 371 (1989).
112. S. K. Basak and M. R. Ladisch, *Anal. Biochem.* 226, 51 (1995).
113. M. A. Survay, D. M. Goodall, S. A. C. Wren, and R. C. Rowe, *J. Chromatogr. A* 636, 81 (1993).
114. F. Nyberg, M. D. Zhu, J. L. Liao, and S. Hjerten, in C. Schafer-Nielsen, Ed., *Electrophoresis '88*, VCH, Weinheim, 1988, pp. 141–150.
115. S. Hjerten, L. Valtcheva, K. Elenbring, and D. Eaker, *J. Liq. Chromatogr.* 12, 2471 (1989).
116. Z. Deyl, V. Rohlicek, and R. Struzinsky. *J. Liq. Chromatogr.* 12, 2515 (1989).
117. P. D. Grossman, J. C. Colburn, and H. H. Lauer, *Anal. Biochem.* 179, 28 (1989).
118. M. Castagnola, L. Cassiano, R. Rabino, D. V. Rossetti, and F. A. Bassi, *J. Chromatogr.* 572, 51 (1991).
119. P. J. Pennino, *BioPharm.* 9, 41 (1989).

7
Separation of DNA by Capillary Electrophoresis

András Guttman *Genetic BioSystems, Inc., San Diego, California*
Kathi J. Ulfelder* *Beckman Instruments, Inc., Fullerton, California*

I.	INTRODUCTION	302
II.	THEORY	303
III.	METHODS	306
	A. Buffer and Gel Systems	306
	B. Injection and Sample Preparation	310
	C. Detection	312
	D. Fraction Collection	316
	E. Quantitation	316
	F. Sizing Techniques	318
IV.	SELECTED APPLICATIONS	319
	A. Short Oligomers	319
	B. Antisense DNA	321
	C. DNA Sequencing	323

Current affiliation: Caliper Technologies Corp., Palo Alto, California.

301

	D. Restriction Digests	326
	E. Polymerase Chain Reaction Products	326
	F. Conformational Polymorphisms	329
	G. Large DNA	331
V.	FUTURE DIRECTIONS	333
	ACKNOWLEDGMENTS	334
	REFERENCES	334

I. INTRODUCTION

Electrophoresis (from the Greek words *elektron*=electron and *phoresis*=carrying) is a separation method in which charged molecules migrate under the influence of an electric field at various rates depending on their charge-to-mass ratio. In the middle of the 1930s, Arnie Tiselius introduced the moving-boundary electrophoresis method to separate human serum into albumin, α-globulin, β-globulin, and γ-globulin [1]. He demonstrated the potential of electrophoresis as a research tool for his pioneering work, and was awarded a Nobel Prize 10 years later. In the 1960s, when electrophoresis-based separation techniques began using some kind of supporting media such as polyacrylamide or agarose gels, this separation method became a standard technique for nucleic acid and protein analysis. The use of the gel matrix had two roles: it acted as an anticonvective medium and as a molecular sieve. Gels reduce convective transport and diffusion, therefore the separated sample components remain positioned in sharp zones during the run. In the meantime, the sieving matrix separates biopolymers according to their size. Classical gel electrophoresis (rod or slab) is often time-consuming and labor-intensive; additionally; it is only qualitative and quantitation is often difficult.

In its short history of a little more than a decade, capillary electrophoresis (CE) has shown great applicability in modern bioanalytical and biopharmaceutical analysis [2,3]. As a relatively new, powerful separation technique, CE is ideally suited for handling very small sample volumes, which is a demand in bioanalytical research, e.g., in biotechnology and in various clinical, diagnostic, genetic and forensic applications. Capillary electrophoresis based analysis of single- and double-stranded DNA molecules has led to considerable advances in molecular biology. Synthetic single-stranded oligonucleotides (ss DNA) are used in a variety of applications: as probes for gene isolation and diagnostics; as primers in the polymerase chain reaction (PCR); and in work dealing with cloning, gene alteration, and antisense therapeutics. The analysis of double-stranded DNA (ds DNA) fragments, such as those produced by restriction digests and PCR [4,5] has led to direct detection and quantification of viruses, tracking of inheritance patterns in a family, diagnosis of numerous genetic diseases, identification of individuals in forensic applications, and aid in mapping the human genome.

Capillary electrophoresis is quite similar to high-performance liquid chromatography (HPLC) in its ease of use, high resolution, speed, on-line detection, and full automation capability and is actually viewed as an automated and instrumental approach to classical electrophoresis [6]. Several recent textbooks cover CE theory, instrumentation, and applications [7–9]. It can be expected that CE will replace classical electrophoretic techniques for DNA analysis, e.g., in various clinical, diagnostic, genetic, forensic, and biotechnology-related applications.

The first papers on the use of CE for DNA separation were published in the late 1980s by Brownlee and co-workers [10] and Karger and co-workers [9]. These publications addressed the application areas of DNA separations by free-solution methods and gel-filled capillaries. Since then, it is more and more evident that almost all methods developed for slab gel electrophoresis separation of DNA molecules can easily be transferred to a capillary format, with the advantages of fast analysis with high resolution, full automation, and online data acquisition and storage capability. Additionally, multiple injections can be made from only microliters of sample, enabling easy validation of analytical methods. High-sensitivity detection systems, such as laser-induced fluorescence (LIF) opened new, so far unimaginably low, detection limits in the zeptomole range [11].

This chapter describes the main principles of the various CE methods for DNA analysis and their applications for oligonucleotides and ds DNA fragments.

II. THEORY

When a uniform electric field (E) is applied to a polyion, in this instance the DNA molecule, with a net charge of Q, the electrical force (F_e) is

$$F_e = QE \tag{1}$$

However, a frictional force (F_f) acts in the opposite direction, especially in viscous media such as crosslinked gel or noncrosslinked gel/polymer network solutions:

$$F_f = f\left(\frac{dx}{dt}\right) \quad (2)$$

where f is the translational friction coefficient and dx and dt are the distance and time increments, respectively. Thus, the motion of the charged solute mediated by the applied electric field can be expressed as

$$m\left(\frac{d^2x}{dt^2}\right) = QE - f\left(\frac{dx}{dt}\right) \quad (3)$$

Equation (3) shows that the product of the mass (m) and acceleration (d^2x/dt^2) is equal to the difference of the electrical and frictional forces. When the force from the applied electric field on the charged solute is counterbalanced by the frictional force, the solute will move with a steady state velocity (v):

$$v = \frac{dx}{dt} = \frac{QE}{f} f \quad (4)$$

The translational friction coefficient is influenced by the absolute temperature (T), especially in noncrosslinked, low-viscosity polymer networks:

$$f = C_1 \exp\left(\frac{1}{T}\right) \quad (5)$$

where C_1 is a constant for a given shape solute.

The electrophoretic mobility (μ) is defined as the velocity per unit field strength [4]:

$$\mu = \frac{v}{E} \quad (6)$$

Shape, size, or net charge variances lead to different electrophoretic mobilities and provide the basis of electrophoretic separation. The retardation of the DNA molecules in capillary gel electrophoresis is a function of the separation polymer concentration (P) and the retardation coefficient (K_R):

$$\mu = \mu_0 \exp(-K_R P) \tag{7}$$

where μ is the apparent electrophoretic mobility and μ_0 is the free-solution mobility of the analyte [12,13].

Sieving, in the classical sense, only exists when the average pore size of the matrix is similar to that of the hydrodynamic radius of the migrating analyte [14]. In constant polymer concentrations, the retardation coefficient (K_R) is proportional to the molecular mass of the analyte [15] (Fig. 1, Ogston sieving regime):

$$\mu \sim \exp(-MW) \tag{8}$$

In this case, the logarithm of the solute's mobility is a linear function of the molecular mass. Then the plots of mobility versus gel concentration (Ferguson plots [16]) are linear and cross each other at zero gel concentration. The Ogston theory [14] assumes that the migrating particles behave as unperturbed spherical objects having similar size as the pores of the sieving matrix. However, it has been reported that large, flexible-chain biopolymer molecules, such as DNA, still migrate through polymer networks with significantly smaller pore sizes [17].

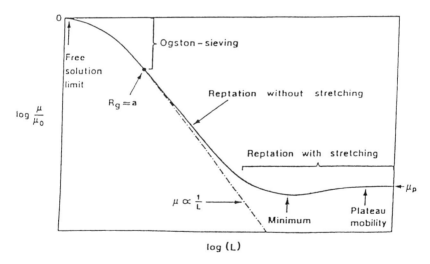

Fig. 1 Schematic representation of the relationship between the logarithmic normalized electrophoretic mobility (μ/μ_0) and the molecular size (L). (Reprinted with permission from Noolandi, *Annu. Rev. Phys. Chem.*, *43*, 237 (1992).)

This phenomenon is explained by the reptation model (Fig. 1, reptation without stretching regime) which describes the migration of the polyelectrolyte as "headfirst, snakelike" motion through the pores of a gel matrix [17–20]. The reptation model suggests an inverse relationship between the mobility of the migrating analyte and its molecular mass [19]:

$$\mu \sim \frac{1}{MW} \tag{9}$$

By the application of higher electric field strengths, reptation with stretching prevails (Fig. 1, reptation with stretching regime) and the mobility of the analyte is described by

$$\mu \sim \left(\frac{1}{MW} + bE^a \right) \tag{10}$$

where a can be between unity [21] and 2 [15], and b is a function of the mesh size of the sieving matrix, and the charge and segment length [15] of the migrating polyions. Since physical gels contain entangled polymers that are not bonded chemically, they can disentangle, allowing motions usually restricted in a crosslinked gel. The theory of constraint release [20] predicts that transient pore size leads to matrix independent mobility of large DNA chains. This effect and the reptation with stretching puts a size limit of DNA separations at about 20 kilobase pair (kbp) at low field strengths. A common way to recognize the Ogston, the reptation, or the reptation with stretching regimes is to plot the log solute mobility versus log solute's molecular mass curves [18,22,23].

III. METHODS

A. Buffer and Gel Systems

The buffer system is one of the most important variables in the CE separation of nucleic acids. Separation of nucleosides, nucleotides, and smaller DNA molecules, such as single-stranded oligonucleotides of less than 10 bases in length, have been demonstrated employing open tubular format by capillary zone electrophoresis [9] or micellar electrokinetic chromatography [24] (Table 1).

In the capillary zone electrophoresis (CZE) separation mode the

Table 1 Different CE Separation Modes

Mode	Separation mechanism	Applications
Free solution		
Capillary zone Electrophoresis (CZE)	Charge-to-mass ratio	Bases, nucleosides, nucleotides, small oligonucleotides, DNA damage
Micellar electrokinetic Chromatography (MEKC)	Charge-to-mass ratio, Partitioning into micelles	Same as CZE
Capillary isotachophoresis (cITP)	Moving-boundary/ displacement	Preconcentration technique for CZE, MEKC
Capillary isoelectric Focusing (cIEF)	Isoelectric point	Proteins
Gel, polymer network		
Capillary gel electrophoresis (CGE)	Molecular sieving, reptation	Oligonucleotides, primers, probes, antisense DNA, PCR products, large ds DNA, point mutations, DNA sequencing

analyte molecules are separated in free solution, with no gel or sieving matrix in the capillary. Here the separation is based on the differences in the charge-to-mass ratio of the solute molecules, and selectivity is attained from the mobility differences of the analytes. Due to the very similar charge density of common size DNA molecules, CZE is only appropriate for the separation of bases, nucleosides and nucleotides. Since the pH of the buffer system determines the degree of ionization of the analytes, selection of the appropriate buffer pH is crucial for their mobilization. The characteristics of the inner surface of the capillary column (uncoated vs. coated) is also very important in the separation of nucleotides. Different buffer systems should be employed for DNA separations by CZE when bare fused silica or surface-modified (coated) capillaries are used.

Micellar electrokinetic chromatography (MEKC) takes advantage of differential partitioning of analytes into a pseudostationary phase of micelles [24]. Both, anionic and cationic surfactants have been utilized as buffer additives for the separation of DNA molecules [25]. Under neutral pH conditions, the bases and nucleosides are not ionized and their separation is governed by the differential partitioning in the micelles. Addition of 1.0 M glucose to micellar buffer systems increases the apparent selectivity by decreasing the partitioning coefficient of the analytes with hydrophilic functional groups, and decreases their residence time in the micellar phase. Methanol can be added to MEKC buffer systems in order to increase the separation time window. In addition to the use of anionic surfactants such as SDS, certain cationic surfactants such as cetyltrimethymmonium bromide (CTAB) and dodecyltrimethylammonium bromide (DTAB), have also been utilized as buffer additives, albeit they change the direction of the electroosmotic flow (EOF)[26].

For CE separations of larger ss DNA and ds DNA fragments, a sieving matrix is required in the buffer system to obtain the separation of the almost equal charge-to-mass ratio DNA molecules. Denaturants, such a urea, formamide, etc. are usually used in ss DNA separations, in DNA sequencing gels and in some instances for ds DNA separations [27]. Nondenaturing conditions are used in most ds DNA separations (DNA restriction fragments, PCR products) and in cases when point mutation detection in ss DNA molecules is required [28]. In this latter case, differences based on the shape, size and charge of the analytes are exploited.

Gels may vary from viscous fluids to solids [29], and the two approaches that have been utilized for DNA analysis by sieving employ chemical gels and physical gels. The chemical or crosslinked gels have well defined pore structure determined by the amount of crosslinker added during the polymerization. These gels are usually chemically bound to the inner wall of the capillary in order to prevent EOF initiated disruption of their structure. Since the pore size of the crosslinked gels is set by the amount of the monomer and crosslinker used, their pore size cannot be varied after polymerization. Also, these gels are extremely heat sensitive and only a minor perturbation, originating, e.g., from the injected samples, can damage them (bubble formation). However, the use of chemical gels ensures very high resolving power for lower base number ss DNA. Because of their extremely high viscosity, crosslinked gels are not replaceable in the capillary; therefore, they should be polymerized within, prior to use. Most commonly acrylamide monomer is used, crosslinked with N,N'-methylene bisacrylamide (BIS) to form chemical gels for capillary gel electrophoresis (CGE) separations.

The other type of gel media that recently become widely used in CGE is the so-called physical gel or polymer network. These are hydrophilic, noncrosslinked, linear polymers with a flexible dynamic pore structure. Usually, they are not attached to the inside wall of the capillary, and their pore size can be easily modified, even during the actual separation by changing, e.g., the column temperature. Noncrosslinked polymer networks are not heat sensitive. Any artifacts originating from injection or developed during the separation can be easily removed by simply replacing the matrix in the capillary. These replaceable gels are very popular for the separation of large molecular weight ds DNA fragments [30]. In the early days, linear (noncrosslinked) polyacrylamides (PA) and various cellulose derivatives (methyl, ethyl, hydroxypropyl, etc.) were used to separate larger DNA restriction fragments [31]. Agarose, which is very popular in conventional slab gel separations of large DNA fragments, can also be filled into capillaries and used as separation media [32]. Poly(N-acryloyl aminoethoxyethanol) (poly[AAEE]), was also employed as sieving polymer and as capillary coating material [23]. Polythylene oxide (PEO) was successfully applied to DNA separations, particularly to DNA sequencing [33]. Interestingly, extremely dilute polymer solutions, with concentrations less than 0.001% could also be applied to separations of larger ds DNA fragments, suggesting that an entangled polymer network is not a prerequisite for separation of polyionic biopolymers [34]. Important to note, that the use of noncrosslinked hydrophilic polymer solutions allows the applicability of the quantitative pressure injection mode, in contrast to the crosslinked gels where, due to their extremely high viscosity, the electrokinetic injection mode is the only viable method.

Good reproducibility of DNA separations with gel or polymer network filled capillary columns is almost always assured with the use of coated capillaries. In fact, a good coating is essential with the use of replaceable polymer matrices. These coatings should be stable around the separation pH of 7–9, minimize EOF, and prevent the interaction of DNA molecules with the silanol groups on the inner wall surface. Different coatings have been utilized in CGE, including crosslinked and noncrosslinked polyacrylamide, dextran, polyethylene glycol, polyvinyl alcohol, and various cellulosic derivatives [35–39].

B. Injection and Sample Preparation

The two basic injection techniques in CE are the hydrodynamic injection, which can be mediated by pressure, vacuum, or buffer vial level differences, and the electrokinetic injection. The first injection method can be easily applied in open tubular low-viscosity physical polymer network filled columns, but cannot be applied to high-viscosity polymers or crosslinked gel-filled capillaries. Sample preparation, such as desalting, is not necessary in this instance. The electrokinetic injection mode in CGE is simply accomplished by starting the electrophoresis process from the sample vial. This injection mode often yields superior peak efficiency compared to pressure injection, especially when the injection is made from low ionic strength solutions or from water. In this instance, the sample components are stacked against the relatively viscous gel matrix. This injection method can result in a sampling bias for small molecules, since the different mobility sample components move into the capillary with different velocities. This is not the case for DNA molecules, since with similar or very close charge-to-mass ratios, no injection bias occurs. It is important to note that quantitation is very difficult during electrokinetic injections without the use of an internal standard (Table 2).

Employing electrokinetic injection, the separation performance in CGE is usually highly dependent on the sample matrix, especially when a high amount of salt is present in the sample [40]. To prevent this so called salt effect, labor-intensive ultrafiltration or desalting is recommended, but it can be avoided by special sampling techniques. In both pressure and electrokinetic injection modes, sampling can be made by the simple injection of the analyte into the capillary, preceded by preinjection of a different buffer system or water, to increase precision [41]. With crosslinked high-viscosity gels, preinjection of water highly increased peak area reproducibility and also slightly increased the injected amount. Preinjection of a low-viscosity buffer plug helped to obtain high-efficiency separation with sharp peaks from high-salt-containing PCR products [42]. Recently, van der Schans et al. [42] studied sample matrix effects in PCR product analysis using the pressure injection mode and replaceable gels. With the increase of the sample plug the separation efficiency decreased and fronting peaks were observed. By simply injecting a plug of low resistance, e.g., $0.1\ M$ Tris-acetate prior to sample injection, sharpening of the peaks was obtained.

Co-injection of PCR products with a standard mixture is a convenient method of verifying the identity of the sample peaks. However, it

Table 2 Peak Area Precision in Capillary Gel Electrophoresis (RSD) with Replaceable Gels

	Hydrodynamic	Electrokinetic
(A) Internal standard		
Adjusted area	8.0%	6.0%
Height	8.1%	3.0%
Migration time	0.1%	0.1%
(B) No internal standard		
Adjusted area	8.4%	28%
Height	8.5%	23%
Migration time	0.2%	0.3%
(C) Internal standard		
Adjusted area	3.0%	
Height	6.7%	
Migration time	0.07%	

(A) 100-bp DNA compared to 200-bp DNA internal standard for 10 runs. Both inlet and outlet run buffer vials were changed after run 5. (B) Without using internal standard. (C) Same as (A) but outlet buffer vial was changed after every run.

Source: Reprinted with permission from Butler et al., *J. Chromatogr. B*, 658, 271 (1994).

was found that co-injection of a DNA standard with the PCR sample can lead to sharpening of the sample peaks or the standard peaks, depending on the order in which the plugs were loaded on the capillary [42]. When a high-salt PCR sample is injected first, followed by a DNA standard, the PCR peak is broadened while the standard peaks are sharp. The opposite is seen when the injection order is reversed. The broadened peaks are due to salt migrating from the PCR sample into the plug containing the DNA standards. During its migration through the capillary, the back end of the standard zone will migrate at a slower velocity relative to the front as here the field strength is lower than the front end of the plug.

Guttman and Schwartz [41] reported two injection-related artifacts in conjunction with CGE. One occurred with consecutive injections from the same, low-volume (0.01–0.1 mL) sample vial, in which case progressively smaller peak heights were obtained with each consecutive injection. An electrochemical processes during the electric field mediated sample introduction was found responsible for this effect. Performing an electrokinetic injection from a water vial prior to the actual sample injection appeared to be a simple solution to solve the problem. The second artifact was related to the capillary edge (at the injection side). Oblique-edge capillaries resulted in poor peak performance compared to straight-edge capillaries [41,43].

C. Detection

Detection of CE based DNA separations is usually obtained by UV absorbance or laser-induced fluorescence (LIF), and very recently by on-line mass spectrometry [44]. All commercially available CE instruments are equipped with a UV detection system, which is universal for many types of analytes and has adequate sensitivity for most applications. However, determination of trace amounts of analytes in the presence of many other sample components makes detection problematic. Stacking or other (e.g., isotachophoretic) preconcentration methods and optimized optics may help to improve detection limits. Alternatively, with fluorescence and especially with LIF detection methods, far lower detection limits can be obtained compared to those based on UV absorbance. Fluorescence-based assays have the advantages of offering both excellent selectivity and very high sensitivity. One of the great advantages of the laser-induced detection systems is that they focus the excitation beam on a very small, nanoliter scale detection volume. With various laser sources and tailor-made instrumentation, it is possible to examine the nucleic acid contents of single human cells or even detect single molecules of stained DNA. Besides the high-cost low-wavelength lasers of He-Cd (325 nm) and Ar-ion (488 nm), low-cost higher-wavelength semiconductor lasers (635–850 nm range) can also be used with CE. Of note is that background fluorescence from biological sample matrices is strongly reduced at these longer wavelengths. In order to extend the applicability of the red diode lasers to CE, new labeling reagents, such as thiazine-, oxazine-, and cyanine-type compounds are being developed [45]. Chen et al. [46] used an oligonucleotide labeled with a cyanine dye (Cy5) in DNA hybridization and sequencing. One of the most straightforward LIF detection schemes of double stranded nucleic acid molecules involves fluorescent intercalators [47–49] which are usually added to the buffer system and/or sample. The intercalator dye then inserts itsel between the two strands of the DNA molecule (Fig. 2). Note

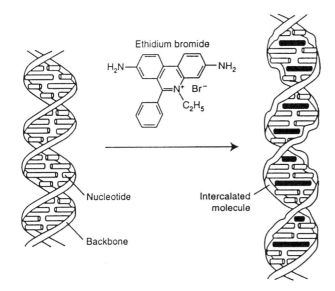

Fig. 2 The intercalation of ethidium bromide into a DNA molecule. (Reprinted with permission from Watson et al., *Molecular Biology of the Gene*, Menlo Park, CA, Benjamin Cummings Publishing Co., 1987.)

that intercalation changes the persistence length of the DNA molecule, as well as its conformation and charge, resulting in a decrease in electrophoretic mobility. As can be seen in Fig. 3, intercalation may result in enhanced resolution. The use of fluorescent intercalating dyes leads to two to three orders of magnitude enhanced sensitivity when compared to UV detection [48,50–53]. The gain in detectability when fluorescent labeling is used is illustrated in Fig. 4 for a 53-base pair (bp) RT-PCR product, Sabin 3, co-injected with a ΦX-174/*Hae*III digest, which also compares the noise level of the UV trace to the LIF trace with no visible noise. The same DNA concentration was used for both the UV detection and that of the LIF. A variety of monomeric and dimeric dyes have been used in recent CE-LIF applications [54]. Besides the easy way of intercalation, which works primarily for ds DNA molecules, direct labeling of DNA molecules with a suitable fluorophore is another recommended approach. Fluorescently labeled

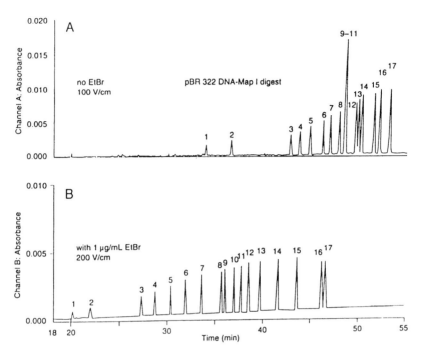

Fig. 3 Effect of ethidium bromide on the CGE separation of a pBR322 DNA/*Msp*I restriction digest. (A) No ethidium bromide; (B) 1 μg/mL ethidium bromide in the gel-buffer system. Peaks (base pairs): 1=24, 2=26, 3=67, 4=76, 5=90, 6=110, 7=123, 8=147, 9=147, 11=160, 12=160, 13=180, 14=190, 15=201, 16=217, 18=238, 19=242. Conditions: (A) 100 V/cm, (B) 200 V/cm. (Reprinted permission from Guttman and Cooke, *Anal. Chem.*, *63*, 2038 (1991).)

probes and primers are used in many molecular biology applications involving hybridization, PCR and sequencing. DNA primers and probes are usually synthesized with a fluorescent label attached to the 5'-end of the molecule, or with postsynthesis attachment of the dye using commercial DNA labeling kits. For subsequent use of the DNA in PCR leading to fluorescent DNA products, conditions can be optimized such that primer purification is not necessary. In DNA sequencing, fluorescent-labeled primers and terminators based on fluorescein, modified fluoresceins, Texas Red, and tetramethylrhodamine are routinely used [55]

While mass spectrometry in conjunction with capillary electrophoresis (CE/MS) is not yet routinely used, special interfaces allow

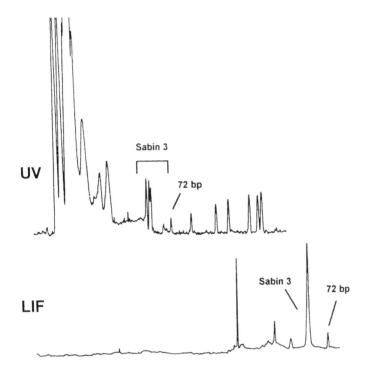

Fig. 4 UV versus LIF detection of the CE separation of a 53 base pair RT-PCR product from the RNA of the polio virus vaccine, Sabin 3. A *Hae*III digested ΦX-174 DNA marker was co-injected with the PCR product for size determination. The same Sabin 3 concentration was used for each analysis, whereas the marker total DNA concentration varied from 200 μg/mL for UV analysis, to 10 μg/mL for LIF analysis. Full scale: UV detection, 0.005 AU; LIF detection, 10 RFU. (Reprinted with permission from Schwartz et al., J. Cap. Elec., 1, 41 (1994).)

CE to be coupled to MS, and even large polyionic biomolecules, such as DNA, are actually amenable to MS. It has been demonstrated that free-solution techniques such as CZE and MEKC can be coupled to MS by employing either fast atom bombardment (CE-FABMS) or electrospray (CE-ESMS) techniques. An example of the former was described by Wolf et al. [56] for the characterization of deoxynucleoside-polyaromatic hydrocarbon adducts. The analysis of antisense oligonucleotides was reported by Bayer et al. [57] using CE-ESMS. Since the exact mass of the double helix can be detected, the CE-ESMS method allows discrimination between specific and nonspecific interactions; therefore the method can be implemented in hybridization reaction studies.

D. Fraction Collection

Although CE is primarily an analytical tool, it can also be used for micropreparative applications. In an early CGE paper, Guttman et al. [58] demonstrated programming the electric field during the collection ("slowing the field down"), simplifying the collection process. Their approach of field programming for fraction collection is demonstrated in Fig. 5. A 47-mer oligonucleotide from a $p(dA)_{40-60}$ mixture was collected. During the micropreparative run (trace A) the field was maintained at 300 V/cm until just before the 47-mer reached the detector. At that point, the field was decreased 10-fold and the fraction was collected for 60 s (the calculated peak width under the low-field conditions was 45 s). Trace B shows the reinjected collected fraction: only the 47-mer is visible together with the internal standard, a 20-mer. Typically, the micropreparative runs yield broader peaks than their corresponding analytical runs. The sample size injected for the micropreparative run resulting in the "overloaded" profile (trace A) was six times higher than the analytical run (trace C). More recently, Kuypers et al. [59] used CGE (replaceable gel, 4% linear PA) to collect multiple peaks corresponding to denatured DNA from a mutated, 372 bp PCR product. The collected fraction were reamplified by PCR and subsequently analyzed again by CGE. It was found that the different peaks corresponded to different gene sequences.

E. Quantitation

There have been a number of attempts to use CE to quantitate therapeutic oligonucleotides and PCR amplified DNA [60–68]. The key to proper quantitation with this application is control of the injection. Given the large potential for undesirable sample matrix effects, electrokinetic injections should be practiced with caution, although with the use of internal standard calibration, it is possible to obtain reproducible results. In most cases hydrodynamic injections are more reproducible [61,65], with longer injection times helping to ensure reproducibility by minimizing any pressure variation effects. It is also important to choose an effective internal standard for quantita-

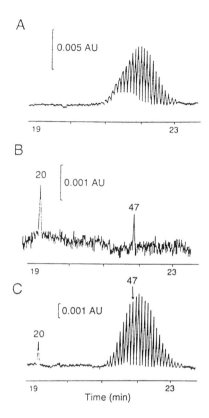

Fig. 5 (A) Micropreparative CGE separation of a polyadenylic acid test mixture, $p(dA)_{40-60}$; (B) analytical run of the isolated $p(dA)_{47}$ spiked with $p(dA)_{20}$; and (C) analytical run of $p(dA)_{40-60}$ spiked with $p(dA)_{20}$. (Reprinted with permission from Guttman et al., *Anal. Chem.*, 62, 137 (1990).)

tion. Several manufacturers now produce individual DNA fragments of known concentration which can be added to the DNA prior to analysis (e.g., GenSura Laboratories, Inc., Del Mar, CA; Bio Ventures, Inc., Murfreesboro, TN). There are a number of advantages associated with using CE for the quantitation of DNA. Unlike traditional techniques such as spectrophotometry or hybridization, impurity levels can be examined, allowing for more accurate and informative assays [61].

F. Sizing Techniques

For applications such as the analysis of genetic variation via the PCR, it is important to produce precise estimates of DNA fragment length. This is a particular problem in CE analysis since, unlike PA slab gels, only a single sample can be analyzed at a time. Sizing standards such as restriction digests can be analyzed prior to sample injection, but due to run-to-run variations in migration time, the precision of this technique is not always adequate. As an alternative, internal markers may be added to or co-injected with the sample to allow fragment size estimation [42,64,65,69]. In such situations, it is important to know the relationship between base-pair size and migration time. In the case of physical gels, a linear relationship exists from 100 to 400 bp [70] while at fragment sizes above 400 bp, the relationship becomes nonlinear due to reptation. For samples sizes under 400 bp, determination of fragment size can be performed by bracketing the peaks of interest between two sizing standards and interpolating [61,71]. Using this technique, PCR products sizes have been determined to a precision of 0.3 bp. Alternatively, precise size determinations can be performed using multichannel fluorescence detection and multiple dyes. In this technique the sample is labeled in one color, the standard is labeled in another, and the two are analyzed simultaneously. This technique is particularly useful for analyzing genetic variation, since a standard consisting of known variants can be run simultaneously with an unknown. Careful attention to the types and locations of the derivatizing dyes is necessary to avoid migration shifts between the standard and sample dyes [72].

Of importance in the use of restriction fragments in CE as sizing standards for PCR analysis is the issue of anomalous migration of certain fragments [73–75]. There are two important factors to consider which account for this problem. The first is the mechanism by which the enzyme cuts the DNA. Certain restriction enzymes such as *Hin*fI produce "sticky" ends with overhanging sections of ss DNA. Even when enzymes such as *Hae*III, which produces blunt-end fragments, are used, the resulting DNA fragments contain terminal phosphates which affect DNA migration [75]. Improved size estimates can be made from "PCR-like" standards which do not contain terminal 5′-phosphates. The second factor to consider is the presence of conformational effects resulting from sequence differences among the fragments. Such effects appear to be a major problem in analyses carried out using linear PA [74] and PEO [73] and to a lesser extent in cellulose polymers [76]. The addition of intercalating dyes appears to eliminate the problem, presumably by unraveling the DNA, minimizing conformational effects [74].

IV. SELECTED APPLICATIONS
A. Short Oligomers

With applications used for the analysis of oligomers used as cloning linkers, for mutagenesis, and as pharmaceuticals, it is essential that the level of impurity be determined following synthesis. Many investigators routinely purify their oligonucleotides regardless of their intended use due to the presence of DNA fragments resulting from failures in the synthesis process. Incomplete deprotection and unwanted chemical modification of the oligomer result in a mixture containing the synthetic oligomer as well as similar molecular weight contaminants known as failure sequences [77]. Separation of these species to determine impurity levels in a crude synthetic mixture can be accomplished using CE. Accordingly, a host of laboratories have published CE applications for the analysis of short oligomers. Whereas small oligonucleotides (<10 bases) can easily be resolved by CZE and MEKC methods, as the number of nucleotide units increase, the separation becomes more difficult in free solution; CGE is suitable for the separation of the larger oligomers. For routine CGE of oligonucleotides, resolution of nucleotides from 10 to 150 bases is typical. The majority of these separations have been performed on chemical PA gels due to their commercial availability and the need for high resolving power. The effect of various operating parameters on oligomer resolution, most notably temperature [78,79], pH [80], and electric field [81] have been described. In the latter, the use of field programming, that is, the use of increasing, decreasing, or step-voltage gradients, was proposed to improve resolution for DNA. Similarly, temperature programming [82] and urea concentration effects [83] were also investigated while using a gel containing the affinity ligand, poly(9-vinyladenine), for base-specific recognition of oligomers. One of the more common anomalies which can affect the mobility of short synthetic oligomers is base composition effects. As a result, not all oligomers of the same size will migrate with the same mobility [77]. The formation of secondary structure, which is not completely eliminated by thermal or chemical denaturants, can also induce mobility shifts [84,85].

Recently, work describing oligonucleotide separations using physical gels has been demonstrated. In order to achieve the resolution required for short oligomer analysis, highly viscous solutions (8–10%T linear PA) are required and are difficult to inject into a narrow bore capillary [86]. Alternatively, these solutions can be polymerized in situ, with unreacted monomers removed by cysteine scavenging [87]. In search of a true replaceable gel system for oligomers, Heller and Viovy [88] reported the use of linear PA with an average molecular mass of 80 kD. A 10% solution in 1 × TBE produces a matrix with chains long enough to be highly entangled at the concentrations needed for oligomer separation, but short enough to allow automated matrix replacement in the coated capillary (low to medium viscosity). Figure 6 shows an example of single base resolution for a p(dA)$_{40-60}$ using a commercially available replaceable PA gel under denaturing conditions.

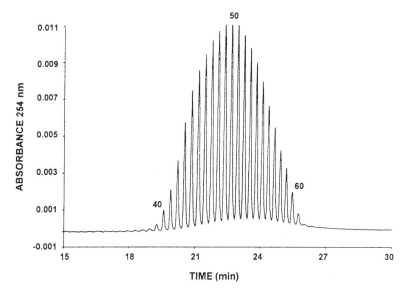

Fig. 6 Size separation of a pd(A)$_{40-60}$ oligonucleotide test mixture using a replaceable PA matrix under denaturing conditions. Single-base resolution is achieved for the <40- to >60-mers. Conditions: 27-cm coated capillry; 2-s electrokinetic injection of sample at 10 kV; field strength, 300 V/cm, 30°C. (Electropherogram courtesy of Dr. P. Shieh, Beckman Instruments, Incorporated, Fullerton, CA.)

Figure 7 exhibits the separation of a synthetic 17-mer oligonucleotide with the trityl group attached to the 5′-end, performed using a replaceable linear PA gel and a coated capillary column of 20 cm and 7 cm effective length, respectively. Mixed PEO polymer solutions may also be used in this application and have a resolving power similar to that of linear PA with reduced viscosity and without the need for a coated capillary [33].

B. Antisense DNA

Antisense therapeutics are synthetic oligonucleotides whose base sequence is complementary to a target sequence on either a messenger RNA (mRNA) which encodes for disease-causing proteins, or the ds DNA from which the mRNA was transcribed [89]. The complementary nature of the antisense molecule allows it to hydrogen bond and inac-

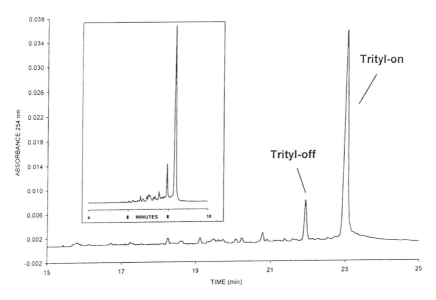

Fig. 7 Separation of a synthetic 17-mer oligonucleotide with the trityl group attached to the 5′-end, performed using a replaceable linear PA gel and a coated capillary column (20 cm effective length). Inset shows high-speed separation of the same oligonucleotide accomplished using only 7-cm effective length capillary.

tivate the genetic message, inhibiting gene translation. Unfortunately, oligonucleotides with a phosphodiester backbone are very susceptible to cellular nuclease degradation. As a result, much interest has been directed toward DNA analogs with phosphorus-modified backbones (e.g., phosphorothioates and methylphosphonates), due to their increased resistance to these nucleases. Another type of antisense, peptide nucleic acids (PNA), in which the entire deoxyribose-phosphate backbone is exchanged for a peptide backbone composed of (2-aminoethyl)glycine units, shows promise as a potent antisense molecule [90]. Because of the therapeutic nature of these antisense DNAs, purity control is essential.

Capillary electrophoresis of these relatively short (10 to 25 bases) oligonucleotides using crosslinked or linear PA gel filled capillaries under denaturing conditions gives single-base resolution of phosphodiester antisense from its failure sequences, but rather unsatisfactory results (broad peaks resulting in poor resolution) with backbone-modified antisense DNA. Phosphorothioate oligonucleotides separations were shown by Warren and Vella [91], using crosslinked PA gels, but with less resolution than that seen by anion-exchange HPLC. Cohen et al. [92] found that chromatographic separation methods worked well for small phosphorothioates, but CGE using linear PA at high density (13–18% T) yielded the optimal resolution. These gels are most often used for DNA sequencing. DeDionisio [93] used a modified formulation of a commercially available gel-filled capillary and achieved single-base resolution of a standard phosphorothioate mixture from 22 to 26 bases by increasing buffer pH and using a more rigid gel. However, for typical phosphorothioate DNA commonly synthesized for biological studies, the (N-1) and N-mer could not be baseline resolved. Using a linear PA gel system, Srivatsa et al. [66] validated the use of CGE for routine analysis of drug product formulations (see, e.g., Fig. 8). The linearity, accuracy, selectivity, precision, and ruggedness of the method were evaluated, with excellent migration time and peak area reproducibility demonstrated by using an internal standard. Bourque and Cohen [67] and Leeds et al. [68] likewise applied CGE for quantitative analysis of antisense oligonucleotides in biological matrices such as human plasma.

Rose [94] studied the binding kinetics of specific PNAs with oligonucleotides using CGE to separate and quantitate both free PNA and oligonucleotides, as well as bound heteroduplexes. In this case, both single- and double-stranded molecules were fully separated on the

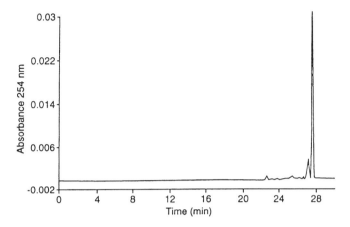

Fig. 8 Purity analysis of a synthetic phosphorothioate 20-mer.

same capillary simultaneously. Vilenchik et al. [95] used CGE as an on-column quantitative Southern analysis tool for monitoring antisense DNA binding (Fig. 9). Detection by LIF monitored the hybridization of a phosphorothioate target with a complementary fluorescein-labeled DNA probe. Finally, Effenhauser et al. [96] described very fast (<1 min) size separations of fluorescent phosphorothioates of 14 to 24 bases using a PA gel in a micromachined glass chip as a CE device.

C. DNA Sequencing

In addition to purity control of synthetic oligonucleotides, another important application involving CE is DNA sequencing. Several research groups are currently exploring the use of CE as an alternative to slab gel electrophoresis for automated DNA sequence determination [55,97]. The large surface-area-to-volume ratio of the capillary permits larger electric fields than typically used with slab gels, resulting in very rapid and efficient separation of sequencing reaction products. Additionally, the capillary format is readily adaptable to automated sample loading and online data collection. With CE, detection of the separated DNA sequencing fragments is performed by LIF. The sensitivity of the LIF detection allows sequencing reactions to be performed on the same template and reagent scale as that of manual DNA sequencing with autoradiographic detection. The identity of the terminal base of each

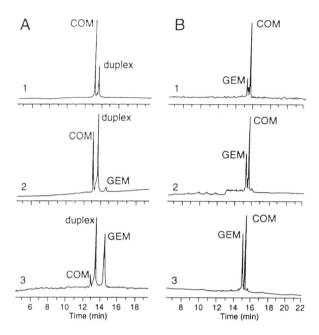

Fig. 9 The separation of GEM, COM, and duplex by CGE. Electropherograms 1, 2, and 3 show different amounts of GEM with constant COM concentration. (A) Nondenaturing conditions. (B) Denaturing conditions. Conditions: (A) 9% linear PA gel, l=20 cm, E=200 V/cm, (B) 13% linear PA gel, l=15 cm, E=400 V/cm. (Reprinted with permission from Vilenchik et al., *J. Chromatogr. A, 663*, 105 (1994).)

DNA sequencing fragment is encoded in the wavelength and/or the intensity of the fluorescent emission.

Sample throughput is a major concern for high-volume sequencing applications. Automation of sample preparation, sequence reactions (including electrophoresis), and data interpretation are necessary in order to achieve the ambitious goal of sequencing the entire human genome (~3 billion bp). With CE, the samples are loaded one at a time. Slab gels, on the other hand, can simultaneously be loaded with 24–36 samples. Instrumentation which would allow running of several capillaries in parallel, together with robotics for sample handling, would dramatically increase the desired sample throughput with CE. DNA sequencing has already been demonstrated in arrays of multiple (20–100) capillaries [98–100]. It can easily be envisioned that this type of instrumentation can also be incorporated in other applications, e.g., screening for genetic diseases, forensic DNA typing, etc.

A large obstacle to the development of commercial CE-based DNA sequencers has been the stability of gel-filled capillaries. While they can provide extremely high resolving power, the crosslinked gels typically last only a few runs when sequencing reactions are loaded, after which time the entire column must be replaced. Recent developments in the CGE column technology (in particular the replaceable gels) should eliminate the time-consuming and laborious procedures of the preparation and alignment of the capillaries [101]. With the replaceable matrix, it is possible to load a sequencing reaction, rapidly separate the DNA fragments at high field, and then reload the gel on the capillary prior to the next run. Figure 10 shows the CGE separation of a single-terminator, Sanger-Coulson reaction using a replaceable linear PA gel. A "red" diode-laser was used for excitation of the fluor (Cy5)-labeled DNA fragments [46,102]. The relatively low viscosity (6% T), gel matrix of these type of capillaries provides reproducible and fast separation of DNA fragments with sequence information extending to at least 400 bases. For DNA sequencing applications, typically a CGE gel buffer containing formamide and urea is used. A denaturing buffer of 30% formamide, 3.5 M urea has a lower viscosity than the typical 7 M urea buffer, and is therefore advantageous to use in a replaceable CGE formulation. In addition, increased decompression of sequences with secondary structures is obtained.

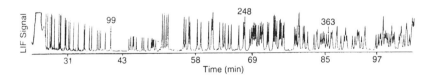

Fig. 10 Electropherogram of fluorescently labeled (Cy5–20 primer) DNA fragments generated enzymatically (M13mp18 template) using ddG terminator. A LaserMax (Rochester, NY) red diode laser (639-nm excitation) was employed with an in-house-built CE-LIF instrument, using linear PA gel. The pattern is recognizable beyond 400 bases. (Electropherogram courtesy of Dr. S. L. Pentoney, Beckman Instruments, Incorporated, Fullerton, CA.)

D. Restriction Digests

A large body of work has been published concerning the application of CE to the analysis of ds DNA in restriction digests, and has been the subject of a number of reviews [6,103,104]. CE separations have been demonstrated on ds DNA from 60 to 20,000 bp in size using both physical and chemical gel systems, with the bulk of this work performed on physical gels due to their convenience and superior column lifetimes. Among the parameters to be considered in the separations is the type of polymer, intercalator, and buffer system. Perhaps the most extensive work in this area was carried out by Nathakarnkitkool et al. [62], who studied the effect of field strength, polymer concentration, pH, ionic strength, and intercalator on the separation of several different restriction digests. Both coated and uncoated capillaries were examined and the polymer used was an HEC of moderate viscosity (300 cP). It was found that a field strength of approximately 200 V/cm produced the best separation and that additives such as salts and intercalating dyes could greatly affect resolution and migration time. Other additives which have been found to be effective in improving DNA separations have included urea, glycerol, and 9-aminoacridine [52,105,106]. With respect to the use of physical gels employing noncellulosic matrices, major work on the analysis of restriction digests using linear PA has been carried out by Pariat et al. [107] and Chiari et al. [87], among others. Of major importance is the removal of excess monomer from the PA. Attempts have also been made to reduce viscosity and to improve column lifetime [23]. Analyses of ds DNA have also been carried out in modified PA, agarose, and PEO [23,73,108].

E. Polymerase Chain Reaction Products

The PCR is a method of amplifying a DNA molecule, allowing the production of millions of copies from a few fragments of template DNA. Applications of PCR range from genetic analysis to disease diagnosis and forensic identification [109]. PCR product analysis is a large potential application for CE, since it provides the ability to rapidly quantitate the products of the reaction [110]. Concurrently, the results from CE analysis enable rapid optimization of the PCR reaction conditions themselves.

Initial work in the analysis of PCR products by CE was hindered by poor separation efficiency and the incompatibility between the high salt content of the buffer and the CE injection [10]. The development of improved sample dialysis techniques has minimized these problems [31,75,111,112] and fluorescence detection makes it possible to analyze samples directly simply by diluting the PCR product with water [53,69]. A further advance in the analysis of PCR-amplified DNA has been the development of internal standards for molecular weight determination and quantitation [64,69,75,113,114].

A wide range of applications has been examined using CE analysis of PCR products. An early example was the use of CE with the PCR in the screening of blood samples for HIV. An HIV *gag* gene of 115 bp in size was coamplified with HLA-DQA [31]. This system utilized a TBE buffer with 0.5% HPMC and the results indicated a certain amount of interference due to the presence of excess primers and dNTPs. These results and others tend to give a lower fragment size limit for PCR amplified fragments of approximately 100 bp unless extensive sample cleanup is utilized. Subsequent work using LIF and an intercalating dye reveals some of the selectivity advantages of fluorescence when applying CE to PCR analysis [48]. While it is technically possible to produce millions of copies of a DNA sequence with the PCR reaction, in many cases sensitivity of the technique is limited by contamination which inhibits the PCR reaction or by limited quantity of template DNA. Accordingly, samples are often diluted to reduce overall ionic strength [69]. For this reason a large proportion of the work performed on PCR products uses fluorescence detection.

Applications of CE with PCR have included the analysis of PCR products and PCR-restriction fragment length polymorphisms indicative of genetic diseases. Such applications in clinical chemistry represent a large potential for CE/DNA analyses. Using buffer systems including cellulose and linear PA polymers, analyses for X-linked recessive disorders, Kennedy's disease, cystic fibrosis, and acyl coenzyme deficiency have been described [53,113,115,116]. Further studies have demonstrated the possibility of multitarget PCR (Fig. 11) [48,64], the quantitative analysis of DNA and RNA by competitive PCR [63,65, 117,118] and a rapid test for the presence of the hepatitis C viral genome in serum [114].

Another area of interest has been the application of PCR and CE to problems associated with human identification. The increasing use of genetic typing in civil and criminal cases has created a need for more automated procedures. PCR amplified variable number tandem repeats

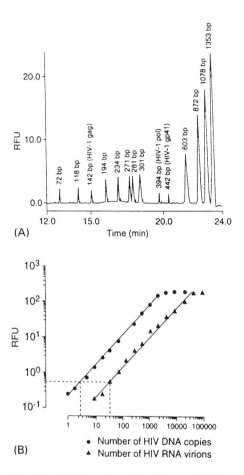

Fig. 11 (A) Detection of multitarget PCR-amplified HIV-1 *gag*, *pol*, and *env* sequences by CE-LIF. (B) Linearity of LIF-CGE analysis of quantitative PCR or RT-PCR products. (Reproduced with permission from Lu et al., *Nature*, *368*, 269 (1994).)

(VNTRs) such as Apo B and D1S80, and smaller repeats (STRs) such as vWa and HUMTH01 have been analyzed using fluorescence detection with intercalating dyes [69,72,75,111,112,119] (Fig. 12). The precision and resolution requirements of the application have resulted in the development of systems using multiple internal standards which have standard errors less than 0.2% for DNA sizing [61].

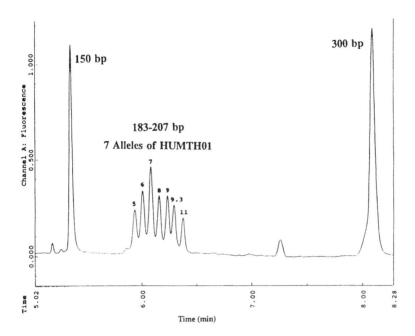

Fig. 12 The HUMTH01 alleles from an individual sample, amplified via the PCR, are separated in less than 7 min using a replaceable polymer matrix and LIF detection. Alleles were sized and compared to the allelic ladder, demonstrating 4 bp resolution. The 150 bp and 300 bp markers allow precise sizing of each allele. Injection: 27 V/cm for 5 s. Voltage programming: 405 V/cm for 4.5 min.; 135 V/cm for 5.5 min. (Reprinted with permission from Butler, et al., *BioTechniques*, *17*, 1062 (1994).)

F. Conformational Polymorphisms

As the individual genes responsible for genetic diseases and cancers become identified, it is increasingly important to be able to analyze them for any aberrations: point mutations, deletions, insertions, or rearrangements, that may result in altered cell behavior. These sequence polymorphisms occur in both strands of a DNA molecule, and may be undetectable by Southern analysis. Recently, various mutation-scanning methods based on differential behavior of the DNA molecules have been adapted from classical slab gel to CE analysis. These methods include single-stranded conformational polymorphism analysis

(SSCP), heteroduplex polymorphism analysis (HPA), and variations on denaturing gradient gel electrophoresis (DGGE).

SSCP was first described by Orita et al. [120]. Under nondenaturing conditions, a ss DNA molecule folds itself into a conformation that is based on its sequence; a point mutation creates a conformational change in the ss DNA that may lead to a shift in mobility during electrophoresis. In SSCP, ds DNA is typically heated to melt and then immediately cooled. The resulting folded ss DNAs are then separated by electrophoresis. Kuypers et al. [59] demonstrated point mutation detection in a known mutation of the p53 gene. A heterozygous individual, having both a normal and mutated p53 gene, showed more than the two normal (wild type) ss DNA fragments using a replaceable polymer network. Cheng et al. [121] enhanced the separation in SSCP by adding glycerol to a HPMC solution; glycerol maintains the stability of the secondary structure of ss DNA. This formulation led to the detection of novel metastable DNA conformational isomers not usually seen by conventional PA slab gel SSCP.

HPA differs from SSCP in that, after the ds DNA is melted, it is allowed to reanneal into its double-stranded form by slow cooling. If a sample contains both a wild type and mutated form of a gene, a heteroduplex will form between the ss DNAs, resulting in mismatched molecules. These ds DNAs are then separated by electrophoresis. Using an entangled polymer solution, Cheng et al. [122] showed differentiation of heteroduplex DNA with a single-nucleotide mismatch from the homoduplex in a PCR product without any sample pretreatment prior to denaturation.

DGGE allows ds DNAs that differ by one or more mutation to be separated in gels having a linear denaturing gradient. The gradient may be formed by chemical means (e.g., urea or formamide), or by temperature. Because a mutation can alter the ds DNA molecule's melting properties, sequence polymorphisms will result in a change in mobility due to conformational differences. Khrapko et al. [123] presented a variation on DGGE for mutational analysis using CE and linear PA gels, termed constant denaturant capillary electrophoresis (CDCE). A constant region of elevated temperature (a denaturing zone) was created over a 10-cm portion of a capillary. The regions outside the heated portion were at ambient temperature. The ds DNA migrating from the ambient to the denaturing zone will partially melt, and its mobility will be based on its conformation due to sequence polymorphism. Another variation to DGGE was reported by Gelfi et al. [124]. Using a linear

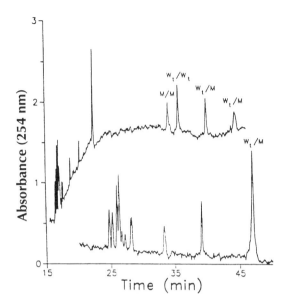

Fig. 13 CE separation of a cystic fibrosis sample exhibiting two polymorphisms on the same chromosome using a temporal thermal gradient in the capillary; M=mutant strands, W_t=wild-type strands. A 60-cm-long coated capillary was filled with 8% poly(AAEE) in TBE under nondenaturing conditions. Lower trace: constant temperature run at 45°C. (Upper trace) temperature gradient run from 45 to 49°C, with increments of 0.2°C/min. (Reprinted with permission from Gelfi et al., *Electrophoresis*, *15*, 1506 (1994).)

poly(AAEE) gel, a temporal temperature gradient could be created by selecting a buffer of proper conductivity and manipulating the applied voltage during the run. Temperature control parameters were determined using a dedicated in-house software program. Figure 13 demonstrates the separation of a cystic fibrosis double polymorphism using temperature gradient versus constant temperature conditions.

G. Large DNA

In this section, large DNA is defined as ds DNA >2000 bp. Attempts to extend the upper size separation limit of DNA fragments in CE have dealt primarily with replaceable polymer matrices using constant fields, due to their relative simplicity. Barron et al. [125] have demonstrated that it is possible to use high-voltage, steady fields in dilute HEC solutions with or without any capillary coating to separate large DNA fragments 2 to 23 kbp in size. Presumably, this separation mechanism is based on entanglement interactions between the cellulose polymer and the DNA. As with conventional slab gels, the demonstrated limit for separations of high resolution by CE is about 20,000 bp [107,126], although low-resolution separation up to 48.5 kbp using linear PA has also been shown [127]. Using a similar gel system, Hebenbrock et al. [128] demonstrated the usefulness of CE for the determination of plasmid copy number in the monitoring of recombinant bacterial cultivation where linearized plasmids up to 10 kbp analyzed, Izumi et al [129]

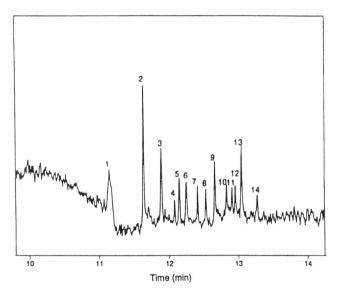

Fig. 14 Electrophoretic separation of a Mbp DNA standard by frequency modulated electric field CE. Run buffer: 0.5× TBE containing 0.00375% HEC and 0.0020% PEO. Electric field: 100 V/cm + 14Hz square wave, 250% modulation. DNA sample at 10 ng/mL injected electrokinetically at 1 kV for 15 s. Peak legend: (1) 0.21, (2) 0.28, (3) 0.35, (4) 0.44, (5) 0.55, (6) 0.60, (7) 0.68, (8) 0.75, (9) 0.79, (10) 0.83, (11) 0.94, (12) 1.10, (13) 1.12, (14)1.6 Mbp. (Reprinted with permission from Kim et al., *Anal. Chem.*, 67, 784 (1995).)

separated a Bluescript II SK(+) plasmid and its *DdeI* restriction digest using entangled glucomannan solutions.

The separation of large DNA fragments also has been applied to agarose solutions [32,130,131]. Due to the light-scattering and absorbing properties of agarose gels, clear agarose solutions which facilitate detection have been used to separate DNA fragments up to 12 kbp in length by using molten low melting agarose heated and maintained in a coated capillary at 40°C Alternatively, urea (up to 7 *M*) may be added to agarose solutions solely to prevent gelling at room temperature by disrupting hydrogen bond information [132].

One way to extend the size limit of DNA fragments in CE sieving beyond 50 kbp is by application of pulsed or frequency modulated electric fields. Using CE, only the field inversion arrangement is possible. Both Sudor and Novotny [133] and Morris' group [134–136] have demonstrated that DNA >23 kbp can be separated in dilute polymer solutions using pulsed or frequency modulated electric fields. Sudor and Novotny used a dilute (0.4% T) linear PA solution to separate DNA up to 1 megabase pair (Mbp). similarly, Kim and Morris [136] separated DNA as large as 1.6 Mbp, but used an ultra-dilute mixture of HEC and PEO to resolve the Mbp-size DNA fragments in 14 min. Figure 14 illustrates this separation. Conventional pulsed-field slab gel electrophoresis often takes one day for such Mbp separations.

V. FUTURE DIRECTIONS

It is clear that DNA applications involving CE have begun the evolution from feasibility to routine use. Whereas slab gel electrophoresis was considered the standard technique, CE is fast becoming a viable alternative. This new-found technology affords high throughput, robust methodology, and complete automation. Dedicated DNA analyzers using CE technology show prospect for this objective. Already, multiplexed arrays of 20 to 100 capillaries simultaneously undergoing DNA electrophoresis have been demonstrated [72,99,100,137]. It has been suggested that even larger arrays should pose no difficulty in their monitoring, especially by fluorescent channel imaging through the use of a charge-coupled device (CCD) camera. Ultrathin slab gel systems with fluorescence detection are also a very promising alternative. Furthermore, micromachined capillary electrophoresis chips have been used successfully for very fast (<45 s) separation and collection of oligonucleotides with excellent migration time reproducibility [96,138]. Coupled with other analytical techniques (e.g., CE-mass spectrometry), chip-based technology would be a valuable tool for increasing selectivity. Based on these new advances, CE is destined to play an essential part in future diagnostic, biomedical, forensic, and biotechnology requisites.

ACKNOWLEDGMENTS

The authors would like to thank Dr. Ágnes Cseh for her aid in the preparation of this manuscript. They also appreciate Beckman Instruments, Incorporated for facilities and time used in preparation of this chapter.

REFERENCES

1. A. Tiselius, *Trans. Faraday. Soc.*, *33*, 524 (1937).
2. S. F. Y. Li, *Capillary Electrophoresis*, Elsevier, Amsterdam, The Netherlands, 1993.
3. *CRC Handbook of Capillary Electrophoresis: Principles, Methods, and Applications*, J. P. Landers, Ed. CRC Press, Boca Raton, FL, 1993.
4. R. K. Saiki, S. J. Scharf, F. Faloona, K. B. Mullis, G. T. Horn, H. A. Erlich, and N. Arnheim, *Science*, *230*, 1350 (1985).
5. K. B. Mullis and F. Faloona, *Meth. Enzymol.*, 155, 335 (1987).
6. B. L. Karger, A. S. Cohen, and A. Guttman, *J. Chromatogr.*, *492*, 585 (1989).
7. S. Hjerten, *J. Chromatogr.*, *270*, 1 (1983).
8. A. S. Cohen and B. L. Karger, *J. Chromatogr.*, *397*, 409 (1987).
9. B. L. Karger, A. Paulus, and A. S. Cohen, *Chromatographia*, *24*, 15 (1987).
10. R. G. Brownlee, F. J. Sunzeri, and M. P. Busch, *J. Chromatogr.*, *533*, 87 (1990).
11. K. J. Ulfelder and P. Shieh, Poster T-211, *Fifth International Symposium on High Performance Capillary Electrophoresis*, Orlando, FL, 1993.
12. A. T. Andrews, *Electrophoresis*, Clarendon Press, Oxford, 2nd ed., 1986.
13. A. Chrambach, *The Practice of Quantitative Gel Electrophoresis*, VCH, Deerfield Beach, FL, 1985.
14. A. G. Ogston, *Trans. Faraday Soc.*, *54*, 1754 (1958).

15. P. D. Grossman, S. Menchen, and D. Hershey, *GATA*, *9*, 9 (1992).
16. K. A. Ferguson, *Metab. Clin. Exp.*, *13*, 985 (1964).
17. O. J. Lumpkin, P. Dejardin, and B. H. Zimm, *Biopolymers*, *24*, 1573 (1985).
18. P. G. de Gennes, *Scaling Concept in Polymer Physics*, Cornell University Press, Ithaca, NY, 1979.
19. L. S. Lerman and H. L. Frisch, *Biopolymers*, *21*, 995 (1982).
20. J. L. Viovy and T. A. J. Duke, *Electrophoresis*, *14*, 322 (1993).
21. T. A. J. Duke, A. N. Semenov, and J. L. Viovy, *Phys. Rev. Lett.*, *69*, 3260 (1992).
22. G. W. Slater and J. Noolandi, *Biopolymers*, *28*, 1781 (1989).
23. M. Chiari, M. Nesi, and P. G. Righetti, *Electrophoresis*, *15*, 616 (1994).
24. A. S. Cohen, S. Terabe, J. A. Smith, and B. L. Karger, *Anal. Chem.*, *59*, 1021 (1987).
25. J. Liu, J. F. Banks, and M. Novotny, *J. Microcol. Sep.*, *1*, 136 (1989).
26. T. Tsuda, G. Nakagawa, M. Sato, and K. Yagi, *J. Appl. Biochem.*, *5*, 330 (1983).
27. A. Guttman, in *Handbook of Capillary Electrophoresis*, J. P. Landers, Ed., CRC Press, Boca Raton, FL, p. 129 (1994).
28. B. K. Clark, C. L. Nickles, K. C. Morton, J. Kovak, and M. J. Sepaniak, *J. Microcol. Sep.*, *6*, 503 (1994).
29. T. Tanaka, *Sci. Am.*, *244*, 124 (1981).
30. A. Guttman, *US Patent* 5,332,481 (1994)
31. H. E. Schwartz, K. J. Ulfelder, F. J. Sunzeri, M. P. Busch, and R. G. Brownlee, *J. Chromatogr.*, *559*, 267 (1991).
32. P. Bocek and A. Chrambach, *Electrophoresis*, *12*, 1059 (1991).
33. E. Fung and E. S. Yeung, *Anal. Chem.*, *67*, 1913 (1995).
34. A. Barron, H. W. Blanch, and D. S. Soane, *Electrophoresis*, *15*, 597 (1994).
35. D. K. Schmalzing, C. A. Piggee, F. Foret, E. Carriho, and B. L. Karger, *J. Chromatog. A*, *652*, 149 (1993).
36. S. Hjerten, and K. Kubo, *Electrophoresis 14*, 390 (1993).
37. J. Townes, J. Bao, and F. Regnier, *J. Chromatog.*, *599*, 227 (1992).
38. P. Shieh, A. Hedeyati, N. Cooke, W. Goetzinger, and B. L. Karger, Poster 161, *Eighth International Symposium on High Performance Capillary Electrophoresis*, Orlando, FL, 1996.
39. J. L. Liao, J. Abramson, and S. Hjerten, *J. Cap. Elec.*, *2*, 191 (1995).

40. H. E. Schwartz and A. Guttman, *Primer: Separation of DNA*, Beckman Instruments, Inc., Fullerton, CA, 1995.
41. A. Guttman and H. E. Schwartz, *Anal. Chem.*, *67*, 2279 (1995).
42. M. J. van der Schaans, J. K. Allen, B. J. Wanders and A. Guttman, *J. Chromatogr. A*, *680*, 511 (1994).
43. N. Cohen and E. Gruschka, *J. Chromatogr.*, *684*, 323 (1994).
44. S. L. Pentoney and J. V. Sweedler, in *Handbook of Capillary Electrophoresis*, J. P. Landers, Ed., CRC Press, Boca Raton, FL, 1994, p. 147.
45. M. Jansson, J. Roeraarde, and F. Laurell, *Anal. Chem.*, *65*, 2766 (1993).
46. F-T. A. Chen, A. Tusak, S. L. Pentoney, K. Konrad, C. Lew, E. Koh, and J. Sternberg, *J. Chromatogr. A*, *652*, 355 (1993).
47. H. E. Schwartz, K. J. Ulfelder, F-T. A. Chen, and S. L. Pentoney, *J. Cap. Elec.*, *1*, 36 (1994).
48. H. E. Schwartz and K. J. Ulfelder, *Anal. Chem.*, *64*, 1737 (1992).
49. A. N. Glazer and H. S. Rye, *Nature*, *359*, 859 (1992).
50. K. J. Ulfelder, *Application Information Bulletin A-1748A*, Beckman Instruments, Inc., Fullerton, CA, 1993.
51. E. F. Rossomando, L. White, and K. J. Ulfelder, *J. Chromatogr. B*, *656*, 159 (1994).
52. H. Zhu, S. M. Clark, S. C. Benson, H. S. Rye, A. N. Glazer, and R. A. Mathies, *Anal. Chem.*, *66*, 1941 (1994).
53. H. Arakawa, K. Uetanaka, M. Maeda, Y. Tsuji, M. Matsubara, and K. Narisawa, *J. Chromatogr. A*, *680*, 517 (1994).
54. K. J. Ulfelder and B. R. McCord, in *Handbook of Capillary Electrophoresis*, 2nd ed., J. P. Landers, Ed., CRC Press, Boca Raton, FL, p. 347 (1997).
55. N. Dovichi, in *Handbook of Capillary Electrophoresis*, J. P. Landers, Ed., CRC Press, Boca Raton, FL, 1994, p. 369.
56. S. M. Wolf, P. Vouros, C. Norwood, and E. Jackim, *J. Am. Soc. Mass. Spectrom.*, *3*, 757 (1992).
57. E. Bayer, T. Bauer, K. Schmeer, K. Bleicher, M. Maier, and H. J. Gaus, *Anal. Chem.*, *66*, 3858 (1994).
58. A. Guttman, A. S. Cohen, D. N. Heiger, and B. L. Karger, *Anal. Chem.*, *62*, 137 (1990).
59. A. W. Kuypers, P. M. Willems, M. J. van der Schans, P. C. Linssen, H. M. Wessels, C. H. de Bruijn, F. M. Everaerts, and E. J. Mensink, *J. Chromatogr.*, *621*, 149 (1993).

60. K. J. Ulfelder, H. E. Schwartz, J. M. Hall, and F. J. Sunzeri, *Anal. Biochem.*, *200*, 260 (1992).
61. J. M. Butler, B. R. McCord, J. M. Jung, M. R. Wilson, B. Budowle, and R. O. Allen, *J. Chromatogr. B*, *658*, 271 (1994).
62. S. Nathakarnkitkool, P. J. Oefner, G. Bartsch, M. A. Chin, and G. K. Bonn, *Electrophoresis*, *13*, 18 (1992).
63. A. W. Kuypers, J. P. Meijerink, T. F. Smetsers, P. C. Linssen, and E. J. Mensink, *J. Chromatogr. B*, *660*, 271 (1994).
64. W. Lu, D. S. Han, J. Yuan, and J. M. Andrieu, *Nature*, *368*, 269 (1994).
65. M. J. Fasco, C. P. Treanor, S. Spivack, H. L. Figge, and L. S. Kaminsky, *Anal. Biochem.*, *224*, 140(1995).
66. G. S. Srivatsa, M. Batt, J. Schuette, R. H. Carlson, J. Fitchett, C. Lee, and D. L. Cole, *J. Chromatogr. A*, *680*, 469(1994).
67. A. J. Bourque and A. S. Cohen, *J. Chromatogr. B*, *662*, 343 (1994).
68. J. M. Leeds, M. J. Graham, L. A. Truong, and L. L. Cummins, *Anal. Biochem.*, *235*, 36 (1996).
69. J. M. Butler, B. R. McCord, J. M. Jung, J. A. Lee, B. Budowle, and R. O. Allen, *Electrophoresis*, *16*, 974 (1995).
70. R. L. Chien and D. S. Burgi, *J. Chromatogr.*, *559*, 141 (1991).
71. J. M. Butler, B. R. McCord, J. M. Jung, and R. O. Allen, *BioTechniques*, *17*, 1062 (1994).
72. Y. Wang, J. Ju, B. A. Carpenter, J. M. Atherton, G. F. Sensabaugh, and A. Mathies, *Anal. Chem.*, *67*, 1197 (1995).
73. H. T. Chang and E. S. Yeung, *J. Chromatogr. B*, *669*, 113 (1995).
74. J. Berka, Y. F. Pariat, O. Müller, K. Hebenbrock, D. N. Heiger, F. Foret, and B. L. Karger, *Electrophoresis*, *16*, 377 (1995)
75. P. E. Williams, M. A. Marino, S. A. Del Rio, L. A. Turni, and J. M. Devaney, *J. Chromatogr. A*, *680*, 525 (1994).
76. D. A. McGregor and E. S. Yeung, *J. Chromatogr.*, *652*, 67 (1993).
77. J. W. Efcavitch, in *Gel Electrophoresis of Nucleic Acids, A Practical Approach*, 2nd ed., D. Rickwood and B. D. Hames, Eds., Oxford University Press, Oxford, 1990.
78. A. Guttman and N. Cooke, *J. Chromatogr.*, *559*, 285 (1991).
79. D. Demorest and R. Dubrow, *J. Chromatogr.*, *559*, 43 (1991).
80. A. Guttman, A. Arai, and K. Magyar, *J. Chromatogr.*, *608*, 175 (1992).
81. A. Guttman, B. Wanders, and N. Cooke, *Anal. Chem.*, *64*, 2348 (1992).

82. Y. Baba, M. Tsuhako, T. Sawa, and M. Akashi, *J. Chromatogr.*, *632*, 137 (1993).
83. Y. Baba, M. Tsuhako, T. Sawa, and M. Akashi, *J. Chromatogr.*, *652*, 93 (1993).
84. T. Satow, T. Akiyama, A. Machida, Y. Utagawa, and H. Kobayashi, *J. Chromatogr.*, *652*, 23 (1993).
85. Y. Cordier, O. Roch, P. Cordier, and R. Bischoff, *J. Chromatogr.*, *680*, 479 (1994).
86. J. Sudor, F. Foret, and P. Bocek, *Electrophoresis*, *12*, 1056 (1991).
87. M. Chiari, M. Nesi, M. Fazio, and P. G. Righetti, *Electrophoresis*, *13*, 690 (1992).
88. C. Heller and J. L. Viovy, *Appl. Theor. Electrophoresis*, *4*, 39 (1994).
89. W. S. Marshall and M. H. Caruthers, *Science*, *259*, 1564 (1993).
90. P. E. Neelsen, M. Egholm, R. H. Berg, and O. Buchardt, *Science*, *254*, 1497 (1991).
91. W. J. Warren and G. Vella, *BioTechniques*, *14*, 598 (1993).
92. A. S. Cohen, M. Vilenchik, and J. L. Dudley, *J. Chromatogr.*, *638*, 293 (1993).
93. L. DeDionisio, *J. Chromatogr.*, *652*, 101 (1993).
94. D. J. Rose, *Anal. Chem.*, *65*, 3545 (1993).
95. M. Vilenchik, A. Belenky, and A. S. Cohen, *J. Chromatogr.*, *663*, 105 (1994).
96. C. S. Effenhauser, A. Paulus, A. Manz, and H. M. Widmer, *Anal. Chem.*, *66*, 2949 (1994).
97. S. L. Pentoney, K. D. Konrad, and W. Kaye, *Electrophoresis*, *13*, 467 (1992).
98. R. A. Mathies and X. C. Huang, *Nature*, *359*, 167 (1992).
99. K. Ueno and E. S. Yeung, *Anal. Chem.*, *66*, 1424 (1994).
100. S. Takahashi, K. Murakami, T. Anazawa, and H. Kambara, *Anal. Chem.*, *66*, 1021 (1994).
101. M. C. Ruiz-Martinez, J. Berka, A. Belenki, F. Foret, A. W. Miller, and B. L. Karger, *Anal. Chem.*, *65*, 2851 (1993).
102. R. A. Evangelista, M-S. Liu, S. Rampal, and F-T. A. Chen, *Anal. Biochem.*, *235*, 89 (1993).
103. J. P. Landers, R. P. Oda, T. C. Spelsberg, J. A. Nolan, and K. J. Ulfelder, *BioTechniques*, *14*, 98 (1993).
104. A. E. Barron and H. W. Blanch, *Separation and Purification Methods*, *24*, 1 (1995).
105. A. Paulus and J. I. Ohms, *J. Chromatogr.*, *507*, 113 (1990).

106. J. Cheng and K. R. Mitchelson, *Anal. Chem.*, *66*, 4210 (1994).
107. Y. F. Pariat, J. Berka, D. N. Heiger, T. Schmitt, M. Vilenchik, A. S. Cohen, F. Foret, and B. L. Karger, *J. Chromatogr.*, *652*, 57 (1993).
108. K. Kleparnik, S. Fanali, and P. Bocek, *J. Chromatogr.*, *638*, 283 (1993).
109. H. A. Erlich (Ed.), *PCR Technology—Principles and Applications for DNA Amplification*, Stockton Press, New York, 1989.
110. K. J. Ulfelder, in *High Performance Liquid Chromatography in Enzymatic Analysis*, 2nd ed., E. Rossomando, Ed., Wiley, New York, in press.
111. B. R. McCord, J. M. Jung, and E. A. Holleran, *J. Liq. Chromatogr.*, *16*, 1963 (1993).
112. K. Srinivasan, S. C. Morris, J. E. Girard, M. C. Kline, and D. J. Reeder, *Appl. Theor. Electrophoresis*, *3*, 235 (1993).
113. M. Nesi, P. G. Righetti, M. C. Patrosso, A. Ferlini, and M. Chiari, *Electrophoresis*, *15*, 644 (1994).
114. T. A. Felmlee, P. S. Mitchell, K. J. Ulfelder, D. H. Persing, and J. P. Landers, *J. Cap Elec.*, *2*, 125 (1995).
115. D. Del Principe, M. P. Iampieri, D. Germani, A. Menichelli, G. Novelli, and B. Dallapiccola, *J. Chromatogr.*, *638*, 277 (1993).
116. C. Gelfi, P. G. Righetti, V. Brancolini, L. Cremonesi, and M. Ferrari, *Clin. Chem.*, *40*, 1603 (1994).
117. J. M. Kolesar, J. D. Rizzo, and J. G. Kuhn, *J. Cap. Elec.*, *2*, 287 (1995).
118. S. J. Williams, C. Schwer, A. S. M. Krishnarao, C. Heid, B. L. Karger, and P. M. Williams, *Anal. Biochem.*, *236*, 146 (1996).
119. B. R. McCord, D. L. McClure, and J. M. Jung, *J. Chromatogr.*, *652*, 75 (1993).
120. M. Orita, H. Iwahana, H. Kanazawa, K. Hayashi, and T. Sekiya, *Proc. Natl. Acad. Sci. USA*, *86*, 2766 (1989).
121. J. Cheng, T. Kasuga, N. D. Watson, and K. R. Mitchelson, *J. Cap. Elec.*, *2*, 24 (1995).
122. J. Cheng, T. Kasuga, K. R. Mitchelson, E. R. Lightly, N. D. Watson, W. J. Martin, and D. Atkinson, *J. Chromatogr. A*, *677*, 169 (1994).
123. K. Khrapko, J. S. Hanekamp, W. G. Thilly, A. Belenkii, F. Foret, and B. L. Karger, *Nucleic Acids Res.*, *22*, 364 (1994).
124. C. Gelfi, P. G. Righetti, L. Cremonesi, and M. Ferrari, *Electrophoresis*, *15*, 1506 (1994).
125. A. E. Barron, W. M. Sunada, and H. W. Blanch, *Electrophoresis*, *16*, 64 (1995).

126. M. Strege and A. Lagu, *Anal. Chem.*, *63*, 1233 (1991).
127. T. Guszczynski, H. Pulyaeva, D. Tietz, M. M. Garner, and A. Chrambach *Electrophoresis*, *14*, 523 (1993).
128. K. Hebenbrock, K. Schügerl, and R. Freitag, *Electrophoresis*, *14*, 753 (1993).
129. T. Izumi, M. Yamaguchi, K. Yoneda, T. Isobe, T. Okuyama, and T. Shinoda, *J. Chromatogr.*, *652*, 41 (1993).
130. P. Bocek and A. Chrambach, *Electrophoresis*, *13*, 31 (1992).
131. P. Bocek and A. Chrambach, *Electrophoresis*, *12*, 620 (1991).
132. S. Rampal, M-S. Liu, R. A. Evangelista, and F-T. A. Chen, P-304, *Seventh International Symposium on High Performance Capillary Electrophoresis*, Würzburg, Germany, 1995.
133. J. Sudor and M. V. Novotny, *Anal. Chem.*, *66*, 2446 (1994).
134. M. J. Navin, T. L. Rapp, and M. D. Morris, *Anal. Chem.*, *66*, 1179 (1994).
135. Y. Kim and M. D. Morris, *Anal. Chem.*, *66*, 3081 (1994).
136. Y. Kim and M. D. Morris, *Anal. Chem.*, *67*, 784 (1995).
137. X. C. Huang, M. A. Quesada, and R. A. Mathies, *Anal. Chem.*, *64*, 2149 (1992).
138. C. S. Effenhauser, A. Manz, and H. M. Widmer, *Anal. Chem.*, *67*, 2284 (1995).

Index

"Activated"-PAD (APAD), 223
Adsorption-desorption process, kinetic model for, 55
Affinity capillary electrophoresis (ACE), 236
Amikacin, 227
Aminoglycosides, analysis of, 196-209
   PAD waveform, 198-199
   review of applications, 205-209
   voltammetry, 196-198
   waveform optimization, 199-205
Amoxicillin, 227
Ampicillin, 227
Antibiotic analysis, PED for, 189-232
   aminoglycosides, 196-209
      PAD waveform, 198-199
      review of applications, 205-209
      voltammetry, 196-198
      waveform optimization 199-205

[Antibiotic analysis, PED for]
   background, 190-191
   electrocatalysis at noble metal electrodes, 193-196
   future research, 223-225
      capillary electrophoresis, 224-225
      future applications, 225
   historical perspective of pulsed electrochemical detection, 191-193
   penicillins and cephalosporins, 209-223
      IPAD waveform, 212-213
      IPAD waveform optimization, 214-216
      review of applications, 217-223
      voltammetry, 210-212
   summary of all antibiotic applications by PED, 227
Antisense DNA, CE separation of, 321-323

Apramycin, 227
A-term dispersion, separation of C-term dispersion and, 31-36

Band spreading in chromatography, 1-49
　anomalous band broadening in gradient elution, 171-173
　column structure, 2-4
　early theories, 5-11
　eddy diffusion coupling, radial dispersion, and the infinite diameter effect, 15-21
　exclusion chromatography, 36-39
　in the electroseparation systems, 41-45
　nonequilibrium between stationary and moving zones or phases, 21-31
　obstructed diffusion and the B-term, 12-15
　reduced parameters, 11-12
　separation of the A-term and the C-term, 31-36
　slow chemical equilibrium and wide peaks, 39-41
B-term dispersion, 12-15
Buffer systems:
　effect of electrolysis on buffer composition, 288-290
　in CE separation of nucleic acids, 306-308

Capillary electrochromatography, 44-45
Capillary electrophoresis (CE), 224-225
　separation of DNA by, 301-340
　　future directions, 333-334
　　methods, 306-319
　　selected applications, 319-333
　　theory, 302-306

Capillary gel electrophoresis (CGE), 236
　separation mode of, 307
Capillary isoelectric focusing (CIEF), 236
　separation mode of, 307
Capillary isotachophoresis (CITP), separation mode of, 307
Capillary zone electrophoresis (CZE), 233-300
　dispersion, 258-265
　　alternative geometry, 263-264
　　broadening during electrophoresis, 260-263
　　introduction, injection, and detection effects, 258-260
　　resolution, 264-265
　effect of electrolysis on buffer composition, 288-290
　electroosmotic flow, 256-258
　　charge separation, 256
　　electroosmotic mobility and zeta potential, 256-258
　electrophoresis in capillaries, 252-256
　　efficiency and speed, 254-256
　　experimental setup, 252-253
　　limitations to field strength, 256
　interaction of ions, 266-288
　　intensity of system peaks, 287-288
　　Kohlrausch regulating function (KRF), 276-282
　　moving boundary equations, 275-276
　　mutual interaction of ions, 274-275
　　nonlinear transport, 266-267
　　non-self-sharpening boundary, 269-271
　　numerical approaches, 282-287

[Capillary zone electrophoresis (CZE)]
   steep, self-sharpening
      boundary, 267-269
   triangular zones, 271-274
 ions in equilibrium, 243-252
   correlation of mobility with
      molecular structure, 248-252
   influence of ionic strength on
      equilibrium, 247-248
   mobility not proportional to
      charge, 246-247
   protolysis, dependence of
      mobility on pH, 244-246
   reactions and equilibrium, 243-244
 mobilities, 237-243
   definitions, 237-238
   models describing mobility of
      ions, 239-243
 peak integrals, 290-292
 separation mode of, 307
Carbon dioxide (CO_2)-modified LC, 78
Cephalexin, 227
Cephalosporins, analysis of, 209-223
 IPAD waveform, 212-213
 IPAD waveform optimization, 214-216
 review of applications, 217-223
 voltammetry, 210-212
Cephapirin, 227
Characteristic function (CF) method in stochastic theory, 64-70
 stochastic model of chromatography investigated by, 70
Charge-coupled device (CCD) camera, 333
Chromatographic column, structure of, 2-4
Clindamycin, 227
Cloxacillin, 227

Column efficiency in SGC, 83-91
 comparision with HPGC, 86-87, 89
Computer simulation of LSS model of gradient elution, 157-160
C-term dispersion, separation of A-term dispersion and, 31-36

Denaturing gradient gel electrophoresis (DGGE), 330
2-Deoxystreptamine, 227
Dicloxacillin, 227
Dihydrostreptomycin, 227
Dispersion during electrophoretic transport, 258-265
 alternative geometry, 263-264
 broadening during electrophoresis, 260-263
 introduction, injection, and detection effects, 258-260
 resolution, 264-265
DNA separation by capillary electrophoresis, 301-340
 future directions, 333-334
 methods, 306-319
  buffer and gel systems, 306-309
  detection, 312-316
  fraction collection, 316
  injection and sample preparation, 309-312
  quantitation, 316-317
  sizing techniques, 317-319
 selected applications, 319-333
  antisense DNA, 321-323
  conformational polymorphisms, 329-331
  DNA sequencing, 323-325
  large DNA, 331-333
  polymerase chain reaction products, 326-328
  restriction digests, 326
  short oligomers, 319-321
 theory, 302-306

Eddy diffusion coupling, 16-18, 21
Edgeworth-Cramér expansion theory, 62-63, 68-70
Electrocatalysis at noble metal electrodes, 193-196
Electrokinetic injection, 310
Electroosmotic flow, 256-258
  charge separation, 256
  electroosmotic mobility and zeta potential, 256-258
Electrophoresis in capillaries, 252-256
  efficiency and speed, 254-256
  experimental setup, 252-253
  limitations to field strength, 256
Electrophoresis mobilities, 237-243
  definitions, 237-238
  models describing mobility of ions, 239-243
Electroseparation systems, band spreading in, 41-45
Exclusion chromatography, 36-39

Flow transport in SGC, 79-82
Foscarnet, 227
Four-step PAD, 223
Free-solution capillary electrophoresis (FSCE), 236

Gas chromatography (GC):
  LSS model of gradient elution applied to, 173-174
  similarities of and differences between liquid chromatography and, 75-76
  variations in velocity between mobile-phase and stationary phase of, 21-31
Gel systems in CE separation of nucleic acids, 306-309
Gentamycin, 227

Glucosamine, voltammetric response of, 196-198
Gradient elution, 116-120, 121-122
  See also Linear-solvent-strength (LSS) model of gradient elution

Height equivalent to a theoretical plate (HETP), 5
Heteroduplex polymorphism analysis (HPA), 330
High-performance anion-exchange chromatography (HPAEC), PAD joined with, 192
High-performance gas chromatography (HPGC):
  packed-column, 77-79
  SGC column efficiency compared with column efficiency in, 86-87, 89
High-performance liquid chromatography (HPLC), 3, 5
  See also Pulsed electrochemical detection (PED) for analysis of antibiotics
High-temperature LC (HTLC), 78
Hydrodynamic injection, 310
Hydrophobic interaction chromatography (HIC), 167
Hygromycin B, 227

Infinite diameter effect, 18
Initial gradient separation, design and use of, 154-157
Injection techniques in CE, 310-312
Integrated pulsed amperometric detection (IPAD), 192
  IPAD waveform for penicillins and cephalosporins, 212-214
  waveform optimization, 214-216
Ion-exchange LC (IELC), 163-166
  retention in, 122

Isocratic elution, 120-121
  comparing separations by gradient elution or, 131-138
    bandwidth and sensitivity, 132
    resolution, 133-135
  mass-overloaded separations, 145-147

Kanamycin, 227
Kinetic model for adsorption-desorption process, 55
Kohlrausch regulating function (KRF), 276-282

Large DNA, CE in separation of, 331-333
Lincomycin, 227
Linear-solvent-strength (LSS) model of gradient elution, 115-187
  application of LSS model to gas chromatography, 173-174
  liquid chromatographic basics, 120-123
    gradient elution, 121-122
    isocratic elution, 120-121
    retention vs. mobile-phase composition, 122-123
  LSS model, 123-150
    comparing separation by gradient or isocratic elution, 131-138
    gradient delay and elution after the gradient, 144-145
    large molecular effects, 138-144
    mass-overloaded separations, 145-150
    quantitative relationships, 123-130
  nonideal effects in gradient elution, 160-173

[Linear-solvent-strength (LSS) model of gradient elution]
  anomalous band broadening in gradient elution, 171-173
  instrumental effects, 160-163
  nonequilibrium effects, 167-168
  non-LSS conditions, 163-167
  potential accuracy of retention predictions from LSS model, 170-171
  retention reproducibility, 168-170
  optimizing gradient elution, 150-160
    computer simulation, 157-160
    design and use of an initial gradient separation, 154-157
    gradient method development, 151-154
Liquid chromatography (LC):
  packed-column, 77-79
  similarities of and differences between gas chromatography and, 75-76
Liquid-liquid partition chromatography, 5

Mass-overloaded separations, 145-150
  gradient elution, 147-150
  isocratic elution, 145-147
Methicillin, 227
Micellar electrokinetic chromatography (MEKC), 236, 307-308
  separation mode of, 307
Mobile phase of GC, variations in velocity between stationary phase and, 21-31

Nafcillin, 227
Neomycin B, C, 227

Noble metal electrodes,
    electrocatalysis at, 193-196
Nonequilibrium effects in gradient
    elution, 167-168
Nonequilibrium theory, 21-31
Nonideal effects in gradient elution,
    160-173
  anomalous band broadening in
    gradient elution, 171-173
  instrumental effects, 160-163
  nonequilibrium effects, 167-168
  non-LSS conditions, 163-167
  potential accuracy of retention
    predictions from LSS
    model, 170-171
  retention reproducibility, 168-170
Normal-phase LC (NP-LC), 166-167

Oligonucleotides, CE analysis of,
    319-321
Optimizing gradient elution, 150-160
Oxacillin, 227

Peak integrals in CZE, 290-292
Peak shape analysis, applications of
    stochastic theory to, 63-64
Penicillin G, 227
Penicillin V, 227
Penicillins, analysis of, 209-223
  IPAD waveform, 212-213
  IPAD waveform optimization,
    214-216
  review of applications, 217-223
  voltammetry, 210-212
Plate theory of band spreading, 5-8
Polarity in SGC, 101-111
Polymerase chain reaction products,
    CE in, 326-328
Potential sweep-pulsed coulometric
    detection (PS-PCD), 192
Pulsed amperometric detection
    (PAD), 193

[Pulsed amperometric detection (PAD)]
  disadvantages for oxide-catalyzed
    detections, 214
  PAD waveform for
    aminoglycosides, 198-199
  waveform optimization, 199-205
Pulsed coulometric detection (PCD),
    192
Pulsed electrochemical detection
    (PED) for analysis of
    antibiotics, 189-232
  aminoglycosides, 196-209
    PAD waveform, 198-199
    review of applications, 205-209
    voltammetry, 196-198
    waveform optimization 199-205
  background, 190-191
  electrocatalysis at noble metal
    electrodes, 193-196
  future research, 223-225
    capillary electrophoresis, 224-
      225
    future applications, 225
  historical perspective of pulsed
    electrochemical detection,
    191-193
  penicillins and cephalosporins,
    209-223
    IPAD waveform, 212-213
    IPAD waveform optimization,
      214-216
    review of applications, 217-223
    voltammetry, 210-212
  summary of all antibiotic
    applications by PED, 227

Radial dispersion in packed beds,
    19-20
Restriction digests, CE in analysis of
    ds DNA in, 326
Retention ratio in chromatography,
    58-59

Retention vs. mobile-phase composition, 122-123
"Reversed"-PAD (RPAD), 223
Reversed-phase gradient elution, 151-154
Reversed-phase liquid chromatography (RP-LC), 119
  characteristics of large-molecule vs. small-molecule RP-LC, 138, 139
  LSS gradient for, 123-130
  retention in, 122
Ribavirin, 227

Separation speed in SGC, 91-101
Sequence polymorphisms in DNA molecules, CE analysis of, 329-331
Short oligomers, CE analysis of, 319-321
Single-stranded conformational polymorphisms (SSCP) analysis, 329-330
Slow chemical equilibration leading to band broadening, 39-41
Solvating gas chromatography (SGC), 75-113
  efficiency in, 83-91
  flow transport in, 79-82
  packed-column HPGC, SGC, SFC, and LC, 77-79
  polarity and solvating power in, 101-111

[Solvating gas chromatography (SGC)]
  separation speed in, 91-101
Spectinomycin, 227
Stationary phase of GC, variations in velocity between mobile phase and, 21-31
Stochastic theory of chromatography, 51-74
  characteristic function method in, 64-70
  fundamentals and basic achievements of, 56-63
  some applications of stochastic theory to experimental peak shape analysis, 63-64
  structural stochastic concepts, 55-56
  theoretical approaches in chromatography, 52-54
Streptomycin, 227
Supercritical fluid chromatography (SFC), packed-column, 77-79

Theoretical investigation in chromatography, 52-54
Tobramycin, 227

Van Deemter equation, 54

Zeta potential, electroosmotic mobility and, 256-258